Climate Mathematics

Theory and Applications

This unique text provides a thorough, yet accessible, grounding in the mathematics, statistics, and programming that students need to master for coursework and research in climate science, meteorology, and oceanography. Assuming only high school mathematics, it presents carefully selected concepts and techniques in linear algebra, statistics, computing, calculus, and differential equations within the context of real climate science examples. Computational techniques are integrated to demonstrate how to visualize, analyze, and apply climate data, with R code featured in the book and both R and Python code available online. Exercises are provided at the end of each chapter with selected solutions available to students to aid self-study and further solutions provided online for instructors only. Additional online supplements to aid classroom teaching include datasets, images, and animations. Guidance is provided on how the book can support a variety of courses at different levels, making it a highly flexible text for undergraduate and graduate students, as well as researchers and professional climate scientists who need to refresh or modernize their quantitative skills.

Samuel S. P. Shen is Distinguished Professor of Mathematics and Statistics, San Diego State University, and Visiting Research Mathematician at Scripps Institution of Oceanography, University of California–San Diego. Formerly, he was McCalla Professor of Mathematical and Statistical Sciences at the University of Alberta, Canada, and President of the Canadian Applied and Industrial Mathematics Society. He has held visiting positions at the NASA Goddard Space Flight Center, the NOAA Climate Prediction Center, and the University of Tokyo.

Shen holds a B.Sc. degree in Engineering Mechanics from Nanjing University of Science and Technology, and M.A. and Ph.D. degrees in Applied Mathematics from the University of Wisconsin–Madison.

Richard C. J. Somerville is Distinguished Professor Emeritus at Scripps Institution of Oceanography, University of California–San Diego. He is a fellow of three scientific societies: the American Association for the Advancement of Science (AAAS), the American Geophysical Union (AGU), and the American Meteorological Society (AMS). He has received awards from the AMS for his research and his popular book *The Forgiving Air: Understanding Environmental Change.* From the AGU, he has received two major honors, the Climate Communication Prize (2015), and the Ambassador Award (2017).

Somerville holds a B.Sc. degree in Meteorology from Pennsylvania State University, and a Ph.D. degree in Meteorology from New York University.

Climate Mathematics

Theory and Applications

SAMUEL S. P. SHEN
San Diego State University

RICHARD C. J. SOMERVILLE
Scripps Institution of Oceanography, University of California, San Diego

CAMBRIDGE
UNIVERSITY PRESS

University Printing House, Cambridge CB2 8BS, United Kingdom

One Liberty Plaza, 20th Floor, New York, NY 10006, USA

477 Williamstown Road, Port Melbourne, VIC 3207, Australia

314–321, 3rd Floor, Plot 3, Splendor Forum, Jasola District Centre, New Delhi – 110025, India

79 Anson Road, #06–04/06, Singapore 079906

Cambridge University Press is part of the University of Cambridge.

It furthers the University's mission by disseminating knowledge in the pursuit of education, learning, and research at the highest international levels of excellence.

www.cambridge.org
Information on this title: www.cambridge.org/9781108476874
DOI: 10.1017/9781108693882

First published 2019

Printed in Singapore by Markono Print Media Pte Ltd, 2019

A catalogue record for this publication is available from the British Library.

Library of Congress Cataloging-in-Publication Data
Names: Shen, Samuel S.P., author. | Somerville, Richard C.J., author.
Title: Climate mathematics : theory and applications / Samuel S.P. Shen (San Diego State University), Richard C.J. Somerville (Scripps Institution of Oceanography, University of California, San Diego).
Description: Cambridge ; New York, NY : Cambridge University Press, 2019. | Includes bibliographical references and index.
Identifiers: LCCN 2019009275 | ISBN 9781108476874 (hardback : alk. paper)
Subjects: LCSH: Climatology–Mathematical models–Problems, exercises, etc. | Climatology–Mathematics–Problems, exercises, etc. | R (Computer program language)
Classification: LCC QC981 .S52275 2019 | DDC 551.60285/5133–dc23
LC record available at https://lccn.loc.gov/2019009275

ISBN 978-1-108-47687-4 Hardback

Additional resources for this publication at www.cambridge.org/climatemathematics

"I hear, and I forget. I see, and I remember. I do, and I understand."

– Confucius

Contents

Preface

How should students in climate science, meteorology, and physical oceanography learn mathematics? What topics in mathematics should they learn, and in what order? What changes are necessary in mathematics education for climate science students in order to meet the needs of climate science education and research in the big data era?

In the last several decades, a typical undergraduate student enrolled in an atmospheric, oceanic, or climate science major might take six university-level mathematics and computing courses: Calculus I, II, and III, Linear Algebra and Differential Equations, Basic Statistics, and Computer Programming. This mathematics curriculum has several shortcomings, and two of them seem to us especially important: much of the mathematics that is taught has little relevance to a major such as climate science, and the pace of mathematics learning in this curriculum is too slow to meet the needs of the major courses. The textbooks used by climate science students in calculus, linear algebra, and statistics courses are usually written by mathematics and statistics professors, and much of the material in these books is simply not very relevant to practical climate science problems. The traditional approach to teaching mathematics treats calculus, linear algebra, statistics, differential equations, and computer programming as isolated and disconnected subjects. Consequently, this traditional teaching methodology not only makes these mathematics topics disconnected from each other, but also makes them even further disconnected from climate science. In reality, a typical climate science research problem does not present itself accompanied by a helpful statement as to what mathematical tools are needed to solve it. The tools can easily involve calculus, linear algebra, statistics, computer programming, or a combination of all these.

Therefore, climate science students have difficulty in seeing why mathematics is taught in this manner and in perceiving how mathematics is used in solving actual climate science problems. For example, a typical linear algebra course would never discuss the significance of eigenvalues in climate science and would never show how to compute the eigenvalues of a large climate data matrix. This pedagogical problem has puzzled many climate science educators for many years. Students are required to spend lots of time learning what they regard as irrelevant and peculiar topics in mathematics, such as using l'Hôpital's rule (named for the seventeenth-century French mathematician Guillaume de l'Hôpital) to evaluate a limit, and a half-angle substitution to calculate an antiderivative, but these students then find themselves unable to cope with the mathematical demands of upper division and graduate-level climate science courses. As a result, climate science professors often complain about their students' weak mathematics backgrounds, and the students in turn are frustrated with the inappropriate and abstract mathematical knowledge that they have acquired from mathematics professors who are unfamiliar with the climate science applications of mathematics. Conversely, the skills and techniques of data analysis and visualization, such as empirical orthogonal functions (EOFs), that are needed by virtually all climate science graduate students are simply not taught in the typical current sequence of mathematics classes that these students are required to take. In brief, conventional mathematics courses are not well suited to the mathematical needs of problems in climate science. Climate science students suffer from a serious mismatch between tasks and tools.

The question then arises as to how best to address the challenges and shortcomings of the current approach for climate science students to learn mathematics, an approach which is both disconnected and time-demanding. We have written this book to answer that question. Our book tries to overcome the above problems by blending the most climate-relevant materials of the six courses (Calculus I, II, and III, Linear Algebra and Differential Equations, Basic Statistics, and Computer Programming) with key climate science examples into one unified course, called Climate Mathematics. This single course may be taught in from one semester up to three semesters, depending on the qualifications and objectives of the students. Climate Mathematics presents the mathematical methods from the point of view of climate science applications. Climate Mathematics prepares climate science students with a sufficient mathematical background to enable them to take not only the upper division courses, but also most graduate-level climate science courses. It also prepares them to conduct research involving analysis and visualization of climate datasets.

Every important mathematical formula and each aspect of theory included in this book is presented together with at least one climate science example. At the same time, we have omitted many of the less relevant topics found in the typical array of six mathematics courses, such as limits in Calculus I, infinite series and integration techniques in Calculus II, determinants in Linear Algebra, and series solutions in differential equations. These omitted materials from the conventional mathematics curriculum are simply not used in typical current climate courses. Even if they are needed in some special cases, they can be more efficiently taught by the climate science instructor if his or her course needs them. We have also included some important modern topics that are not taught in the traditional six courses, such as the singular value decomposition (SVD) method to compute EOFs from a rectangular space–time data matrix.

The second defining feature of this book is the extensive use of computer codes. We have selected the programming language R for these codes. We provide an introduction to R for students to whom this remarkable language is completely new, and we include advanced R graphics suitable for application to global climate model (GCM) datasets. We also present R analysis and visualization techniques appropriate for observational climate datasets characterized by missing data.

The R codes are included in the book and are also available as an open source from the book website www.cambridge.org/climatemathematics. Also available from the book website are the equivalent computer codes written in Python. In recent years, Python has become a very popular computer language in science and engineering, and has begun to become popular among climate scientists. We have supplied a Python equivalent code for every R code in this book. The corresponding Python codes are available as an open source in the Jupyter Notebook format. They may be freely used by anyone.

The third distinct feature of the book is that we have also updated some topics based on recent advances in science and technology, such as the GPS-based radiosonde, a new hypsometric equation derived without the usual isothermal assumption, and the GPS-planimeter as a smartphone app to measure the area of a region on the Earth's surface based on Green's theorem.

We recognize that our book can be quite demanding for an instructor, who will ideally be familiar with all the mathematical tools we discuss, as well as the history of calculus, linear algebra, statistics, and computing. The instructor should also be conversant with the important applications of these tools to climate science, such as the SVD approach to EOFs and principal components, and techniques for the efficient analysis and visualization of Reanalysis GCM data. Perhaps at this moment not many instructors will be fully prepared to teach all the material in this book. However, this situation is changing rapidly. Currently, more Ph.D. students probably use SVD rather than a covariance matrix to compute EOFs. Many students are already skillful at employing data visualization and signal extraction using R, Python, and MATLAB packages. These graduate students can now use our book as a research reference and toolbox, and they will be well qualified to teach from our book when they become faculty members themselves. Thus, we in climate science and allied fields may learn from biology and psychology, which today are among the most popular majors in large public universities in the United States. Years ago, biology and psychology programs had begun to teach their own statistics courses, often staffed by biologists and psychologists. Engineering colleges had also started to teach their own linear algebra and differential equations courses. Calculus for Life Sciences and Calculus for Business are taught today on many campuses. We pose the following question: Should climate science departments or the geoscience college now begin to teach their own mathematics, statistics, and computing courses, using a book like this one, in the era of big data and artificial intelligence? Should climate science departments and mathematics departments make joint faculty appointments to teach Climate Mathematics? Or should the mathematics department first hire (if necessary) and then assign a faculty member who is thoroughly familiar with climate science to teach Climate Mathematics?

A closely related and important question is this: How should we teach the necessary mathematical methods to students who have decided to change their fields from humanities to climate science? These students typically come to climate science lacking a calculus background, and they may have no intention of employing the advanced mathematics needed for, say, climate modeling, but they can certainly contribute to interdisciplinary fields employing climate science, such as policy-making for climatic change. In fact, this book originated from lecture notes in a course created to teach exactly this type of student at Scripps Institution of Oceanography, University of California, San Diego. The course was called Climate Mathematics (SIOC 290). This course was taught by one of us (S.S.P.S.) and consisted of 45 hours of instruction in a special summer session of five weeks in 2015, with nine instruction hours per week. The course was designed for the students in a newly established program called Masters of Advanced Studies in Climate Sciences and Policy. These students would need to understand, explain, and present the results from climate models and observations. The students used SIOC 290 to prepare themselves to take graduate courses such as SIO 210 (Physical Oceanography), SIO 217A (Atmospheric Thermodynamics), and SIO 260 (Marine Chemistry). In addition to strictly mathematical topics, the students taking this course would also learn sufficient graphics and visualization techniques to enable them to present and describe climate data from both observations and models. SIOC 290 was taught again in 2016 and a third time in 2017.

In choosing the title Climate Mathematics, we wished to emphasize that this book is not aimed at a mathematics class for mathematics majors. Several in-depth mathematics courses on the topics we cover, such as linear algebra, can and should be taught for mathematics majors and other students who love mathematics for its own sake. Also, in order to limit this book to a reasonable size, we have not included several aspects of mathematics that are useful in climate science. Numerical methods for ordinary and partial differential equations, for example, are essential for climate modeling, but they are beyond the scope of this book. We hope to include them in a future book on Advanced Climate Mathematics.

Suggestions on the Usage of the Book

Who Will Use this Book?

This book can be used as both a textbook and a reference manual for the following five groups of students as illustrated in Fig. 0.1.

Group A. This group includes undergraduate students majoring in climate science, meteorology, or oceanography. The book can be used as a textbook for three semesters of Climate Mathematics I (Appendices, Chapters 1–4), II (Chapters 5–8), and III (Chapters 9–11) to replace the traditional six semesters of courses: Calculus I, II, and III, Linear Algebra and Differential Equations, Basic Statistics, and Computer Programming. Climate Mathematics I contains an accelerated introduction to calculus including the divergence theorem and line integrals, and it prepares students to take courses in mechanics and atmospheric or oceanic dynamics and thermodynamics. When students complete all the three Climate Mathematics courses, they should have sufficient quantitative analysis background to take upper division undergraduate courses in climate science, such as the courses based on the book by Wallace and Hobbs entitled *Atmospheric Science: An Introductory Survey*, that by Talley et al. entitled *Descriptive Physical Oceanography: An Introduction*, and that by Curry and Webster entitled *Thermodynamics of Atmospheres and Oceans*. Students should have also mastered sufficient skills for data analysis and data mappings for course work and research projects. Because students complete the mathematics training early in their undergraduate programs,

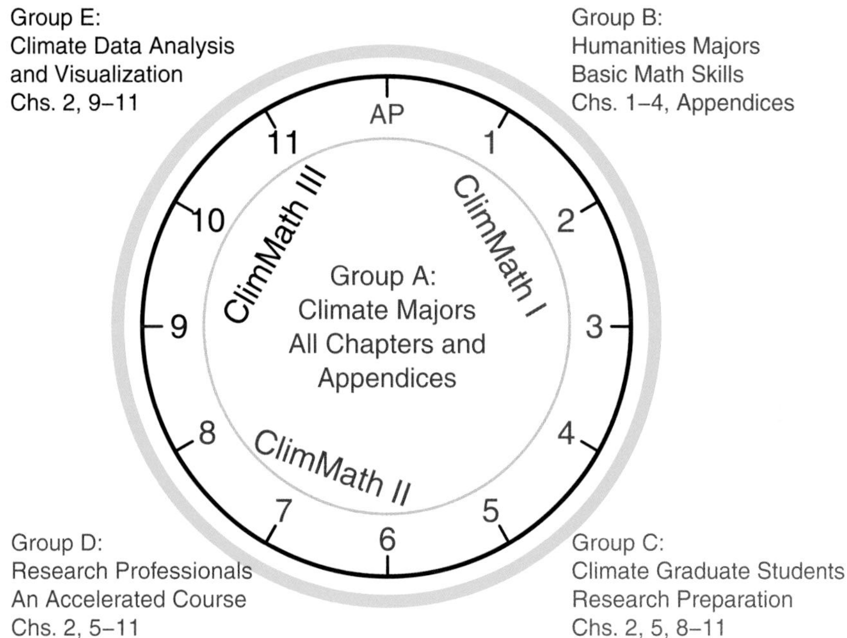

Fig. 0.1 Schematic diagram illustrating how this book might be used in several different courses for students having different backgrounds and motivations. ClimMath I, II, and III stand for three semesters of Climate Mathematics course based on this book. ClimMath I covers the basics of Calculus in the Appendices (denoted by AP in the 12 o'clock position), R Program in Chapter 2, Statistics in Chapter 3, and Linear Algebra in Chapter 4. ClimMath II covers Chapters 5–8 and focuses on the methods of climate modeling as applications of the mathematical methods described in ClimMath I. ClimMath III covers Chapters 9–11 plus one or two research topics selected by an instructor, and focuses on climate data analysis and visualization.

they will have more time to take newly created courses, such as machine learning in climate science, geographic information systems (GIS), satellite meteorology, climate data visualization, or climate modeling.

Alternatively, a student might choose to accelerate the Climate Mathematics series by taking Climate Mathematics II and III simultaneously in the second semester of the freshman year. This would allow the student to have more freedom to choose her/his courses, particularly the big data courses that will dramatically improve the employment opportunities of the climate science majors.

Group B. This group includes those students who originally were humanities majors, but now wish to change their emphasis to climate science or simply to take several upper division or graduate climate science courses. They had not majored in physical sciences and have little background in calculus, statistics, or linear algebra, but they are well motivated to learn climate science. Yet they have no time to go through the full series of Climate Mathematics I, II, and III. They prefer to have the minimum mathematics background necessary to enable them to take a few upper division climate science courses. The background can be achieved by taking one semester simplified Climate Mathematics using selected sections from our book as the text. We suggest the following sections: Appendices A, B, C, and D.1, D.3, D.4.1–2, D.4.5–7, and Chapters 1.1–1.4, 2.1–2, 3.1–3, 5.1–2, 6.4–7, and 7.1–2. This course may be called Climate Mathematics B.

Group C. This group includes the graduate students in climate science who have received the traditional mathematics education and wish to enhance or refresh their modern mathematical skills for research, such as the energy balance models, modern hypsometric equation, R analysis of the GCM data, and data visualization. This can be a one-semester course, called Climate Mathematics C, which uses Chapters 2 and 5, Section 7.2, and Chapters 8–11. It is basically Climate Mathematics III plus some selected topics to prepare for climate science research. Because the students have already taken at least the six traditional mathematics courses, Climate Mathematics C can be taught at an accelerated pace compared to Climate Mathematics III.

Group D. This group includes research scientists and professionals who wish to modernize their skills in modern mathematics, statistics, climate data analysis, and data visualization. They may take a one-semester selected topics course with the topics selected from Chapters 2 and 5–11. This is Climate Mathematics D.

Group E. This group includes people who are mainly interested in climate data analysis and visualization, such as climate science professionals in government laboratories or businesses. They might spend a week-long effort with this book in preparation for working on R analyses and graphics for climate data. They can study topics from Chapters 2 and 9–11 to learn how to analyze and visualize climate data for presentations or research, such as R plotting for climate data and a comprehensive R analysis for the gridded United States surface temperature data. One of us (S.S.P.S.) taught such a short course in 2017 for scientists at the National Centers for Environmental Information in the United States, and also at the Chinese Academy of Sciences in Beijing.

Because this book contains numerous R codes for climate data analysis and graphics, the book can also be used as a reference manual for a climate scientist. The book's website www.cambridge.org/climatemathematics provides updated computer codes in both R and Python to support all users of this book.

How to Use this Book

We have tried to make this book "user-friendly" for both students and instructors. An instructor can simply follow the book when lecturing and can assign the exercise problems. Complete solutions for a few selected problems are presented at the end of the book as Appendix E to help train students in writing about mathematics. A solutions manual with answers for all the problems is available through the publisher to qualified instructors.

This book emphasizes the value of practice in acquiring quantitative analysis skills. The book contains many R codes, which can also be found at the book's website www.cambridge.org/climatemathematics showing details of our R codes and their output. The relevant climate data used in the R codes can also be downloaded directly from the website.

The exercise problems that require R codes are marked with a computer symbol. Problems that do not have this symbol can be solved analytically without using a computer.

When doing computations, we strongly recommend that readers of conventionally published books should use the R codes (or their Python equivalents) that can be found on the book's website, www.cambridge.org/climatemathematics. There is no need to re-type the computer codes. The open-source online codes will be maintained and will include corrections, updates, and improvements to the original codes. Although e-book readers may be able to copy and paste the R code from a pdf file to RStudio, some special symbols, such as ~ and ^, may become altered in this copy-and-paste procedure and may have to be corrected in RStudio.

The websites cited in this book may be updated or changed in the future. Readers can find the updated website addresses on the book website www.cambridge.org/climatemathematics.

The datasets used in this book can be downloaded as a data.zip file from the book website: www.cambridge.org/climatemathematics. Updated data in some cases can be downloaded from the data developers' websites provided in the book.

A new mathematics textbook of this size is virtually certain to contain errors, despite the diligent efforts of many people to find them all before publication. We take full responsibility for all remaining errors, and we will post any errata on the book's website, www.cambridge.org/climatemathematics.

Acknowledgements

We thank the students at Scripps Institution of Oceanography (SIO), University of California, San Diego (UCSD), in the SIOC 290 Climate Mathematics courses from 2015 through 2017. They made many suggestions based on their experience with successive versions of the course lecture notes from which this book originated. In particular, Larry Serra, Jr. proofread early drafts of the first nine chapters. Minyang Wang also proofread part of the book and helped with typesetting the index.

Early versions of the material in Chapters 9–11 on R codes for the analysis and visualization of climate data were first taught in May 2017 at the National Centers for Environmental Information in the United States, and again in June 2017 at the Third Pole Environment of the Chinese Academy of Sciences in Beijing, China.

We are also deeply indebted to our colleague Professor Lynn M. Russell, the founding director of the SIO Masters of Advanced Studies in Climate Science and Policy program. She worked with one of us (SSPS) in the development of SIOC 290 in 2014–2015. It was her vision then that it would be worthwhile for the students in the program to learn R. This book has also benefited greatly from Professor Russell's long collaboration with one of us (RCJS) in both climate research and teaching.

Dr. Jean-Pierre Williams of the University of California, Los Angeles provided the Diviner satellite lunar temperature data to us, together with relevant background material that was invaluable in developing Chapter 5.

We have benefited from exceptionally thoughtful reviews of an early version of the manuscript by Professor Donald J. Wuebbles of the University of Illinois and Professor Gerald R. North of Texas A&M University, as well as by several anonymous reviewers. We have also benefited greatly from discussions with Professor Geoffrey K. Vallis of the University of Exeter, who shares our conviction that numerical models and codes can usefully be combined with conventional text and equations in a book such as this one. We quote Professor Vallis at length on this topic in the introduction to Chapter 2.

We thank Jon Billam and Irene Pizzie for excellent copy-editing. At Cambridge University Press, Nicola Chapman, Ilaria Tassistro, and Heather Brolly have provided us with extremely helpful and detailed guidance in preparing our manuscript for publication. As our editor throughout the process of conceiving and developing this book, Dr. Susan Francis has been especially generous in advising us and has been an invaluable source of expertise and wisdom. Finally, we thank our colleagues, friends, and families for their support and tolerance during the many intensive periods of writing and revision, when we devoted considerable time to the book.

Main Symbols and Acronyms

Notation	Description
$\|\|\cdot\|\|$	Length or magnitude of a vector
a, A	Acceleration; Wave amplitude; Intercept of a linear regression a or b_0; Data matrix $A = [a_{ij}], B = [b_{ij}]$; Area A
A^t	Transpose of matrix A
A^{-1}	Inverse matrix of A
α	Albedo, dimensionless constant in dimensional analysis
AMO	Atlantic Multi-decadal Oscillation
$\beta, b, \hat{b}, b_1, \hat{b}_1$	Slope for linear regression; Rossby parameter $\beta = 2\Omega\cos\phi/R$; Solar zenith angle β
c	Speed of light $c \approx 3\times10^5$ [km s^{-1}]; Speed of shallow water waves $c = \sqrt{gH}$
C, c, c_p, c_v	Specific heat equal to energy per unit mass per unit kelvin
Coriolis force	$\mathbf{F}_c = -2m\mathbf{\Omega}\times\mathbf{v}$ points to the right of the mass m's velocity $\mathbf{v}$ in the Northern Hemisphere, and to the left in the Southern Hemisphere
$corr(x,y), r_{xy}, \rho_{xy}$	Correlation between x and y
$cov(x,y), C_{xy}, \Sigma_{xy}$	Covariance between x and y
CSOI	Cumulative SOI; Also see SOI, WSOI, and CWSOI
CWSOI	Cumulative WSOI
D	Thermal diffusivity with dimension L^2T^{-1}
DA pair	(f', f) such that $f(x) - f(0) = I[f', 0, x]$
dH	Differential of enthalpy $dH = dU + pdV + Vdp$
dU	Internal energy increment $dU = dQ + dW$ is equal to the heat increment dQ plus the work on the system dW
dW	Work done due to gas expansion $dW = pdV$
$\frac{D\rho}{Dt}$	Total derivative of ρ: $\frac{D\rho}{Dt} = \frac{\partial\rho}{\partial t} + u\frac{\partial\rho}{\partial x} + v\frac{\partial\rho}{\partial y} + w\frac{\partial\rho}{\partial z}$
ΔS_{AB}	Entropy increment from state A to state B $\Delta S_{AB} = \int_A^B \frac{dq}{T}$
E	Energy
E_{bb}	Blackbody radiation power intensity $E_{bb} = \sigma T^4$ [W m^{-2}]
EBM	Energy balance model
$E[\lambda, T]$	Spectral flux of radiation power [(W m^{-2}) m^{-1}]

Notation	Description
EM	Error margin
ENSO	El Niño–Southern Oscillation
EOF	Empirical orthogonal function
ϵ	Surface emissivity; Random error in a statistical model
$\exp(x) = e^x$	Exponential function
f	Coriolis frequency $f = 2\Omega \sin\phi$; Force
$f', \dot{f}, df/dx$	Derivative of function f
F	Force; Newton's second law of motion $F = ma$
g	Gravitational acceleration; the Earth's gravitational acceleration is $g \approx 9.8$ [m s^{-2}]
G	Universal gravitational constant
γ_3	Skewness
γ_4	Kurtosis
GCM	Global climate model; General circulation model
GIS	Geographic Information System
GPCP	Global Precipitation Climatology Project
GPS	Global Positioning System
h	Planck constant $h = 6.626070040 \times 10^{-34}$ [J sec]
H_0	Null hypothesis
H_1, H_a	Alternative hypothesis
i	Imaginary unit $\sqrt{-1}$; Electric current
IEEE	Institute of Electrical and Electronics Engineers
k	Wave number $k = 1/\lambda$
κ	Thermal conductivity, unit [W m^{-1} K^{-1}]
k_b	Boltzmann constant $k_b = 1.38064852 \times 10^{-23}$ [J K^{-1}]
L	Length dimension $[l] = L$; Wavelength
λ_k	The kth eigenvalue: $\mathbf{C}\mathbf{u}_k = \lambda_k \mathbf{u}_k, k = 1, 2, \ldots, N$
$L(x)$	Linear approximation $L(x) = f(a) + f'(a)(x-a)$
M	Mass dimension $[m] = M$; Molecular weight of dry air $M = 28.97$ [g mol^{-1}]
μ	Expected value; Mean
MVT	Mean value theorem in the integral form $I[f,a,b] = f(c)(b-a)$, or in the differential form $f'(\xi) = [f(b) - f(a)]/(b-a)$
n	Sample size
$\nabla, \nabla p$	Gradient differential operator $\nabla = \left(\frac{\partial}{\partial x}, \frac{\partial}{\partial y}, \frac{\partial}{\partial z}\right)$; Gradient of pressure field ∇p
NASA	National Aeronautics and Space Administration, USA
NCAR	National Center for Atmospheric Research, USA
NCEI	National Centers for Environmental Information, NOAA, USA
NCEP	National Centers for Environmental Prediction, NOAA, USA
$N(\mu, \sigma^2)$	Normal distribution with mean μ and standard deviation σ
NOAA	National Oceanic and Atmospheric Administration, USA

Notation	Description
ω	Wave frequency $\omega = 1/\tau$
Ω	Angular speed of the Earth's rotation $\Omega \approx 7.3 \times 10^{-5}$ [radian sec^{-1}]
p	Pressure
P	Power
PC	Principal component
PGF	Pressure gradient force $\mathbf{F}_p = -\nabla p/\rho$, force per unit mass
ϕ	Angle, dimensionless, unit: radian or degree; Latitude
Q	Total amount of heat energy
R	Specific gas constant for dry air $R = R^*/M = 287$ [J K^{-1}(kg)$^{-1}$]; Radius of Earth; Radius of a nuclear shock wave
R^*	Universal gas constant $R^* = 8.314$ [J K^{-1}mol^{-1}]
Ro	Rossby number Ro$= U/Lf$
ρ	Mass density
s, S	Standard deviation; Solar constant S
$s, s_y, \hat{s}_y, \hat{s}$	Standard deviation
SE	Standard error $SE(\bar{x}) = S/\sqrt{n}$
σ	Standard deviation; Stefan–Boltzmann constant $\sigma = 5.670367 \times 10^{-8}$ [W m^{-2} K^{-4}]
SLP	Sea level pressure
SSE	Sum of squared errors
SOI	Southern Oscillation Index; Also see CSOI, WSOI, and CWSOI
$\int_a^b f(x)dx, I[f,a,b]$	Integration from a to b for function f
SVD	Singular value decomposition $A = UDV^t$ for spatial patterns U, "energy" D, and temporal patterns V
t	Time; Random variable for the t-distribution
T	Time dimension $[t] = T$; Temperature; Total time involved
τ	Time; Period of an oscillation
Θ	Temperature dimension $[T] = \Theta$
θ	Angle, dimensionless, unit: radian or degree; Longitude
$U, u, v, w, \mathbf{v}$	Velocity
UCAR	University Corporation for Atmospheric Research, USA
WSOI	Weighted SOI; Also see SOI, CSOI, and CWSOI
$X = [x_{it}]_{N \times Y}$	Space–time data matrix with rows for space and columns for time
$\bar{x}$	Mean of a data string: $\bar{x} = \frac{\sum_{i=1}^n x_i}{n}$
z	The vertical coordinate; z-score for a normal distribution
ζ	Vorticity $\boldsymbol{\zeta} = \nabla \times \mathbf{u}$, where $\mathbf{u} = (u, v, w)$ is velocity vector
$\zeta^{(z)}$	Vertical vorticity $\zeta^{(z)} = v_x - u_y$
ζ_a	Absolute vorticity $\zeta_a = f + \zeta^{(z)}$
ζ_p	Potential vorticity $\zeta_p = \zeta_a/H$

1 Dimensional Analysis for Climate Science

"There is something fascinating about science. One gets such wholesale returns of conjecture out of such a trifling investment of fact." That perceptive comment was written in *Life on the Mississippi* (1883) by Samuel Clemens (1835–1910), known by his pen name, Mark Twain, who has been called "the father of American Literature," and who wrote *The Adventures of Tom Sawyer* and *The Adventures of Huckleberry Finn*.

In a similar vein, but more relevant to a mathematical textbook like the one you are reading now, is a remark by the physicist Eugene Wigner. In his paper, "The Unreasonable Effectiveness of Mathematics in the Natural Sciences," Wigner (1960) wrote, "it is important to point out that the mathematical formulation of the physicist's often crude experience leads in an uncanny number of cases to an amazingly accurate description of a large class of phenomena."

Dimensional analysis, the subject of this introductory chapter, illustrates this phenomenon well. Consider a simple pendulum, consisting of the Earth's gravity g acting on a mass m suspended on a string of length l. How does the period of the pendulum depend on m, l, and g? In this chapter, we show how to solve this problem using only dimensional analysis, without any physics or calculus.

"The principal use of dimensional analysis is to deduce from a study of the dimensions of the variables in any physical system certain limitations on the form of any possible relationship between those variables. The method is of great generality and mathematical simplicity." (Bridgman 1969).

1.1 Dimension and Units

In the physical sciences, measurements and parameters are often expressed as a number plus a dimensional unit. Examples are 10 meters, 30 seconds, 15 kilograms, and 100 degrees Celsius. These are examples of the dimensions of length, time, mass, and temperature, respectively. Length may be called the dimension of a line, and it is denoted by L. Length can be measured in SI Units, such as meters, or in Imperial units, such as feet. The abbreviation "SI" comes from the French term "Système International d'Unités," i.e., the International System of Units. The SI system is also known as the metric system. The commonly used SI length units include meters, centimeters, millimeters, and kilometers. Similarly, SI time units include seconds and microseconds, and SI mass units include grams and kilograms. The corresponding Imperial system units are feet, seconds, and pounds. The United States customary units developed from the British Imperial units system, and the two systems are similar but not identical. The SI system in its present form was established in 1960. It is now the most prevalent units system used in science and engineering around the world. The United States is the only major country that still prefers a form of Imperial units in everyday usage and in some aspects of engineering. However, most scientific publications in the US have adopted the SI or metric system. The United Kingdom had adopted the metric system in the 1960s.

Any given climate parameter has a dimension and units. The dimension is an intrinsic property of the parameter, such as wind speed with dimension $[LT^{-1}]$. This dimension is applicable to speeds of all kinds of variables, such as the speed of a hurricane wind, the speed of the California Current, and the speed of a river flow. The units, on the other hand, are a description of the dimension appropriate for given scales, such as using kilometers to describe the horizontal size of the Gulf Stream, but using meters to describe the amplitude of oceanic surface waves. Thus, a given dimension can have many different units. All the units can be converted to each other with fixed formulas, such as one kilometer = 1,000 meters.

The systematic use of units is very important. Misuse can have serious consequences. In 1999, a Mars Climate Orbiter spacecraft costing hundreds of millions of dollars was lost because of human error confusing the SI system with the Imperial system:
https://en.wikipedia.org/wiki/Mars_Climate_Orbiter

The units have cultural origins and are external properties. The dimension and the laws of nature are intrinsic properties and should be independent of units. Newton's second law of motion $F = ma$ is valid in both Imperial and metric systems. Thus, it is critical that a law of nature must be expressed in a single units system, not in a mixture of two systems.

1.2 Fundamental Dimensions: *LMT*θ*I*-class

The fundamental dimensions used in climate science are the five listed in Table 1.1 for length, time, mass, temperature, and electric current. The dimensions of most other physical quantities can be derived from the five. For example, speed is the displacement in a unit time and has the dimension LT^{-1}. Table 1.2 shows the dimensions of some of the most commonly used climate parameters.

Table 1.1 Fundamental dimensions: *LMT*Θ*I*-class

Notation	Meaning	Dimension	Units
$[l]$	Length	L	m
$[t]$	Time	T	sec or s
$[m]$	Mass	M	kg
$[\theta]$	Temperature	Θ	K
$[I]$	Electric current	I	amp or A

Table 1.2 Dimensions of derived physical quantities

	Meaning	Dimension	SI Units
$[v]$	Velocity	LT^{-1}	m s^{-1}
$[a]$	Acceleration	LT^{-2}	m s^{-2}
$[F]$	Force ($F = ma$)	MLT^{-2}	N = 1.0 kg m s^{-2}
$[\rho]$	Mass density	ML^{-3}	kg m^{-3}
$[p]$	Pressure (force per area)	$MLT^{-2}L^{-2} = ML^{-1}T^{-2}$	Pa = N m^{-2}
$[E]$	Energy or Work or Heat	ML^2T^{-2}	joule = 1.0 N m
$[P]$	Power (work per time)	ML^2T^{-3}	watt = 1.0 joule s^{-1}
$[J]$	Momentum (mass times velocity)	MLT^{-1}	kg m s^{-1}
$[S]$	Entropy (energy per K)	$ML^2T^{-2}\Theta^{-1}$	joule K^{-1}
$[Q]$	Electric charge	IT	coulomb: C = 1.0 A s
$[E]$	Electric field (force per coulomb)	$NC^{-1} = MLT^{-3}I^{-1}$	kg m/(A s^3)
$[B]$	Magnetic field	$N(IL)^{-1} = MT^{-2}I^{-1}$	kg/(A s^2)
$[\phi]$	Angle	1 (dimensionless)	radian

We normally use square brackets $[\cdot]$ to denote dimension. If v is velocity, then $[v] = LT^{-1}$ as shown in Table 1.2.

Example 1.1 Dimensional analysis of the geopotential energy: The geopotential energy E of an air parcel of mass m at height h is defined as

$$E = mgh, \tag{1.1}$$

where g is the acceleration of gravity.

Following Table 1.2, we have

$$[E] = [m][g][h] = M(LT^{-2})L = ML^2T^{-2}. \tag{1.2}$$

The last expression can be further organized into $M(LT^{-1})^2$, hence mass times speed squared, which has a clear physical meaning: kinetic energy, defined as $(1/2)mv^2$. This simple analysis links the potential energy and kinetic energy, and helps one to think about the conversion of potential energy into kinetic energy, such as a wind caused by geopotential differences or air pressure differences.

Hence, a simple algebraic manipulation of a dimension formula may lead to a new physical insight, such as the realization of the relationship between potential energy and kinetic energy. Mathematical manipulations such as this are regarded as being in the category of dimensional analysis, which may lead to discoveries of physical laws and relationships.

Example 1.2 Dimensional analysis of π: The constant π is the ratio of circumference to diameter $\pi = C/D$ for any circle. The dimensional form of this equation is

$$[\pi] = [C]/[D] = L/L = 1. \tag{1.3}$$

Thus, π is dimensionless, or non-dimensional. The angle of π radians is equivalent to that of $180°$; so the angle degrees are also dimensionless. Any angle can be measured by radians or degrees, and is thus dimensionless. The trigonometric functions, logarithmic functions, and exponential functions can only be applied to dimensionless quantities, such as 0.5π, or 2.3, or 1.0. These are pure numbers, but can also be regarded as radians, measuring an angle. However, radian is not a dimension. Of course, the ranges of trigonometric functions and logarithmic functions are also dimensionless. In the expressions $y = \sin x, y = \ln x, y = \exp(x)$, both x and y are dimensionless. In $\sin(\pi/6) = 0.5$, $\pi/6$ radians is considered dimensionless since radian is dimensionless, and 0.5 is also dimensionless.

Because of this common dimensionless feature of trigonometric functions, logarithmic functions, and exponential functions, one may think that these functions should be related. Yes, they are. The exponential function and trigonometric functions are related by

$$e^{i\theta} = \cos\theta + i\sin\theta, \tag{1.4}$$

where $i = \sqrt{-1}$ is the imaginary unit. This is usually called Euler's formula (Leonhard Euler, 1707–1783, Swiss mathematician), illustrated by Fig. 1.1. Physics Nobel laureate Richard Feynman called Euler's equation "the most remarkable formula in mathematics." This equation can help in expressing and understanding numerous physical properties, such as the wave function in quantum mechanics, as well as in analyzing oceanic waves and alternating electric current.

The length of the arc of an interior angle θ and radius r is

$$s = \theta r. \tag{1.5}$$

The dimension of the above equation is

$$[s] = [\theta][r], \tag{1.6}$$

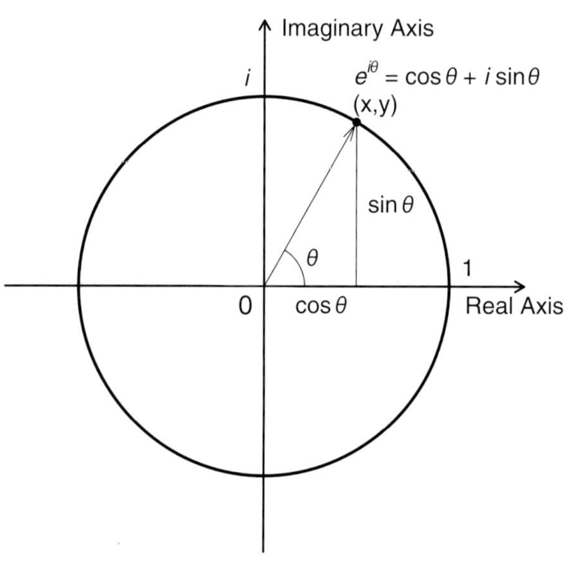

Fig. 1.1 Euler's formula: $e^{i\theta} = \cos\theta + i\sin\theta$.

which is

$$L = [\theta]L. \tag{1.7}$$

Hence, $[\theta] = 1$ is dimensionless. This is another way to illustrate that an angle is dimensionless, although we customarily use radian or degree as units to measure angles.

The logarithmic function is the inverse function of the exponential function. Both its input and output are dimensionless. Other trigonometric functions can be derived from the cosine and sine functions and hence also have dimensionless inputs and outputs.

Example 1.3 Dimensional analysis of pressure: The dimension of pressure is $[p] = ML^{-1}T^{-2}$ from Table 1.2 and can be re-written as $M(LT^{-1})^2L^{-3}$. Because $M(LT^{-1})^2$ is kinetic energy, $M(LT^{-1})^2L^{-3}$ is thus the kinetic energy per unit volume. This is the thermodynamic definition of pressure. The pressure exerted by an ideal gas on its container wall is measured by the magnitude of the kinetic energy contained in a unit volume of the gas.

Thus, the rearrangement of different dimensions can result in revealing very interesting and profound laws of nature. Dimensional analysis provides a powerful tool for discovery. We may say that dimensional analysis is a shortcut for discovery and can lead us to become aware of many useful mathematical formulas for climate science, and for understanding nature in general.

The pressure dimension can be written in another form

$$[p] = ML^{-1}T^{-2} = (MLT^{-2})L^{-2}. \tag{1.8}$$

This provides another physical meaning. Since LT^{-2} is acceleration, MLT^{-2} is thus ma the force, following Newton's second law of motion $F = ma$. Thus, $(MLT^{-2})L^{-2}$ means the force per unit area, which is exactly the physical definition of pressure.

According to this definition of pressure, the atmospheric pressure at sea level can be defined as the total gravitational force per unit area acting on the surface, whether the surface is ocean or land at sea level. The pressure can be expressed as an integral

$$p = \frac{\int_0^\infty g\rho(z,\phi,\theta,t)dzA}{A}, \tag{1.9}$$

where A is the horizontal cross section area of the air column from sea level to infinite height, g is the gravitational acceleration, ϕ is latitude, θ is longitude, t is time, and the atmospheric density ρ is a function of z,ϕ,θ,t. The definition and calculation of an integral, which is a part of calculus, are described in Appendix D.

The cancellation of A in the above equation yields

$$p(\phi,\theta,t) = \int_0^\infty g\rho(z,\phi,\theta,t)dz. \tag{1.10}$$

This is precisely the definition of the sea level pressure (SLP), which is a very important parameter for weather forecasting and is a function of latitude, longitude, and time.

1.3 Dimensional Analysis for a Simple Pendulum

Simple pendulum clocks are based on the concept of the oscillation of a simple pendulum. For such a clock, its most important function is to record time measured as the number of oscillation periods. The period of an idealized pendulum is given by

$$\tau = 2\pi\sqrt{l/g} \tag{1.11}$$

where l is the length of the rod or string and g is the gravitational constant. The rod or string is assumed to be massless for this formula. The formula can be derived using many methods, including an approach involving a second-order ordinary differential equation. Here we provide a simple approach via dimensional analysis.

A pendulum involves three quantities: the mass of the pendulum, the length of the massless rod or string, and the Earth's gravitational acceleration (Fig. 1.2). We may assume that the pendulum's oscillation period depends on these three quantities.

To understand the gravity dependence, one can think of an extreme environment: outer space with zero gravity, where the pendulum will not oscillate because of the absence of gravity. The period is thus infinity. Similarly one may reasonably conclude that the same pendulum oscillates more slowly on the Moon than on the Earth, because the Moon has a smaller gravitational acceleration.

Thus, we can assume that the pendulum's period depends on mass, length, and gravity, written in the following form:

$$\tau = \alpha m^a l^b g^c, \tag{1.12}$$

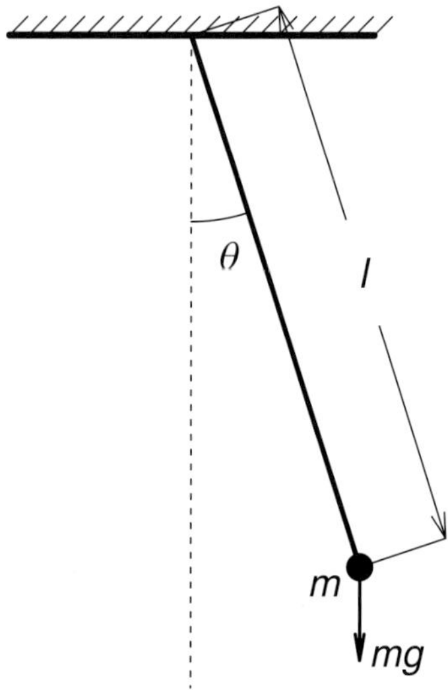

Fig. 1.2 A simple pendulum of mass m and length l.

where the exponents a, b, c are to be determined by reducing this equation to an equation of the five fundamental dimensions L, T, M, Θ, and I, and α is a dimensionless constant to be determined by an experiment or by another mathematical constraint.

Taking dimensions for both sides of the equation yields

$$[\tau] = [\alpha][m]^a[l]^b[g]^c. \tag{1.13}$$

Because α is dimensionless, $[\alpha] = 1$. Tables 1.1 and 1.2 provide the dimensions of τ, m, l, and g:

$$[\tau] = T, \ [m] = M, \ [l] = L, \ [g] = LT^{-2}.$$

The equation of dimensions (1.13) becomes

$$T = M^a L^b (LT^{-2})^c \tag{1.14}$$

or

$$M^0 L^0 T^1 = M^a L^{b+c} T^{-2c}. \tag{1.15}$$

The exponents of each fundamental dimension on both sides of the equation must be equal, which leads to

$$a = 0, \tag{1.16}$$

$$b + c = 0, \tag{1.17}$$

$$-2c = 1. \tag{1.18}$$

These three equations have three unknowns a, b, and c and the following solution

$$a = 0, \ c = -1/2, \ b = 1/2. \tag{1.19}$$

The period is thus proportional to $m^0 l^{1/2} g^{-1/2}$, i.e.,

$$\tau = \alpha m^0 l^{1/2} g^{-1/2} = \alpha\sqrt{l/g}. \tag{1.20}$$

An experiment was conducted in a classroom with a string length equal to 0.88 meters. Two periods were recorded by several groups of students using smartphone stopwatches. An average value was found to be equal to 3.75 seconds. Substitution of these values into the above equation yields that

$$3.75 = 2 \times \alpha\sqrt{0.88/9.8} = 0.60\alpha, \tag{1.21}$$

which leads to

$$\alpha = 3.75/0.60 = 6.25 = 1.99\pi \approx 2\pi. \tag{1.22}$$

This is an easy experiment to carry out. Since the motion is relatively slow when the string is long enough, using a smartphone stopwatch, it is fairly easy to record the period. One can improve the experimental results by making several experiments and using the average period as the final value for α.

A pendulum clock's rod is often made of metal, whose thermal expansion and contraction can change the rod length l. This factor alone can slow down a steel-rod clock in summer by half a minute each day when temperature increases by 10 degrees Celsius. To reduce a pendulum clock's error due to thermal expansion, the British watchmaker George Graham made an invention in 1721 to use mercury to cancel the effect of the rod's expansion, since mercury has a five times larger thermal expansion coefficient.

1.4 Dimensional Analysis for the State Equation of Air

A basic thermodynamical relationship is the equation of state for an ideal gas that relates pressure p, temperature T, volume V, and the amount of gas n [moles]:

$$pV = nR^*T, \tag{1.23}$$

where R^* is called the universal gas constant. The dimension of R^* is

$$\begin{aligned}[R^*] &= [p][V]/([n][T]) \\ &= (MT^{-2}L^{-1})(L^3/[n])(\Theta^{-1}) \\ &= ML^2T^{-2}\Theta^{-1}[n]^{-1} \\ &= M(L/T)^2\Theta^{-1}[n]^{-1}.\end{aligned} \tag{1.24}$$

The factor $M(L/T)^2$ is mass times velocity squared and is hence energy. We can use joule as the unit of energy, kelvin as the unit of temperature, and moles as the unit of gas amount. Thus, the universal gas constant is energy per degree of temperature per mole of mass. The universal gas constant R^* is equal to

$$R^* = 8.314\ [\mathrm{J\ K^{-1} mol^{-1}}]. \tag{1.25}$$

For the same one mole of molecules, different gases have different weights. Thus, for a specific gas, the mass of one mole of molecules of the gas is called molecular weight (or molecular mass) and denoted by M. A unit of the molecular weight may be [g/mol]. The molecular weight of dry air is approximately $M = 28.97$ [g/mol]. Thus, the specific gas constant

$$R = R^*/M. \tag{1.26}$$

For the dry air, the specific gas constant can be computed as follows

$$R = \frac{R^*}{M} = \frac{8.314[\mathrm{J\ K^{-1}\ mol^{-1}}]}{28.97[\mathrm{g\ mol^{-1}}]} = 0.287[\mathrm{J\ K^{-1} g^{-1}}], \tag{1.27}$$

or

$$R = 287[\mathrm{J\ K^{-1}(kg)^{-1}}]. \tag{1.28}$$

1.5 Dimensional Analysis of Heat Diffusion

A point source of heat with Q joules of heat is initially placed at the midpoint of a one-dimensional heat-conducting rod. The midpoint is denoted by $x = 0$. The heat will diffuse as time increases, and the temperature $T(x,t)$ at the location x and time t is given by the following formula:

$$T = \frac{Q}{\rho C \sqrt{2\pi D t}} \exp\left(-\frac{x^2}{2Dt}\right). \tag{1.29}$$

The parameters in this expression are described as follows:

- Q is the heat energy in joules or other units with dimension ML^2T^{-2},
- ρ is the linear density (not the density in a three-dimensional volume) of the one-dimensional rod with dimension ML^{-1},
- C is the specific heat which is the energy per unit mass per temperature degree (i.e., the energy needed to increase the temperature of a unit mass by one degree) with unit joule / (g K) and dimension $ML^2T^{-2}M^{-1}\Theta^{-1}$, and
- D is a constant called thermal diffusivity with dimension L^2T^{-1}.

The dimension of Dt is

$$[Dt] = [D][t] = (L^2T^{-1})T = L^2,$$

which is the same as the dimension of x^2. Thus, the term

$$\frac{x^2}{2Dt}$$

inside the exponential function is dimensionless, which it clearly should be, since exponential, trigonometric, and logarithmic functions can act only on dimensionless quantities.

The thermal diffusivity D measures the diffusion rate, a property of the material in which the heat diffuses. A large value of D means fast heat diffusion. For example, heat diffuses faster in air with $D = 22.39$ [mm^2 sec^{-1}] (at temperature $T = 25$°C and at the usual sea level pressure) than in water with $D = 0.14$ [mm^2 sec^{-1}] (under the same temperature and pressure conditions).

The dimension of the right-hand side is

$$\begin{aligned} &\left[\frac{Q}{\rho C\sqrt{2\pi D t}}\right] \\ &= \frac{[Q]}{[\rho][C]\sqrt{[Dt]}} \\ &= \frac{ML^2T^{-2}}{(ML^{-1})(ML^2T^{-2}M^{-1}\Theta^{-1})\sqrt{L^2}} \\ &= \Theta = [T] \text{ (the dimension of temperature).} \end{aligned} \tag{1.30}$$

This is the same as the dimension of the left-hand side T.

1.6 Dimensional Analysis of Rossby Waves and Kelvin Waves

This section will describe the following parameters from the point of view of dimensional analysis: Rossby number, Rossby parameter, the strength of a Rossby wave, Rossby radius of deformation, phase and energy speeds of a Kelvin wave, wave frequency, and wave number.

1.6.1 Parameters for Rossby Waves

Rossby waves, sometimes called planetary waves, occur in both the atmosphere and the ocean. Their properties depend on the rotation of the Earth. Atmospheric Rossby waves are meanders in high-altitude winds. They can have a strong influence on weather. These waves are associated with atmospheric pressure systems and with the jet stream, as shown in Fig. 1.3. Oceanic Rossby waves move along the thermocline, which is the boundary region between the warm upper ocean and the cold deeper ocean.

The speed U of an idealized Rossby wave depends on the wavelength L, and the Coriolis force described by the Coriolis frequency f:

$$f = 2\Omega \sin\phi \tag{1.31}$$

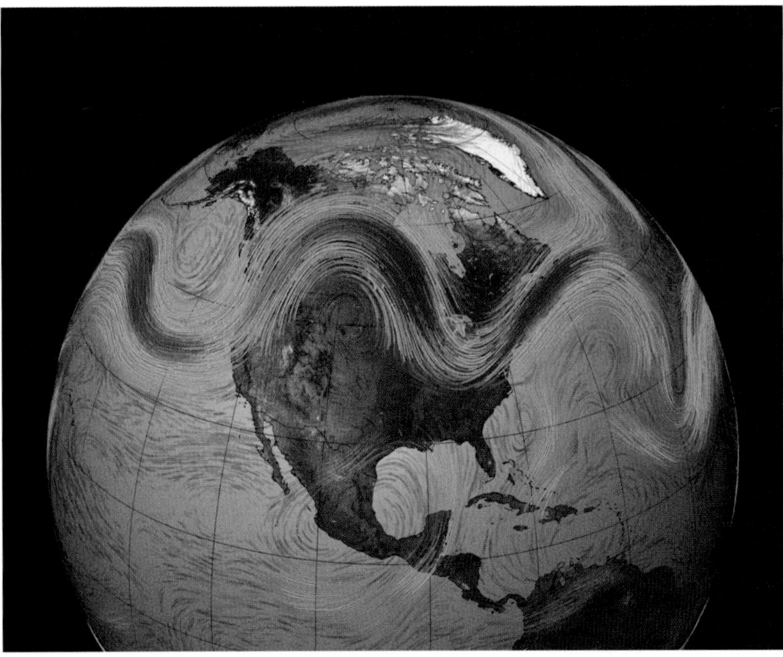

Fig. 1.3 Rossby wave and jet stream over a mid-latitude zone of the northern hemisphere. The colors represent the wind speed: blue and green for slow to red and dark for fast. *Credit: NASA Goddard Space Flight Center*, URL: https://nasaviz.gsfc.nasa.gov/10902

where $\Omega = 7.3 \times 10^{-5}$ [radian sec^{-1}] is the angular speed of the rotation of the Earth, and ϕ is the latitude of the location. The Coriolis force is zero at the equator and is largest at the poles. The wave speed U can help predict when a Rossby wave can arrive. The wavelength L measures the size of a wave. The Coriolis frequency measures the magnitude of the Earth's rotational effect.

Following the general rule of dimensional analysis, we may arrange products and quotients from these three quantities U, L, and f to form a non-dimensional quantity, which may turn out to be important in explaining physical phenomena. The dimensions of the three quantities are

$$[U] = LT^{-1}, [L] = L, \text{and } [f] = T^{-1}. \tag{1.32}$$

Placing them together, we form a dimensionless index

$$[\text{Ro}] = [U]^a[L]^b[f]^c = (LT^{-1})^a L^b (T^{-1})^c = L^{a+b}T^{-a-c}. \tag{1.33}$$

To make Ro dimensionless, we have

$$a + b = 0, \tag{1.34}$$

$$-a - c = 0. \tag{1.35}$$

These two equations yield

$$b = -a, \ c = -a. \tag{1.36}$$

The simplest non-trivial solution is $a = 1, b = -1$, and $c = -1$, which leads to

$$\text{Ro} = U^1 L^{-1} f^{-1}, \quad \text{or} \quad \text{Ro} = \frac{U}{Lf}. \tag{1.37}$$

This is the famous Rossby number, which is the unique dimensionless combination of the three key fundamental quantities: U, L, and f. The Rossby number (Ro) is a geophysical fluid dynamics parameter that measures the influence of the Coriolis effect. It was named for Carl-Gustav Arvid Rossby (1898–1957).

Physically, the Rossby number can be regarded as the ratio of the inertial force per unit mass $(\mathbf{u} \cdot \nabla)\mathbf{u} \sim U^2/L$ to the Coriolis force per unit mass $\mathbf{\Omega} \times \mathbf{u} \sim U\Omega$:

$$\frac{(\mathbf{u} \cdot \nabla)\mathbf{u}}{\mathbf{\Omega} \times \mathbf{u}} \sim \frac{U^2/L}{U\Omega} \tag{1.38}$$

where $\mathbf{u}$ is the flow velocity of the unit mass, ∇ denotes spatial differentiation, $\cdot$ denotes the dot product of two vectors, and $\times$ denotes the cross product of two vectors (see Appendices A and B for the definitions of these mathematical notations).

The above formula can be simplified to

$$\text{Ro} = \frac{U}{Lf}. \tag{1.39}$$

A large Rossby number means that a system is dominated by inertial forces, and the Coriolis force is relatively small. For example, tornadoes have a high air flow speed and a large speed gradient, and hence have a large Rossby number. The Coriolis force due to Earth's rotation is relatively less important than the centrifugal force in the tornado's fast rotation. A tornado eye has a low central pressure. The pressure has a large gradient from the center to the tornado's outer boundary. The pressure gradient is mainly balanced by centrifugal forces, which is called cyclostrophic balance. The Coriolis force in a tornado is negligible.

A small Rossby number indicates relatively small inertial forces. The Coriolis force then plays a critical role in determining the flow. The geophysical force balance for slow large-scale motions of the atmosphere is mainly between the Coriolis force and the pressure gradient, a situation which is called geostrophic balance. This balance makes the air flow parallel to isobars that are the lines of equal pressure.

Here, "strophic" is from the Greek, meaning "turning." "Strophe" originally denoted a movement from right to left made by a Greek chorus. "Geo" is also Greek and means Earth. "Geostrophic" in climate science means atmospheric or oceanic motion characterized by a balance between the pressure gradient force and the Coriolis force due to the Earth's rotation.

Another important geophysical parameter is called the Rossby (or beta) parameter which measures the gradient of the Coriolis force in the meridional (north–south) direction:

$$\beta = \frac{\partial f}{\partial y} = \frac{1}{R}\frac{d}{d\phi}(2\Omega \ \sin\phi) = \frac{2\Omega\cos\phi}{R} \tag{1.40}$$

where Ω is the angular speed of the Earth's rotation [rad/sec], and R is the radius of the Earth. Here, $dy = d(R\phi) = Rd\phi$ designates a small distance increment in the meridional direction. The dimension of the Rossby parameter is $1/(TL)$.

The Rossby or beta parameter describes the variation of the Coriolis force with latitude (hence the latitudinal derivative) and does not depend on atmospheric motions, while the Rossby number measures the strength of inertial forces relative to the Coriolis force and thus strongly depends on atmospheric motions.

For a wave, its amplitude A is important. The following expression

$$\mathrm{So} = \frac{UA}{Lf} \tag{1.41}$$

has the length dimension. This parameter is a measure of the strength of the Rossby wave with respect to the wavelength.

Another measure using the depth D of an atmospheric or oceanic Rossby wave is

$$L_R = \frac{\sqrt{gD}}{f} = \frac{\sqrt{gD}}{2\Omega\sin\phi}. \tag{1.42}$$

This parameter also has the length dimension and measures the length scale at which rotational effects become as important as buoyancy or gravity wave effects. The Coriolis frequency is approximately equal to $f = 1\times 10^{-4}$ $[\mathrm{s}^{-1}]$ at $\phi = 45°$ latitude. If $D = 4{,}000$ [m] from land surface, then $L_R = 2{,}000$ [km] is large. For a very shallow Rossby wave $D = 40$ [m], then $L_R = 200$ [km] is small. L_R is called the Rossby radius of deformation.

1.6.2 Non-Dispersive Properties of Kelvin Waves

A Kelvin wave is an important type of wave in the atmosphere or ocean that is non-dispersive, i.e., the phase speed of the wave crests is equal to the speed of wave energy. Physically, this equality means that the wave's shape is retained when it moves, similar to a solitary wave. Mathematically, this equality means that wave frequency ω [1/time] is proportional to wave number k [1/length]

$$\omega = ak + b. \tag{1.43}$$

The dimension of this formula is

$$T^{-1} = [a]L^{-1}. \tag{1.44}$$

Thus, $[a] = L/T$ has the dimension of velocity, which is both phase speed and energy speed in the case of Kelvin waves.

In a stratified ocean of mean depth H, free waves propagate along coastal boundaries (and hence become trapped in the vicinity of the coast itself) in the form of internal Kelvin waves with a wavelength about 30 km. These ocean water waves are called coastal Kelvin waves, and have propagation speeds of approximately 2 m/s (173 km/day). An example is the Kelvin waves of sea surface height in the Bay of Bengal along the coast of India.

In the ocean, a coastal Kelvin wave will have a phase speed that is equal to the speed of shallow-water gravity waves when the Coriolis force is negligible:

$$c = \sqrt{gH}, \tag{1.45}$$

where H is the mean depth of the coastal region and g is the Earth's gravitational acceleration.

1.7 Estimating the Shock Wave Radius of a Nuclear Explosion by Dimensional Analysis

The instantaneous energy release from a nuclear explosion causes a shock wave in the air, with an inside pressure that is thousands of times greater than the outside pressure. This pressure difference can knock down trees and structures, and destroy many objects. Let us assume the shock wave to be spherical with radius $R(t)$, which depends on time t after the explosion. Given the energy E of the nuclear explosion, we wish to calculate the shock wave radius function $R(t)$, and hence predict the shock wave's arrival time at any location.

A shock wave will occur in the atmosphere due to the supersonic compression of the air from one side, so that the air mass from that side accumulates and cannot escape. Thus, the pressure builds up and a large pressure difference with the other side develops, and hence forms a shock. Two critical elements here are the supersonic speed and the compressibility property of air. Thus, the shock wave radius should be related to the density ρ of compressible air, the total energy of the nuclear explosion E, and time t. Because air is light on the scale of a nuclear explosion, gravity effects can be assumed to be negligible. Thus, we assume the following

$$R = \alpha E^a \rho^b t^c. \tag{1.46}$$

The dimension of the above equation is

$$[R] = [\alpha][E]^a[\rho]^b[t]^c, \tag{1.47}$$

which leads to

$$L = 1 \times (ML^2T^{-2})^a(ML^{-3})^bT^c = M^{a+b}L^{2a-3b}T^{-2a+c}. \tag{1.48}$$

The exponents of both sides of this equation should be equal for each fundamental dimension M, L, and T:

$$a + b = 0, \tag{1.49}$$

$$2a - 3b = 1, \tag{1.50}$$

$$-2a + c = 0. \tag{1.51}$$

These three equations have a unique solution

$$a = 1/5, b = -1/5, c = 2/5. \tag{1.52}$$

Therefore,

$$R = \alpha E^{1/5}\rho^{-1/5}t^{2/5}. \tag{1.53}$$

or

$$R = \alpha\left(\frac{Et^2}{\rho}\right)^{1/5}. \tag{1.54}$$

This makes Et^2 a very special term, which is the fifth power of density times length according to the principle of dimensional equality, where the density ρ is the density of the air before the explosion.

Another expression of the shock radius is

$$R = \alpha\left(\frac{E}{\rho}\right)^{1/5} t^{2/5}. \tag{1.55}$$

The log-plot of this R–t relationship is a straight line with slope 2/5:

$$\ln R = \ln\alpha + \ln\left(\frac{E}{\rho}\right)^{1/5} + \frac{2}{5}\ln t. \tag{1.56}$$

Any of the above three formulas can be used to predict the position $R(t)$ of the shock wave for a given time t, if α is known. Yet, it is not easy to evaluate this α by an experiment since such an experiment is too destructive. An alternative is to derive it from mathematical models. The fluid dynamicist G. I. Taylor (1886–1975) used a mathematical model and estimated that $\alpha \approx 1.0$.

Still another way of writing the above equation is

$$E = \frac{R^5 \rho}{t^2}. \tag{1.57}$$

This allows one to estimate the power of a nuclear bomb from news reports of the shock arrival time t at a given location R.

If a nuclear bomb test is made underground, a seismograph can measure R and t. With the known density of the Earth's crust, one can then estimate the bomb's power. There are about 500 seismograph stations distributed around the world to detect ground-shaking incidents, including earthquakes and nuclear tests.

One can use a similar model to estimate the shock waves caused by supernova explosions. Interested readers are referred to the book *Exploring the X-ray Universe* by Seward and Charles (2010) listed in References and Further Readings.

1.8 Chapter Summary

This chapter on the basic method of dimensional analysis has shown how to derive the dimension of a climate parameter from the five fundamental dimensions L, T, M, Θ, and I. One may think of the five fundamental dimensions as elements. The key points are listed below:

(i) Dimensional analysis can be used to discover useful formulas or laws of physics, such as the formulas describing a free-falling object and the period of a simple pendulum. The approach is to express a climate parameter x in terms of the product of the five elements, each raised to an unknown power, times a dimensionless constant α

$$x = \alpha L^a T^b M^c \Theta^d I^e. \tag{1.58}$$

The values of the unknown exponents a, b, c, d, and e can be determined in three steps:

- Take the dimension for both sides of this equation,
- Compare the exponents for each fundamental dimension L, T, M, Θ, or I to obtain a set of linear equations for a, b, c, d, and e, and
- Solve the linear equations for a, b, c, d, and e.

The dimensionless constant α can be determined from experimental data, or by invoking another constraint, such as the conservation of energy.

(ii) A rearrangement of the dimension formula of a quantity may lead to a new interpretation and hence may show the intrinsic connections among different forms of a physical phenomenon. For example, the dimension formula $M(LT^{-1})^2$ for energy means mass times velocity squared and hence the kinetic energy $(1/2)mv^2$; a rearrangement of this formula $M(LT^{-2})L$ means mass times acceleration times distance, which is the potential energy mgh. The dimension formulas in Table 1.2 show several other examples of the connections among different forms of a physical phenomenon, such as the electric field being connected to the magnetic field times a velocity which illustrates the principle of electric power generation by a conductor moving inside a magnetic field.

(iii) An angle is dimensionless and may have either radian or degree as its unit.

(iv) Some climate parameters involve exponential, logarithmic, or trigonometric functions. The input and output of these functions must be dimensionless. For example, in $y = \sin x$, both x and y must be dimensionless. Another example is the use of the sine function to model the periodic seasonal cycle of a climate parameter, such as temperature,

$$T(t) = A\sin(\omega t + \gamma), \tag{1.59}$$

then ωt must be dimensionless, which makes $[\omega] = T^{-1}$, and γ must also be dimensionless, so it can be regarded as an angle and is called the phase in climate oscillations.

(v) A dimensionless quantity composed of relevant important control parameters, such as the Rossby number, can help provide insight into a climate phenomenon.

(vi) Details of several dimensional analysis examples relevant to climate sciences have been presented, including the period of a simple pendulum, equation of state for air, heat diffusion, Rossby waves, and atmospheric shock waves arising from a nuclear explosion.

The exercises illustrate several applications of the above methods and ideas. Solving these problems can help one to appreciate that the dimensional analysis "method is of great generality and mathematical simplicity," as we quoted from Bridgman (1969) at the beginning of this chapter.

The materials on dimensional analysis included in this chapter are sufficient for most climate science studies and research. However, many advanced topics of dimensional analysis are not covered, such as why it is justified to express a climate parameter x in terms of the product of the five dimensional elements, each raised to an unknown power. Interested readers are referred to the Further Reading list at the end of this chapter.

References and Further Readings

[1] Barenblatt, G. I., 1987: *Dimensional Analysis*. Gordon and Breach Science Publishers, New York.

> Grigory Isaakovich Barenblatt (born 1927) is a Russian mathematician. One of his many awards was given "for seminal contributions to nearly every area of solid and fluid mechanics, including fracture mechanics, turbulence, stratified flows, flames, flow in porous media, and the theory and application of intermediate asymptotics."

[2] Bridgman, P. W., 1969: "Dimensional Analysis" in *Encyclopaedia Britannica*, Vol. 7, pp. 439–449: Encyclopaedia Britannica, Chicago.

Percy William Bridgman (1882–1961) was both an outstanding physicist and a penetrating thinker and prolific writer on many aspects of the philosophy of science. He spent virtually his entire career at Harvard University. Bridgman won the 1946 Nobel Prize in Physics for research on high-pressure physics. A practical man of many talents, he was also good at plumbing, carpentry, gardening, music, and chess. He committed suicide because he suffered from advanced cancer.

[3] Seward, F. D., and P. A. Charles, 2010: *Exploring the X-ray Universe*, 2nd Edition, Cambridge University Press, New York.

This book is a very readable overview of the accomplishments of X-ray astronomy.

[4] Twain, M., 1883: *Life on the Mississippi*. URL: www.gutenberg.org/files/245/245-h/245-h.htm

We open Chapter 1 with a quote from this memoir by the great American writer Mark Twain. It tells of his travels on the Mississippi River. It is a delightful book, although except for that quote, it may be quite irrelevant to science and mathematics.

[5] Wigner, E.,1960: The unreasonable effectiveness of mathematics in the natural sciences. *Communications in Pure and Applied Mathematics*, 13, 1–14.

Eugene Wigner (1902–1995) was a brilliant Hungarian-American mathematician and theoretical physicist. He was a key participant in the Manhattan Project, which developed the first atomic bomb during World War II. Wigner shared the 1963 Nobel Prize in Physics for his fundamental discoveries. His paper on "The Unreasonable Effectiveness of Mathematics in the Natural Sciences" is easily available online and is well worth reading. He concludes it by saying, "The miracle of the appropriateness of the language of mathematics for the formulation of the laws of physics is a wonderful gift which we neither understand nor deserve. We should be grateful for it and hope that it will remain valid in future research and that it will extend, for better or for worse, to our pleasure, even though perhaps also to our bafflement, to wide branches of learning."

Exercises

Exercise 1.10 requires the use of a computer.

1.1 An object of mass m falls freely under the action of only the gravitational force. The air friction is ignored. Let v be the velocity of the object at time t. Assume

$$v = \alpha m^a g^b t^c. \tag{1.60}$$

Use the dimensional analysis method to determine a, b, and c.

1.2 For the same free-fall problem above, let h be the distance the object has traveled in the free fall process. Assume

$$h = \beta m^a g^b t^c. \tag{1.61}$$

Use the dimensional analysis method to determine a, b, and c.

1.3 Conduct experiments to determine the undetermined constant β in the above problem

(a) Perform the experiments on free-fall using a coin or any heavy metal object or a stone to determine the dimensionless constant for distance. You may find that this is a difficult experiment because it happens very quickly, so it is not easy to record the time.

(b) Change the free-fall experiment to a free-roll experiment as Galileo Galilei (1564–1642) did. Place a ball on an inclined plane and let it roll down by gravity. The gravity along the plane is now reduced to $g \sin\phi$ where ϕ is the angle between the plane and a horizontal surface (ideally, the horizontal tangent plane is perpendicular to the Earth's radius). The experiment is now easier to carry out, because it is easier to record the elapsed time. However, the inclined angle should not be too small, to avoid making the friction force non-negligible. As is often the case, one can achieve better accuracy when repeating the experiment many times and using the average result.

1.4 The constant β in Exercise 1.2 can also be determined by a physical constraint formula without experiments. Given that velocity is the derivative of the distance with respect to time

$$v = \frac{dh}{dt}, \tag{1.62}$$

use the law of energy conservation at the time zero when the object begins to fall and at the time T when the object has fallen a vertical distance H to the ground

$$mgH = \frac{1}{2}mv^2 \tag{1.63}$$

to determine the value for β. This approach requires the calculation of a derivative.

1.5 Sine and cosine functions are often used to model oscillations in climate science. Let us model the oscillatory motion of the simple pendulum shown in Fig. 1.2 by the following sine function

$$\theta = A\sin(Bt + C), \tag{1.64}$$

where A, B, and C are the constants to be determined. This model makes the pendulum's angle θ have the periodic dependence on time t. Suppose that we release the pendulum at time zero with the pendulum mass at its highest position, i.e., the maximum angle θ_M to the right. Then, at $t = 0$,

$$\theta_M = A\sin C. \tag{1.65}$$

The maximum value of $\sin C$ is one if $C = \pi/2$, which implies $A = \theta_M$. So, the mathematical model (1.64) becomes

$$\theta = \theta_M \sin(Bt + \pi/2). \tag{1.66}$$

The question for this exercise is to determine B up to a constant using the dimensional analysis method.

Hint: The relevant physical parameters for this pendulum are m, l, and g. Let

$$B = \alpha m^a l^b g^c. \tag{1.67}$$

Taking the dimension for both sides of this equation yields

$$[B] = [\alpha][m]^a[l]^b[g]^c. \tag{1.68}$$

The term Bt is inside a trigonometric function and must be dimensionless. Thus, $[B] = T^{-1}$. From here, the dimensional analysis steps can lead to values of a, b, and c, but this still leaves the dimensionless constant α undetermined.

1.6 Conduct experiments to record the period data of a simple pendulum oscillation and use the average data to determine the dimensionless constant α in the above exercise.

Hint: Since we have assumed that $\theta = \theta_M$ at $t = 0$, the function $\sin(Bt + \pi/2)$ must be decreasing when t increases at first from zero, in order for the pendulum angle θ to decrease. This is possible only when B is negative, which in turn implies $\alpha < 0$. The period of the function $\sin(Bt + \pi/2)$ is $-B\tau = 2\pi$ since the period of $\sin t$ is 2π. Thus,

$$-\alpha m^a l^b g^c \tau = 2\pi, \tag{1.69}$$

or

$$\alpha = -\frac{2\pi}{\tau m^a l^b g^c}. \tag{1.70}$$

1.7 When a derivative is used, the dimensionless constant α in Exercise 1.5 can be determined by the constraint of energy conservation without using the experimental data.

Hint: The law of energy conservation used here is that the potential energy at $\theta = \theta_M$ is all converted into the kinetic energy at $\theta = 0$. The velocity of the pendulum mass is

$$v = l\frac{d\theta}{dt}. \tag{1.71}$$

The pendulum mass has dropped a vertical distance $l(1 - \cos\theta_M)$ when it swings from at $\theta = \theta_M$ to $\theta = 0$. A hidden approximation is required here: θ_M must be fairly small so that $\sin(\theta_M/2) \approx \theta_M/2$ when the unit of θ_M is radian. This approximation is applied to

$$1 - \cos(\theta_M) = 2\sin^2(\theta_M/2) \approx \frac{\theta_M^2}{2}. \tag{1.72}$$

1.8 The law of universal gravitation states that every object in the universe attracts every other object with a force directly along the line connecting the centers of the two objects. The force is given by the following formula

$$F_g = G\frac{m_1 m_2}{r^2}, \tag{1.73}$$

where m_1 and m_2 are the masses of the two objects, r is the distance between the two centers, and G is the universal gravitational constant. Find the dimension of G.

1.9 From Newton's second law of motion $F = ma$, find the dimension of force from the dimensions of mass and acceleration.

1.10 Design an experiment to "discover" Newton's second law of motion: $F = ma$, when assuming the mass does not change and measuring the data of acceleration and force. Make a scatter plot on the (F,a)-plane and plot a regression line for the scatter plot. *Hint: Search the Internet to find instruments to measure force and acceleration. Force is the cause of an acceleration. However, it may be difficult to actually do the experiment using the true instruments. You can design your experiment based on the instrument specifications, and describe your procedures of plotting the scatter plot and regression line and drawing the conclusion that the regression slope is approximately equal to* $1/m$.

1.11 Dimensional analysis example: the heat flux from ocean floor to ocean water. In the ocean, there is a flux of heat from certain regions of the ocean floor. The heat flux is denoted by q_{cond} in units [W cm^{-2}]. This flux is related to the four critical quantities:

(i) Thermal molecular diffusivity: D_H[cm^2 s^{-1}],
(ii) Temperature gradient: ∇T[K cm^{-1}],
(iii) Heat capacity of seawater: C_p[J (g K)$^{-1}$], and
(iv) Density of seawater: ρ[g cm^{-3}].

Do the following:

(a) Find the dimension for each of the five quantities above: $[q_{cond}], [D_H], [\nabla T], [C_p], [\rho]$. Use the given units as a hint.

(b) Find how q_{cond} is related to the other four quantities using the following dimensional analysis equation:

$$[q_{cond}] = [D_H]^a [\nabla T]^b [C_p]^c [\rho]^d. \tag{1.74}$$

Use the results in part (a) to show that

$$a = b = c = d = 1. \tag{1.75}$$

This leads to the following equation of heat diffusion from the higher temperature ocean floor to the lower temperature seawater:

$$q_{cond} = D_H \nabla T C_p \rho. \tag{1.76}$$

(c) From the units of the right-hand side of the above equation:

$$[\text{cm}^2\ \text{s}^{-1}][\text{K cm}^{-1}][\text{J(g K)}^{-1}][\text{g cm}^{-3}], \tag{1.77}$$

show that the unit of q_{cond} is W cm^{-2}.

2 Basics of R Programming

When discussing the future of geophysics education, G. K. Vallis (2016) made this perceptive statement:

> Scientists will always have personal preferences and differing expertise but combining analytical ideas with simple numerical models can be a very powerful tool in both research and education, and modern tools can be used to enable this at an early stage in the classroom. A numerical model transparently coded in 100 lines and run on a laptop can then play a similar role to that of a rotating tank in illustrating phenomena and explaining what equations mean, and the rift between theory, models and phenomena then never opens. Although conventional books will remain important for years to come, the next textbook or monograph in GFD, or really in any similar field, could to great effect be written using a Jupyter Notebook (formerly IPython Notebook), or similar, which can combine numerical models with conventional text and equations (e.g. LaTex markup), figures, and even symbolic manipulation in a single document, enabling interactive exploration of both analytical and numerical GFD concepts. Such an effort would be a major undertaking, so a collaborative effort may be needed, perhaps like the development of open source software, and the end product would hopefully be free like both beer and speech.

Our book is written from this perspective for the future of mathematics education in climate sciences. The book uses R and the R Notebook. This chapter explains the installation of R and RStudio and demonstrates some basic uses of R.

Equivalent Python codes and their Jupyter Notebooks may be found at the book website www.cambridge.org/climatemathematics.

2.1 Download and Install R and RStudio

For Windows users, visit the website
https://cran.r-project.org/bin/windows/base/
to find the instructions for R program download and installations.

For Mac users, visit
https://cran.r-project.org/bin/macosx/

If you experience difficulties, please refer to online resources, Google, or YouTube. A recent 3-minute YouTube instruction for R installation for Windows can be found from the following link:
www.youtube.com/watch?v=Ohnk9hcxf9M

```
R version 3.4.0 (2017-04-21) -- "You Stupid Darkness"
Copyright (C) 2017 The R Foundation for Statistical Computing
Platform: x86_64-apple-darwin15.6.0 (64-bit)

R is free software and comes with ABSOLUTELY NO WARRANTY.
You are welcome to redistribute it under certain conditions.
Type 'license()' or 'licence()' for distribution details.

  Natural language support but running in an English locale

R is a collaborative project with many contributors.
Type 'contributors()' for more information and
'citation()' on how to cite R or R packages in publications.

Type 'demo()' for some demos, 'help()' for on-line help, or
'help.start()' for an HTML browser interface to help.
Type 'q()' to quit R.

[R.app GUI 1.70 (7338) x86_64-apple-darwin15.6.0]

[Workspace restored from /Users/sshen/.RData]
[History restored from /Users/sshen/.Rapp.history]

> 2+3
[1] 5
>
```

Fig. 2.1 R Console window after opening R.

The same author also has a YouTube instruction about R installation for Mac (2 minutes): www.youtube.com/watch?v=uxuuWXU-7UQ

When R is installed, one can open R. The `R Console` window will appear. See Fig. 2.1. One can use `R Console` to perform calculations, such as typing 2+3 and hitting return. However, most people today prefer using RStudio as the interface. To install RStudio, visit www.rstudio.com/products/rstudio/download/
This site allows one to choose Windows, or Mac OS, or Unix.

After both R and RStudio are installed, one can use either R or RStudio, or both, depending on one's interest. However, RStudio will not work without R. Thus, always install R first.

When opening RStudio, four windows will appear as shown in Fig. 2.2: The top left window is called R script, for writing the R code. The green arrow on top of the window can be clicked to run the code. Each run is shown in the lower left R Console window, and recorded on the upper right R History window. When plotting, the figure will appear in the lower right R Plots window. For example, `plot(x,x*x)` renders the eight points in the Plots window, because `x=1:8` defines a sequence of numbers from 1 to 8. `x*x` yields a sequence from 1^2 to 8^2.

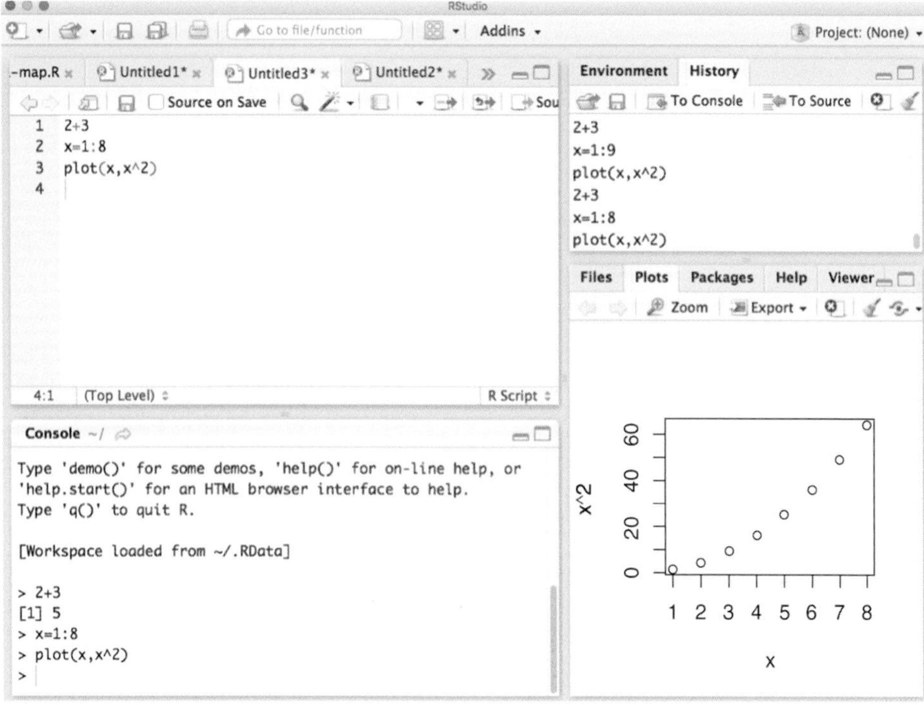

Fig. 2.2 RStudio windows.

2.2 R Tutorial

Many excellent tutorials for quickly learning R programming, using a few hours or a few evenings, are available online and in YouTube, such as
ww2.coastal.edu/kingw/statistics/R-tutorials/.
You can easily perform an Internet search and find your preferred tutorials.

It can be extremely difficult for the beginners of R to navigate through the official, formal, detailed, and massive R-Project documentation:
www.r-project.org/

2.2.1 R As a Smart Calculator

R can be used like a smart calculator that allows more elaborate calculations than those done with ordinary calculators.

```
1+4
#[1] 5  This is the result
2+pi/4-0.8
#[1] 1.985398
x<-1
y<-2
```

```
z<-4
t<-2*x^y-z
t
#[1] -2
u=2         # "=" sign and "<-" are almost equivalent
v=3         # The text behind the "#" sign is comments
u+v
#[1] 5
sin(u*v)     # u*v = 6 in the sine function is considered radian by R
#[1] -0.2794155
```

R programming uses assignment operator `a <- b` to assign b to a. Often the equal operator `a=b` can do the same job or vice versa. The two operators are equivalent in general. However, certain R formulas have specific meanings for `=` and cannot be replaced by `<-`. Most veteran R users use `<-` for assignment and `=` for defined R formulas.

2.2.2 Define a Sequence in R

Directly enter a sequence of daily maximum temperature data at San Diego International Airport (Lat: 32.7336°N, Lon: 117.1831°W) during May 1–7, 2017 [unit: °F].
`tmax <- c(77, 72, 75, 73, 66, 64, 59)`
The data are from the Daily Summary of the Local Climatological Data (LCD), National Centers for Environmental Information (NCEI)
www.ncdc.noaa.gov/cdo-web/datatools/lcd
The command `c( )` is used to hold a data sequence and is named `tmax`. Entering the `tmax` command will render the temperature data sequence:

```
tmax
#[1] 77 72 75 73 66 64 59
```

You can generate different sequences using R, e.g.,
`1: 8 #Generates a sequence 1,2,...,8`
Here the pound sign # begins R comments which are not executed by R calculations. The same sequence can be generated by different commands, such as

```
seq(1,8)
seq(8)
seq(1,8, by=1)
seq(1,8, length=8)
seq(1,8, length.out =8)
```

The most useful sequence commands are `seq(1,8, by=1)` and `seq(1,8, length=8)` or `seq(1,8, len=8)`. The former is determined by a begin value, end value, and step size, and the latter by a begin value, end value, and number of values in the sequence. For example, `seq(1951,2016, len=66*12)` renders a sequence of all the months from January 1951 to December 2016.

2.2.3 Define a Function in R

The function command is

`name <- function(var1, var2, ...) expression of the function.`

For example,

```
samfctn <- function(x) x*x
samfctn(4)
#[1] 16
fctn2 <- function(x,y,z) x+y-z/2
fctn2(1,2,3)
#[1] 1.5
```

2.2.4 Plot with R

R can plot all kinds of curves, surfaces, statistical plots, and maps. Below are a few very simple examples for R beginners. For adding labels, ticks, color, and other features to a plot, you will learn them from later parts of this book, and you may also carry out an Internet search on R plot to find the commands for the proper inclusion of the desired features.

R plotting is based on the coordinate data. The following command plots the seven days of San Diego Tmax data above:

```
plot(1:7, c(77, 72, 75, 73, 66, 64, 59))
```

The resulting graph is shown in Fig. 2.3.

```
plot(sin, -pi, 2*pi)   #plot the curve of y=sin(x) from -pi to 2 pi

square <- function(x) x*x   #Define a function
plot(square, -3,2)   # Plot the defined function

# Plot a 3D surface
```

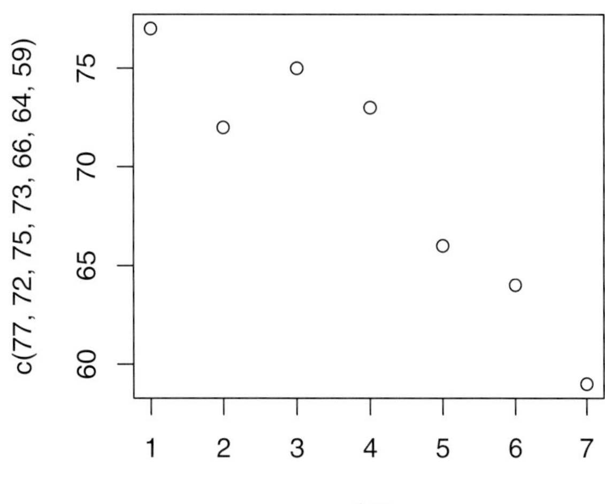

Fig. 2.3 The daily maximum temperature during May 1–7, 2017 at San Diego International Airport.

```
x <- seq(-1, 1, length=100)
y <- seq(-1, 1, length=100)
z <- outer(x, y, function(x, y)(1-x^2-y^2))
#outer (x,y, function) renders z function on the x, y grid
persp(x,y,z, theta=330)
# yields a 3D surface with perspective angle 330 deg

#Contour plot
contour(x,y,z) #lined contours
filled.contour(x,y,z) #color map of contours
```

The color map of contours resulting from the last command is shown in Fig. 2.4.

2.2.5 Symbolic Calculations by R

Many people once thought that R can handle only numbers. Actually R can also do symbolic calculations, such as finding a derivative, although at present R is not the best symbolic calculation tool. One can use WolframAlpha, SymPy, and Yacas for free symbolic calculations or use the paid software packages Maple or Mathematica. Carry out an Internet search on symbolic calculation for calculus to find a long list of symbolic calculation software packages, e.g.,
https://en.wikipedia.org/wiki/List_of_computer_algebra_systems.

```
D(expression(x^2,'x'), 'x')
# Take derivative of x^2 w.r.t. x
2 * x #The answer is 2x
```

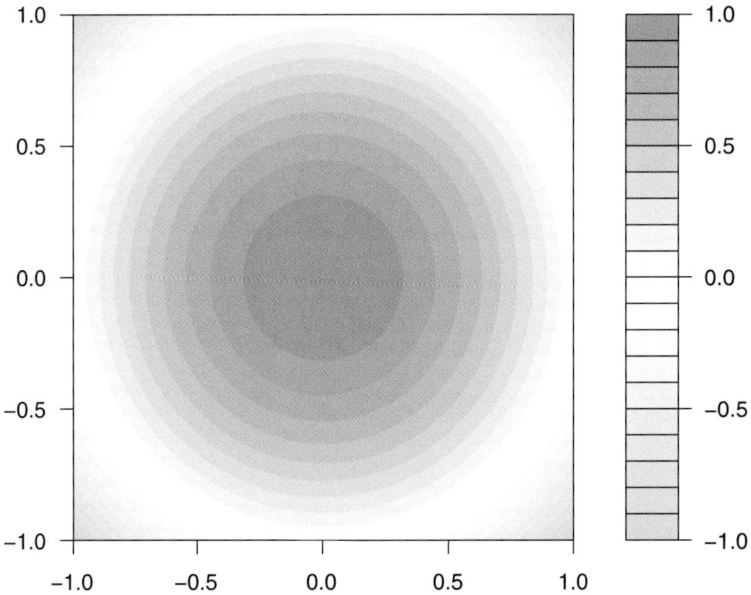

Fig. 2.4 The color map of contours for the function $z = 1 - x^2 - y^2$.

```
fx= expression(x^2,'x')  #assign a function
D(fx,'x') #differentiate the function w.r.t. x
2 * x  #The answer is 2x

fx= expression(x^2*sin(x),'x')
#Change the expression and use the same derivative command
D(fx,'x')
2 * x * sin(x) + x^2 * cos(x)

 fxy = expression(x^2+y^2, 'x','y')
#One can define a function of 2 or more variables
 fxy #renders an expression of the function in terms of x and y
#expression(x^2 + y^2, "x", "y")
D(fxy,'x') #yields the partial derivative with respect to x: 2 * x
D(fxy,'y') #yields the partial derivative with respect to y: 2 * y

square = function(x) x^2
integrate (square, 0,1)
#Integrate x^2 from 0 to 1 equals to 1/3 with details below
#0.3333333 with absolute error < 3.7e-15

integrate(cos,0,pi/2)
#Integrate cos(x) from 0 to pi/2 equals to 1 with details below
#1 with absolute error < 1.1e-14
```

The above two integration examples are for definite integrals. It seems that no efficient R packages are available for finding antiderivatives, or indefinite integrals.

2.2.6 Vectors and Matrices

R can handle all kinds of operations involving vectors and matrices.

```
c(1,6,3,pi,-3) #c() gives a vector, considered a 4X1 column vector
#[1]  1.000000  6.000000  3.000000  3.141593 -3.000000
seq(2,6) #Generate a sequence from 2 to 6
#[1] 2 3 4 5 6
seq(1,10,2) # Generate a sequence from 1 to 10 with 2 increment
#[1] 1 3 5 7 9
x=c(1,-1,1,-1)
x+1 #1 is added to each element of x
#[1] 2 0 2 0
2*x #2 multiplies each element of x
#[1]  2 -2  2 -2
x/2 # Each element of x is divided by 2
#[1]  0.5 -0.5  0.5 -0.5
y=seq(1,4)
x*y  # This multiplication * multiplies each pair of elements
#[1]  1 -2  3 -4
x%*%y #This is the dot product of two vectors and yields
#      [,1]
#[1,]   -2
t(x)  # Transforms x into a row 1X4 vector
#      [,1] [,2] [,3] [,4]
#[1,]     1   -1    1   -1
t(x)%*%y #This is equivalent to dot product and forms 1X1 matrix
#      [,1]
#[1,]   -2
```

```
x%*%t(y) #This column times row yields a 4X4 matrix
#      [,1] [,2] [,3] [,4]
#[1,]    1    2    3    4
#[2,]   -1   -2   -3   -4
#[3,]    1    2    3    4
#[4,]   -1   -2   -3   -4
my=matrix(y,ncol=2)
#Convert a vector into a matrix of the same number of elements
#The matrix elements go by column, the first column, second, etc
#Command matrix(y,ncol=2, nrow=2), or matrix(y, ncol=2)
#or matrix(y,2), or matrix(y,2,2) does the same job
my
#      [,1] [,2]
#[1,]    1    3
#[2,]    2    4
dim(my)  #find dimensions of a matrix
#[1] 2 2

bigM=matrix(1:100, nrow=10) #Generate a 10-by-10 matrix
subM=bigM[4:6,3:7] #Extract a sub-matrix from a big matrix
subM
#      [,1] [,2] [,3] [,4] [,5]
#[1,]   24   34   44   54   64
#[2,]   25   35   45   55   65
#[3,]   26   36   46   56   66

as.vector(my) #Convert a matrix to a vector, again via columns
#[1] 1 2 3 4
mx <- matrix(c(1,1,-1,-1), byrow=TRUE,nrow=2)
mx*my #multiplication between each pair of elements
#      [,1] [,2]
#[1,]    1    3
#[2,]   -2   -4
mx/my #division between each pair of elements
#      [,1]       [,2]
#[1,]  1.0  0.3333333
#[2,] -0.5 -0.2500000
mx-2*my
#      [,1] [,2]
#[1,]   -1   -5
#[2,]   -5   -9
mx%*%my #This is the real matrix multiplication in matrix theory
#      [,1] [,2]
#[1,]    3    7
#[2,]   -3   -7
det(my) #determinant
#[1] -2
myinv = solve(my) #yields the inverse of a matrix
myinv
#      [,1] [,2]
#[1,]   -2  1.5
#[2,]    1 -0.5
myinv%*%my #verifies the inverse of a matrix
#      [,1] [,2]
#[1,]    1    0
#[2,]    0    1
```

```
diag(my) #yields the diagonal vector of a matrix
#[1] 1 4
myeig=eigen(my) #yields eigenvalues and unit eigenvectors
myeig
myeig$values
#[1]  5.3722813 -0.3722813
myeig$vectors
#           [,1]       [,2]
#[1,] -0.5657675 -0.9093767
#[2,] -0.8245648  0.4159736
mysvd = svd(my) #SVD decomposition of a matrix M=UDV'
#SVD can be done for a rectangular matrix of mXn
mysvd$d
#[1] 5.4649857 0.3659662
mysvd$u
#           [,1]       [,2]
#[1,] -0.5760484 -0.8174156
#[2,] -0.8174156  0.5760484
mysvd$v
#           [,1]       [,2]
#[1,] -0.4045536  0.9145143
#[2,] -0.9145143 -0.4045536

ysol=solve(my,c(1,3))
#solve linear equations matrix %*% x = b
ysol  #solve(matrix, b)
#[1]  2.5 -0.5
my%*%ysol #verifies the solution
#     [,1]
#[1,]    1
#[2,]    3
```

2.2.7 Simple Statistics by R

R was originally designed to do statistical calculations. Thus, R has a comprehensive set of statistics functions and software packages. This subsection gives a few basic commands. More will be described in the statistics chapter of this book.

```
x=rnorm(10) #generate 10 normally distributed numbers
x
#[1]  2.8322260 -1.2187118  0.4690320 -0.2112469  0.1870511
#[6]  0.2275427 -1.2619005  0.2855896  1.7492474 -0.1640900
mean(x)
#[1] 0.289474
var(x)
#[1] 1.531215
sd(x)
#[1] 1.237423
median(x)
#[1] 0.2072969
quantile(x)
#         0%        25%        50%        75%       100%
#-1.2619005 -0.1994577  0.2072969  0.4231714  2.8322260
range(x) #yields the min and max of x
```

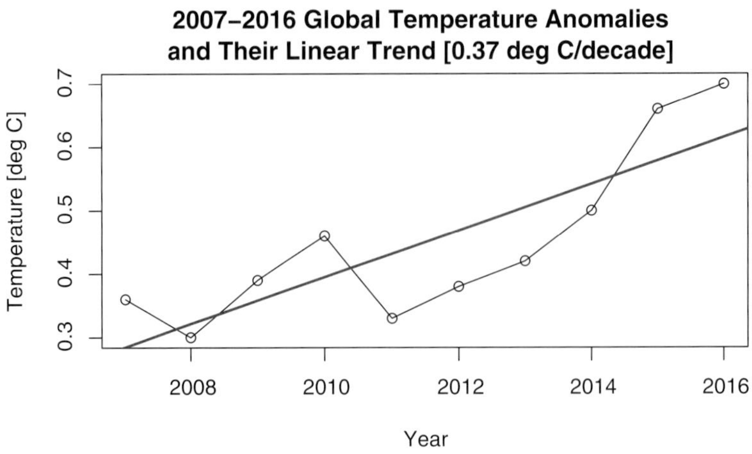

Fig. 2.5 The 2007–2016 global average annual mean surface air temperature anomalies with respect to the 1971–2000 climate normal. The red line is a linear trend line computed from a linear regression model.

```
#[1] -1.261900  2.832226
max(x)
#[1] 2.832226

boxplot(x) #yields the box plot of x
w=rnorm(1000)

summary(rnorm(12)) #statistical summary of the data sequence
#   Min. 1st Qu.  Median    Mean 3rd Qu.    Max.
#-1.9250 -0.6068  0.3366  0.2309  1.1840  2.5750

hist(w)
#yields the histogram of 1000 random numbers with a normal distribution

#Linear regression and linear trend line
#2007-2016 data of the global temperature anomalies
#Source: NOAAGlobalTemp data
t=2007:2016
T=c(.36,.30, .39, .46, .33, .38, .42, .50, .66, .70)
lm(T ~ t) #Linear regression model of temp vs time
#(Intercept)            t
#-73.42691      0.03673
#Temperature change rate is 0.03673  deg C/yr or 0.37 deg C/decade
plot(t,T, type="o",xlab="Year",ylab="Temperature [deg C]",
     main="2007-2016 Global Temperature Anomalies
          and Their Linear Trend [0.37 deg C/decade] ")
abline(lm(T ~ t), lwd=2, col="red") #Regression line
```

The global temperature data from 2007–2016 in the above R code example are displayed in Fig. 2.5, together with their linear trend line

$$T = -73.42691 + 0.03673t. \tag{2.1}$$

2.3 Online Tutorials

Numerous online R tutorials are available. Several are relatively efficient for learning climate mathematics and are recommended below.

2.3.1 YouTube Tutorial: For True Beginners

This is a very good and slow-paced 22-minute YouTube tutorial: Chapter 1. An Introduction to R
www.youtube.com/watch?v=suVFuGET-0U

2.3.2 YouTube Tutorial: For Some Basic Statistical Summaries

This is a 9-minute tutorial by Layth Alwan.
www.youtube.com/watch?v=XjOZQN-Nre4

2.3.3 YouTube Tutorial: Input Data by Reading a csv File into R

An excel file can be saved as csv file: xxxx.csv. This 15-minute YouTube video by Layth Alwan shows how to read a csv file into R. He also shows linear regression.
www.youtube.com/watch?v=QkE8cp0B9gg

R can input all kinds of data files, including xlsx, txt, netCDF, MatLab data, Fortran file, and SAS data. Some commands are below. One can search the Internet to find proper data reading commands for any particular data format.

```
mydata <- read.csv("mydata.csv")
# read csv file named "mydata.csv"

mydata <- read.table("mydata.txt")
# read text file named "my data.txt"

library(gdata)                    # load the gdata package
mydata = read.xls("mydata.xls")   # read an excel file

library(foreign)                  # load the foreign package
mydata = read.mtp("mydata.mtp")   # read from .mtp file

library(foreign)                  # load the foreign package
mydata = read.spss("myfile", to.data.frame=TRUE)

ff <- tempfile()
cat(file = ff, "123456", "987654", sep = "\n")
read.fortran(ff, c("F2.1","F2.0","I2")) #read a fotran file

library(ncdf4)
ncin <- ncdf4::nc_open("ncfname")  # open a NetCDF file
lon <- ncvar_get(ncin, "lon") #read data "lon" from a netCDF file into R
```

Many more details of reading and reformatting of `.nc` files will be discussed later when dealing with NCEP/NCAR Reanalysis data.

Some libraries are not in the R project. For example,

```
library(ncdf4) #The following error message pops up
Error in library(ncdf4) : there is no package called ncdf4
```

You can install the R package by

```
install.packages("ncdf4")
```

After this installation, `library(ncdf4)` will run, and the functions in the ncdf4 package will work.

You only need to install the package once on your computer, but in each new R session you must run `library(package)` in order to activate the package functions. Many examples will be shown in the rest of this book.

The R packages and the datasets used in this book are listed below and can be downloaded and installed first before proceeding to the R codes in the rest of the book.

```
#R packages: animation, chron, e1071, fields, ggplot2, lattice,
#latticeExtra, maps, mapdata, mapproj, matrixStats, ncdf4,
#NLRoot, RColorBrewer, rgdal, rasterVis, raster, sp, TTR

#The zipped data:
#www.cambridge.org/climatemathematics

#To load a single package, such as "animation", you can do
library(animation)

#You can also load all these packages in one shot using
# pacman

install.packages("pacman")
library(pacman)
pacman::p_load(animation, chron, e1071, fields, ggplot2, lattice,
            latticeExtra, maps, mapdata, mapproj, matrixStats, ncdf4,
            NLRoot, RColorBrewer, rgdal, rasterVis, raster, sp, TTR)
```

On your computer, you can create a directory called `climmath` under your user name. The one used in the book is `Users/sshen/climmath`. You unzip the data and move the data folder under the `Users/sshen/climmath` directory. A data folder will be created: `Users/sshen/climmath/data`. The data folder contains about 400 MB of data. Place all the R codes in the directory `Users/sshen/climmath`. Then, you can run all the codes in this book after replacing `sshen` by your user name on your own computer.

2.4 Chapter Summary

This chapter has described the following basic aspects of the R programming language

(i) Installation of both R and RStudio.
(ii) Layout of RStudio panels and functions.

(iii) R commands for defining and computing vectors, matrices, and functions.
(iv) Simple statistics using R: mean, median, standard deviation, variance, histogram, boxplot, scatter plot, and linear regression.
(v) Online resources for R tutorial material.

We suggest that reading and practicing the R basics included in this chapter might require about three hours. You might then use perhaps five to ten hours to do the exercise problems of this chapter. After that introduction to R, we estimate that you should be able to develop and carry out projects using R independently. Familiarizing yourself with additional R commands may require spending some time online with a search engine.

More sophisticated examples of using R for the analysis and visualization of the data from climate models and observations are included in Chapters 9–11. These chapters also provide research-level R-graphics codes for climate sciences, R programs for advanced statistical analysis of climate data, such as empirical orthogonal functions computed from the climate model datasets, and R programming techniques for analyzing datasets with missing data, either in space or time. To develop R projects at a more advanced level, you will need to read these chapters or search for the R code for the specific analysis or graphics tasks described in these chapters. These tasks include examples such as the preparation of data from a global climate model (GCM) for a singular value decomposition (SVD) analysis.

The R programming language was created by statisticians Ross Ihaka and Robert Gentleman in New Zealand and was first released in 1995. R is named partly after the first names of the two original authors of R, and partly as a play on the name of the S programming language for statistics. Currently, R is a popular computer programming language used in almost every field of science and engineering. Among the top programming languages ranked by the Institute of Electrical and Electronics Engineers (IEEE) in 2017, R ranks sixth, following Python, C, Java, C++, and C#. For a climate science student or professional who is not a specialist in computer programming or information technology, R is easy to learn, and it offers numerous public-domain software packages that are free to use. As an alternative to R, equivalent Python codes for our entire book are available online at the book website www.cambridge.org/climatemathematics.

References and Further Readings

[1] King, K. B., 2016: *R Tutorials*, Coastal Carolina University. URL: ww2.coastal.edu/kingw/statistics/R-tutorials/

> This easy-to-learn tutorial is for beginners of R, and does not require any computer programming background. It contains many statistics examples.

[2] Jost, S., 2018: *Introduction to R: An R Tutorial for Data Analysis and Regression.* De Paul University. URL:
http://facweb.cs.depaul.edu/sjost/csc423/

This is a very brief R tutorial from the perspective of computer programming for beginners. One needs only about one hour to go through the entire tutorial.

[3] Vallis, G. K., 2016: Geophysical fluid dynamics: whence, whither and why? *Proceedings of the Royal Society A*, 472: 20160140. http://dx.doi.org/10.1098/rspa.2016.0140

In this stimulating article, Vallis discusses the role of geophysical fluid dynamics in understanding the dynamics of atmospheres and oceans. Geoffrey K. Vallis, a Professor in the Department of Mathematics at the University of Exeter, is an expert in climate dynamics, the circulation of planetary atmospheres, and dynamical meteorology and oceanography. He is the author of *Atmospheric and Oceanic Fluid Dynamics: Fundamentals and Large-Scale Circulation*. This widely praised standard text is a magisterial treatment of geophysical fluid dynamics. The book was first published in 2006, and its second edition in 2017 contains nearly 1,000 pages.

Exercises

All exercises in this chapter require the use of a computer.

2.1 Use R to define a data sequence `t=seq(2015,2018, length=100)`, and then plot the following two functions on the same figure: $y = \sin(2\pi(t-0.1))$ and $y = \cos^2(2\pi t)$.

2.2 (a) Use R to make a contour plot of the function $z = \sin^2 x \cos^2(y-\pi)$ over the domain of $[0,2\pi] \times [0,2\pi]$.

(b) Use R to plot a color contour map for the same function on the same domain.

2.3 Use R to solve the following linear equations:

$$\begin{cases} 9x + 8y = 87 \\ 6x - 20y = 126 \end{cases}$$

2.4 Use R to solve the following linear equations:

$$\begin{cases} -3x + 2y + z = 1 \\ -2x - y + z = 2 \\ 2x + y - 4z = 0 \end{cases}$$

2.5 For some purposes, climatology or climate is defined as the mean state, or normal state, of a climate parameter, and is calculated from data over a period of time called the climatology period (e.g., 1961–1990). Thus the surface air temperature climate or climatology at a given location may be calculated by averaging observational temperature data over a period such as 1961 through 1990. Thirty years are often considered in the climate science community as the standard length of a climatology period. Due to the relatively high density of weather stations in 1961–1990, compared to earlier periods, investigators have often used 1961–1990 as their climatology period, although some may now choose 1971–2000 or 1981–2010. Surface air temperature (SAT) is often defined as the temperature inside a white-painted louvered instrument container or box, known as a Stevenson screen, located on a stand about 2 meters above the ground. The purpose of the Stevenson screen is to shelter the instruments from radiation, precipitation, animals, leaves, etc., while allowing the air to circulate freely inside the box. The daily maximum temperature (Tmax) is the maximum temperature measured inside the screen box by a maximum temperature thermometer within 24 hours. The daily minimum temperature (Tmin) is the minimum temperature within 24 hours. The daily mean temperature (Tmean) is the average of Tmax and Tmin.

Go to the NOAA NCEI website

https://www.ncdc.noaa.gov/cdo-web/datatools/findstation

and download the monthly Tmax, Tmin, and Tmean data of the Cuyamaca station (Station No. 042239) or another station near San Diego, California, USA. Or download the data.zip file of this book at

www.cambridge.org/climatemathematics

Unzip the data and find the data from the file named `CA042239T.csv`

(a) Use R to arrange the monthly Cuyamaca Tmax data from January 1961 to December 1990 as a matrix with each row as year and each column as month.
(b) Do the same for Tmin.
(c) Do the same for Tmean.

2.6 (a) Use R to calculate the August climatology of Tmax, Tmin, and Tmean for the Cuyamaca station according to the 1961–1990 climatology period.
(b) Use R to compute the standard deviation of Tmax, Tmin, and Tmean of the Cuyamaca station for January during the 1961–1990 climatology period.

2.7 (a) Use R to plot the Cuyamaca January Tmin time series from 1951 to 2010 with a continuous curve.
(b) Use R to plot the linear trend lines of Tmin on the same plot as (a) in the following time periods:

(i) 1951–2010,
(ii) 1961–2010,
(iii) 1971–2010, and
(iv) 1981–2010.

(c) Finally, what is the temporal trend per decade for each of the four periods above?

2.8 Use R to plot the time series and its trend line for P. D. Jones' global average annual mean temperature anomaly data: `JonesGlobalT.txt`. This data file can be found from the book's data.zip file downloaded from the book website.

(a) Plot the global average annual mean temperature from 1880 to 2015.

(b) Find the linear trend of the temperature from 1880 to 2015. Plot the trend line on the same figure as (a).

(c) Find the linear trend from 1900 to 1999. Plot the trend line on the same figure as (a).

2.9 Use the gridded NOAA global monthly temperature anomaly data NOAAGlobal-Temp from the following website or another data source

www.ncdc.noaa.gov/data-access/marineocean-data/
noaa-global-surface-temperature-noaaglobaltemp

Or use the `NOAAGlobalT.csv` data file from the book's data.zip file downloaded from the book website. Choose two 5-by-5 degrees lat-lon grid boxes of your interest. Plot the temperature anomaly time series of the two boxes on the same figure using two different colors.

2.10 Using the same NOAAGlobalTemp dataset, choose sufficiently many grid boxes that cover the state of Texas, USA. Compute the average temperature anomalies of these boxes for each month. Then plot the monthly average temperature anomalies as a function of time. Plot a linear trend line on the same figure.

2.11 Choose a 5-by-5 degrees grid box that covers Edmonton, Canada, and another grid box that covers San Diego, USA.

(a) Use R and 30 years of the January NOAAGlobalTemp data from January 1981 to January 2010 to compute the standard deviations for each grid box for January.

(b) Do the same for February, March, ..., December.

(c) Use R to write your standard deviation results in a 12-by-2 matrix with each row for a month, and each column for a grid box ID.

(d) Describe the main differences between the values of the two columns.

3 Basic Statistical Methods for Climate Data Analysis

You may have seen statements such as, "The available evidence rejects the null hypothesis at the 5% significance level." That language certainly sounds "scientific." What such statements really mean, however, is strongly dependent on context. You might be enthusiastic about trying a new restaurant or seeing a new movie, if 19 out of 20 online reviews were favorable. But you would never get on an airplane if you thought the odds were 1 in 20 that it would crash. We use "statistics" to mean a suite of scientific methods for analyzing data and for drawing credible conclusions from data. We provide basic concepts and useful R codes covering commonly used statistical methods in climate data analysis, so that users can arrive at credible conclusions based on the data, together with a given error probability. To interpret the statistical results in a meaningful way, however, knowledge of climate science is essential. The statistical methods in this chapter have been chosen to focus on making credible inferences in climate science, with a given error probability, based on the analysis of climate data, so that observational data can lead to objective and reliable conclusions. We will first describe a list of statistical indices, such as mean, variance, and quantiles, for climate data. We will then take up probability distributions and statistical inferences. You may be surprised to learn that some well-known and powerful statistical techniques were developed by a scientist who spent his entire career working as a beer brewer at the Guinness Brewery in Dublin, Ireland.

3.1 Statistical Indices from the Global Temperature Data from 1880 to 2015

The following link provides data for the global average annual mean surface air temperature anomalies from 1880 to 2015 (NOAA GlobalTemp dataset at NCDC)
www1.ncdc.noaa.gov/pub/data/noaaglobaltemp/operational/
In the data list, the first datum corresponds to 1880 and the last to 2015. These 136 years of data are used to illustrate the following statistical concepts: mean, variance, standard deviation, skewness, kurtosis, median, 5th percentile, 95th percentile, and other quantiles. The anomalies are with respect to the 20th century mean, i.e., the 1900–1999 climatology period. The global average of the 20th century mean is 12.7 °C. The 2015 anomaly was 0.65 °C. Thus, the 2015's global average annual mean temperature is 13.4 °C.

```
  [1] -0.367918 -0.317154 -0.317069 -0.393357 -0.457649 -0.468707
  [7] -0.451778 -0.498811 -0.403252 -0.353712 -0.577277 -0.504825
 [13] -0.556487 -0.568014 -0.526737 -0.475364 -0.340468 -0.367002
 [19] -0.505967 -0.368630 -0.315155 -0.387099 -0.494861 -0.585158
 [25] -0.663492 -0.535226 -0.457892 -0.617208 -0.684107 -0.672176
 [31] -0.624129 -0.675199 -0.570521 -0.558340 -0.379505 -0.308313
 [37] -0.531023 -0.551480 -0.444860 -0.444257 -0.451256 -0.388185
 [43] -0.469536 -0.455500 -0.489551 -0.385962 -0.305391 -0.393436
 [49] -0.416556 -0.538602 -0.339823 -0.316963 -0.360309 -0.486954
 [55] -0.347795 -0.383147 -0.356958 -0.262097 -0.272009 -0.257514
 [61] -0.152032 -0.050356 -0.095295 -0.088983  0.044418 -0.073264
 [67] -0.251405 -0.297744 -0.296136 -0.303984 -0.405346 -0.255647
 [73] -0.218081 -0.146923 -0.358796 -0.377482 -0.441748 -0.194232
 [79] -0.133076 -0.184608 -0.222896 -0.165795 -0.154384 -0.137509
 [85] -0.393492 -0.322453 -0.267491 -0.257946 -0.274517 -0.151345
 [91] -0.207025 -0.322901 -0.216440 -0.080250 -0.316583 -0.241672
 [97] -0.323398 -0.046098 -0.131010 -0.016080  0.021495  0.057638
[103] -0.061422  0.099061 -0.093873 -0.109097 -0.015374  0.125450
[109]  0.129184  0.050926  0.186128  0.159565  0.010836  0.038629
[115]  0.092131  0.211006  0.074193  0.269107  0.384935  0.194762
[121]  0.177381  0.296912  0.351874  0.363650  0.329436  0.408409
[127]  0.362960  0.360386  0.291370  0.385638  0.453061  0.325297
[133]  0.370861  0.416356  0.491245  0.650217
```

Because we have just quoted numbers that purport to be observations of annual mean global mean surface temperatures, this may be a good place to mention an important caveat. The caveat is that observational estimates of the global mean surface temperature are less accurate than similar estimates of year-to-year changes. This is one of several reasons why global mean surface temperature data are almost always plotted as anomalies (such as differences between the observed temperature and a long-term average temperature) rather than as the temperatures themselves. It is also important to realize that the characteristic spatial correlation length scale for surface temperature anomalies is much larger (hundreds of kilometers) than the spatial correlation length scale for surface temperatures. The use of anomalies is also a way of reducing or eliminating individual station biases that are invariant with time. A simple example of such biases is that due to station location, which usually does not change with time. It is easy to understand, for instance, that a station located in a valley in the middle of a mountainous region might report surface temperatures that are higher than an accurate mean surface temperature for the entire region, but the anomalies at the station might more accurately reflect the characteristics of the anomalies for the region. For a concise summary of these important issues, with references, see a thoughtful analysis at this link

www.realclimate.org/index.php/archives/2017/08/observations-reanalyses-and-the-elusive-absolute-global-mean-temperature/

3.1.1 Mean, Variance, Standard Deviation, Skewness, Kurtosis, and Quantiles

We use R to calculate the relevant statistical parameters. The data is read as `tmean15`.

```
setwd("/Users/sshen/climmath")
dat1 <- read.table("data/aravg.ann.land_ocean.90S.90N.v4.0.0.2015.txt")
```

```
dim(dat1)
tmean15=dat1[,2] #Take only the second column of this data matrix
head(tmean15) #The first six values
#[1] -0.367918 -0.317154 -0.317069 -0.393357 -0.457649 -0.468707
mean(tmean15)
#[1] -0.2034367
sd(tmean15)
#[1] 0.3038567
var(tmean15)
#[1] 0.09232888
library(e1071)
#This R library is needed to compute the following parameters
#install.packages("e1071") #if it is not installed on your computer
skewness(tmean15)
#[1] 0.7141481
kurtosis(tmean15)
#[1] -0.3712142
median(tmean15)
#[1] -0.29694
quantile(tmean15,probs= c(0.05,0.25, 0.75, 0.95))
#         5%         25%         75%          95%
#-0.5792472 -0.4228540 -0.0159035  0.3743795
```

The above statistical indices were computed using the following mathematical formulas, described by $x = \{x_1, x_2, \ldots, x_n\}$ as the sampling data for a time series:

$$\text{mean: } \mu(x) = \frac{1}{n}\sum_{k=1}^{n} x_k, \tag{3.1}$$

$$\text{variance by unbiased estimate: } \sigma^2(x) = \frac{1}{n-1}\sum_{k=1}^{n}(x_k - \mu(x))^2, \tag{3.2}$$

$$\text{standard deviation: } \sigma(x) = (\sigma^2(x))^{1/2}, \tag{3.3}$$

$$\text{skewness: } \gamma_3(x) = \frac{1}{n}\sum_{k=1}^{n}\left(\frac{x_k - \mu(x)}{\sigma}\right)^3, \tag{3.4}$$

$$\text{kurtosis: } \gamma_4(x) = \frac{1}{n}\sum_{k=1}^{n}\left(\frac{x_k - \mu(x)}{\sigma}\right)^4 - 3. \tag{3.5}$$

The significance of these indices is as follows. The mean gives the average of samples. The variance and standard deviation measure the spread of samples. They are large when the samples have a broad spread. Skewness is a dimensionless quantity. It measures the asymmetry of samples. Zero skewness signifies a symmetric distribution. For example, the skewness of a normal distribution is zero. Negative skewness denotes a skew to the left, meaning that the long distribution tail is on the left side of the distribution. Positive skewness has a long tail on the right side. Kurtosis is also dimensionless and measures the peakedness of a distribution. The kurtosis of a normal distribution is zero. Positive kurtosis means a high peak at the mean, thus a slim and tall shape for the distribution. This is referred to as leptokurtic. "Lepto" is Greek in origin and means thin or fine. Negative kurtosis means a low peak at the mean, thus a fat and short shape for the distribution, referred to as platykurtic. "Platy" is also Greek in origin and means flat or broad. "Kurtic" and "kurtosis" are Greek in origin and mean peakedness.

For the 136 years of global average annual mean temperature data given above, the skewness is 0.71, meaning skew to the right with a long tail on the right, thus with more extreme high temperatures than low temperatures (see Section 3.2 for a histogram). The kurtosis is −0.37, meaning the distribution is flatter than a normal distribution, also shown in the histogram.

The median is a number characterizing a set of samples, such that 50% of the samples are less than the median, and another 50% are greater than the median. To find the median, sort the samples from the smallest to the largest. The median is then the sample number in the middle. If the number of the samples is even, then the median is equal to the mean of the two middle samples.

Quantiles are defined in the same way by sorting. For example, 25-percentile (also called 25th percentile) is a sample such that 25% of sample values are less than this sample value. By definition, 75-percentile is thus larger than 40-percentile. Here, percentile is a description of quantile relative to 100. Obviously, 100-percentile is the largest sample, and 0-percentile is the smallest sample. Often, a box plot is used to show the typical quantiles (see Section 3.2).

The 50-percentile (or 50th percentile) is called the median. If the distribution is symmetric, then the median is equal to the mean. Otherwise these two quantities are not equal. If the skew is to the right, then the mean is on the right of the median: the mean is greater than the median. If the skew is to the left, then the mean is on the left of the median: the mean is less than the median. Our 136 years of temperature data are right skewed and have mean equal to −0.2034 °C, greater than their median equal to −0.2969 °C.

3.1.2 Correlation, Covariance, and Linear Trend

The following R commands can plot the time series of the global average annual mean surface air temperature anomaly data from 1880 to 2015 with a linear trend (see Fig. 3.1).

```
yrtime15 = seq(1880, 2015)
reg8015 = lm(tmean15 ~ yrtime15)
# Display regression results
reg8015
#Call:
#lm(formula = tmean15 ~ yrtime15)
#Coefficients:
#(Intercept)      yrtime15
#-13.208662      0.006678
# Plot the temperature time series and its trend line
plot(yrtime15,tmean15,xlab="Year",ylab="Temperature deg C",
     main="Global Annual Mean Land and Ocean Surface
     Temperature Anomalies 1880-2015", type="l", lwd=2)
abline(reg8015, col="red")
text(1930, 0.4, "Linear temperature trend 0.67 deg C per century",
     col="red",cex=1.2)
```

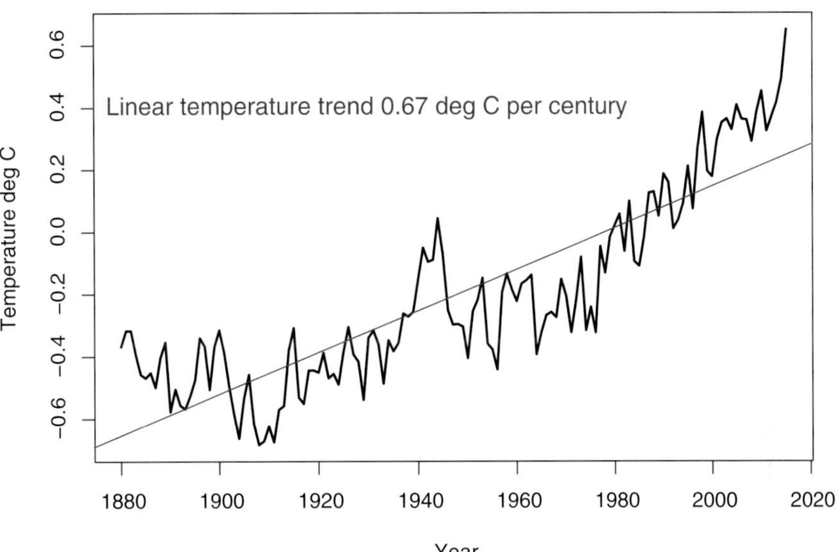

Fig. 3.1 Time series of the global average annual mean temperature with respect to 1900–1999 climatology: 12.7 °C.

The calculation of the linear trend $\bar{b}_1 = 0.67$ °C per century is related to the following two concepts: covariance and correlation coefficient. The covariance between two variables x and y measures how much the two variables vary together. This is in contrast to the variance defined in Section 3.1.1 above, which measures how much a single variable varies. A positive covariance means a positive trend, or a positive linear relationship between two variables, and a negative covariance implies the opposite. The absolute value of the covariance quantifies the magnitude of how much the two variables vary together. Figure 3.1 shows a positive linear trend, and hence the covariance between the temperature anomalies and time is positive. The covariance can be computed by the following R commands

```
 x=yrtime15
 y=tmean15
cov(x, y)
#[1] 10.36856
```

The mathematical expression of the covariance between two datasets $x = (x_1, x_2, \ldots, x_n)$ and $y = (y_1, y_2, \ldots, y_n)$ is

$$\mathrm{cov}(x,y) = \frac{\sum_{i=1}^{n}(x_i - \bar{x})(y_i - \bar{y})}{n-1} \tag{3.6}$$

where $\bar{x} = \mu(x)$ and $\bar{y} = \mu(y)$ are respectively the sample means of x and y defined by formula (3.1).

The dimension of covariance $\text{cov}(x,y)$ between x and y is the product of the dimensions of x and y: $[\text{cov}(x,y)] = [x][y]$. Other commonly used notations for the covariance $\text{cov}(x,y)$ are $C_{xy}, c_{xy}, \Sigma_{xy}, \sigma_{xy}$, or simply C.

A correlation coefficient measures the strength of the linear relationship between two datasets. Its mathematical formula is

$$\text{corr}(x,y) = \frac{\text{cov}(x,y)}{s_x s_y}, \tag{3.7}$$

where $s_x = \sigma(x)$ is the standard deviation of x, and $s_y = \sigma(y)$ is that of y, defined by formula (3.3).

Other commonly used notations for the correlation coefficient corr(x,y) between x and y include r_{xy}, ρ_{xy}, or simply r. Formula (3.7) implies that the correlation coefficient r_{xy} is dimensionless, and its value is between -1 and 1. When $r_{xy} = 0$, the two datasets x and y are totally uncorrelated, i.e., not related to each other at all. If $r_{xy} = 1$, the two datasets x and y are perfectly correlated to each other, and vary together in a completely synchronized way. If $r_{xy} = -1$, the two datasets x and y vary in an exactly opposite way. The sign and magnitude of the correlation coefficient r_{xy} are also displayed in many statistical plots, such as the regression plot and scatter plot to be described in Section 3.2 below.

The correlation coefficient between the temperature data and the time in Fig. 3.1 can be computed by the following R command

```
cor(tmean15, yrtime15)
#[1] 0.8659857
```

This means a positive correlation, implying that the temperature increases with time from 1880 to 2015.

One can also use R and formula (3.7) to verify this correlation result.

```
x=yrtime15
y=tmean15
sx=sd(yrtime15)
sy=sd(tmean15)
cxy=cov(x,y)
rxy=cxy/(sx*sy)
rxy
#[1] 0.8659857 #The same result as cor(tmean15, yrtime15)
```

The concepts of correlation and covariance have many applications in climate science, such as statistical approaches to climate prediction, pattern detection in climate dynamics, and trend analysis in climate change detection and attribution. Chapters 4 and 10 will present more climate science examples of covariance and correlation, including the detection of the spatial and temporal patterns of El Niño.

The linear trend of the temperature increase in Fig. 3.1 is determined by

$$\hat{b}_1 = \frac{\sum_{i=1}^{n}(x_i - \bar{x})(y_i - \bar{y})}{(n-1)s_x^2}. \tag{3.8}$$

Following the covariance formula (3.6) and the correlation formula (3.7), this formula can also be expressed in terms of correlation coefficient and standard deviations:

$$\hat{b}_1 = r_{xy}\frac{s_y}{s_x}, \tag{3.9}$$

or in terms of covariance and variance:

$$\hat{b}_1 = \frac{C_{xy}}{s_x^2}. \tag{3.10}$$

The R command to find $\hat{b}_1$ for the temperature data of this section is
`reg8015$coefficients[2]`,
which outputs 0.006677908 °C/year, approximately equivalent to 0.67 °C/century shown in Fig. 3.1. One can also use R and formula (3.9) to verify this trend result:

```
rxy*sy/sx
#[1] 0.006677908 #verified
cxy/sx^2
#[1] 0.006677908 #verified
```

Further discussion on the calculation and inference of a linear trend is in Section 3.5.

3.2 Commonly Used Statistical Plots

We will use the 136 years of temperature data and R to illustrate some commonly used statistical figures, namely the histogram, boxplot, scatter plot, q q plot, and linear regression trend line.

3.2.1 Histogram of a Set of Data

The histogram of the global average annual mean temperature anomalies data from 1880–2015 is shown in Fig. 3.2, which can be generated by the following R commands.

```
#Plot histogram of the tmean15 data
h = hist(tmean15, main="Histogram of 1880-2015 Temperature Anomalies",
     xlab=expression(paste("Temperature anomalies [", degree, "C]")),
     xlim=c(-1,1), ylim=c(0,50))
xfit<-seq(-1,1, length=100)
areat = sum((h$counts)*diff(h$breaks[1:2]))#Normalization area
#diff(h$breaks[1:2])= h$breaks[2] - h$breaks[1]
#is the histogram's bin width
yfit<-areat*dnorm(xfit, mean=mean(tmean15), sd=sd(tmean15))
#Plot the normal fit on the histogram
lines(xfit, yfit,col="blue",lwd=2)
```

3.2.2 Box Plot

Figure 3.3 is the box plot of the 136 years of global average annual mean temperature data, and can be made from the following R command

```
boxplot(tmean15, ylim=c(-0.8,0.8),
    ylab=expression(paste("Temperature anomalies [", degree, "C]")))
```

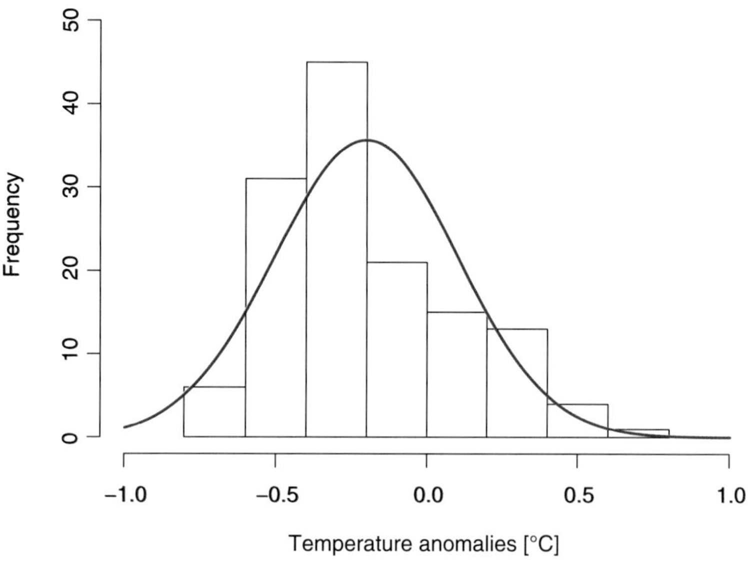

Fig. 3.2 Histogram and its normal fit of the global average annual mean temperature anomalies from 1880–2015.

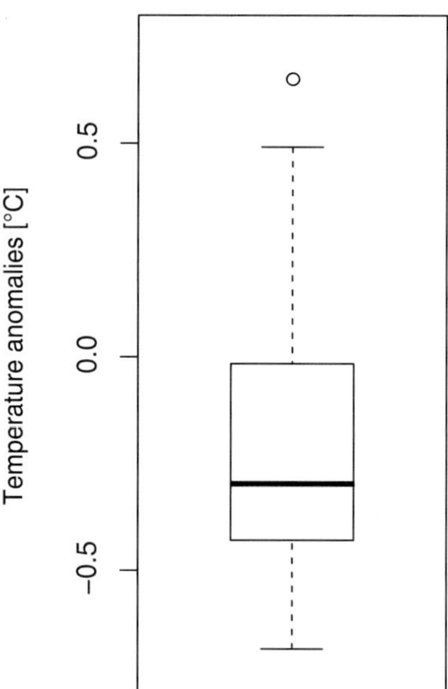

Fig. 3.3 Box plot of the global average annual mean temperature anomalies from 1880–2015.

The rectangular box's mid line indicates the level of the median, which is −0.30 °C. The rectangular box's lower boundary is the first quartile, i.e., 25-percentile. The box's upper boundary is the third quartile, i.e., the 75-percentile. The box's height is the third quartile minus the first quartile, and is called the interquartile range (IQR). The upper "whisker" is the third quartile plus 1.5 IQR. The lower whisker is supposed to be at the first quartile minus 1.5 IQR. However, this whisker would then be lower than the lower extreme. In this case, the lower whisker takes the value of the lower extreme, which is $-0.68\,^\circ$C. The points outside of the two whiskers are considered outliers. Our dataset has one outlier, which is $0.65\,^\circ$C. This is the hottest year in the dataset. It was year 2015.

Sometimes, one may need to plot multiple box plots on the same figure, which can be done by R. One can look at an example in the R-project document
https://www.rdocumentation.org/packages/graphics/versions/3.5.3/topics/boxplot

3.2.3 Scatter Plot

The scatter plot is convenient for displaying whether two datasets are correlated with one another. We use the Southern Oscillation Index (SOI) and the contiguous United States temperature anomalies as an example to describe the scatter plot. The SOI is a standardized index based on the observed sea level pressure (SLP) differences between two southern tropical Pacific locations: Tahiti and Darwin. The data can be downloaded from
www.ncdc.noaa.gov/teleconnections/enso/indicators/soi/
www.ncdc.noaa.gov/temp-and-precip/
See Section 4.6 for further information about the detailed analysis and visualization of the SLP and SOI data.

The following R code can produce the scatter plot shown in Fig. 3.4.

```
#Use setwd("working directory") to work in the right directory
rm(list=ls())
setwd("/Users/sshen/climmath")
par(mgp=c(1.5,0.5,0))
ust=read.csv("data/USJantemp1951-2016-nohead.csv",header=FALSE)
soi=read.csv("data/soi-data-nohead.csv", header=FALSE) #Read data
soid=soi[,2] #Take the second column SOI data
soim=matrix(soid,ncol=12,byrow=TRUE)
#Make the SOI into a matrix with each month as a column
soij=soim[,1] #Take the first column for Jan SOI
ustj=ust[,3] #Take the third column: Jan US temp data
setEPS()
postscript("fig0304.eps", height=7, width=7)
par(mar=c(4.5,5,2.5,1), xaxs = "i", yaxs = "i")
plot(soij,ustj,xlim=c(-4,4), ylim=c(-8,8),
    main="January SOI and the US Temperature Anomalies",
    xlab="SOI [dimensionless]",
    ylab=expression(paste("Temperature Anomalies [", degree, "F]")),
    pch=19, cex.lab=1.3)
# Plot the scatter plot
soiust=lm(ustj ~ soij) #Linear regression
abline(soiust, col="blue", lwd=4) #Linear regression line
dev.off()
```

The correlation between the two datasets is 0. Thus, the slope of the blue trend line is also zero.

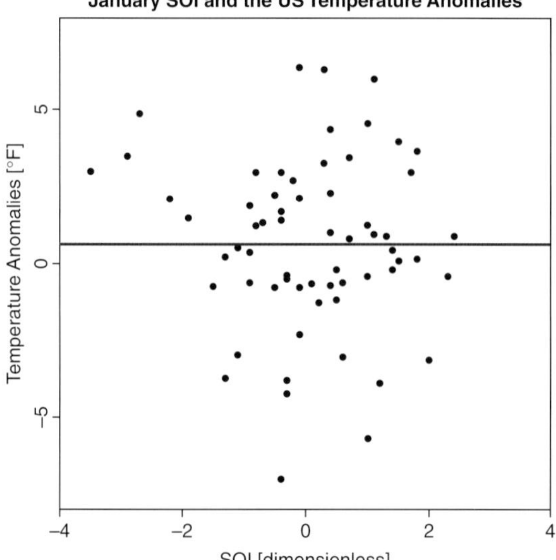

Fig. 3.4 Scatter plot of the January US temperature vs. the January SOI from 1951–2016.

The scatter plot shows that the nearly zero correlation is mainly due to the five negative SOI values, which are El Niño Januaries: 1983 (−3.5), 1992 (−2.9), 1998 (−2.7), 2016 (−2.2), 1958 (−1.9). When these strong El Niño Januaries are removed, then the correlation is 0.2. The slope is then 0.64, compared with 1.0 for perfect correlation.

The R commands to retain the data without the above five El Niño years are below

```
soijc=soij[c(1:7,9:32,34:41,43:47,49:65)]
ustjc=ustj[c(1:7,9:32,34:41,43:47,49:65)]
```

With these data, the scatter plot and trend line can be produced in the same way.

We thus may say that the SOI has some predictive skill for the January temperatures of the contiguous United States, for the non-El Niño years. This correlation is stronger for specific regions of the US. The physical reason for this result has to do with the fact that the temperature field over the US is inhomogeneous, and in different regions, it is related to the tropical ocean dynamics in different ways. This gives us a hint as to how to find the predictive skill for a specific objective field: to create a scatter plot using the objective field, which is being predicted, and the field used for making the prediction. The objective field is called the predicant or predictand, and the field used to make the prediction is called the predictor. A very useful predictive skill would be that the predictor leads the predicant by a certain time, say one month. Then the scatter plot will be made from the pairs between predictor and predicant data with one-month lead. The absolute value of the correlation can then be used as a measure of the predictive skill. Since the 1980s, the US Climate Prediction Center has been using sea surface temperature (SST) and sea level pressure (SLP) as predictors for the US temperature and precipitation via the canonical correlation analysis method (CCA). Therefore, before a prediction is made, it is a good idea to examine the predictive skill via scatter plots, which can help identify the best predictors.

However, the scatter plot approach above for maximum correlation is only applicable for linear predictions or for weakly nonlinear relationships. Nature can sometimes be very nonlinear, which requires more sophisticated assessments of predictive skill, such as neural networks and time-frequency analysis. The CCA and other advanced statistical prediction methods are beyond the scope of this book.

3.2.4 Q–Q Plot

Figure 3.5 shows Quantile–Quantile (Q–Q) plots, also denoted by q–q plots, qq-plots, or QQ-plots.

The function of a Q–Q plot is to compare the distribution of a given set of data with a specific reference distribution, such as a standard normal distribution with zero mean and standard deviation equal to one, denoted by $N(0,1)$. A Q–Q plot lines up the percentiles of data on the vertical axis and the same number of percentiles of the specific reference distribution on the horizontal axis. The pairs of the quantiles $(x_i, y_i), i = 1, 2, \ldots, n$ determine the points on the Q–Q plot. Here, x_i and y_i correspond to the same cumulative percentage or probability p_i for both x and y variables, where p_i monotonically increases from approximately zero to one as i goes from 1 to n. A red Q–Q reference line is plotted as if the vertical axis values are also the quantiles of the given specific distribution. Thus, the Q–Q reference line should be diagonal.

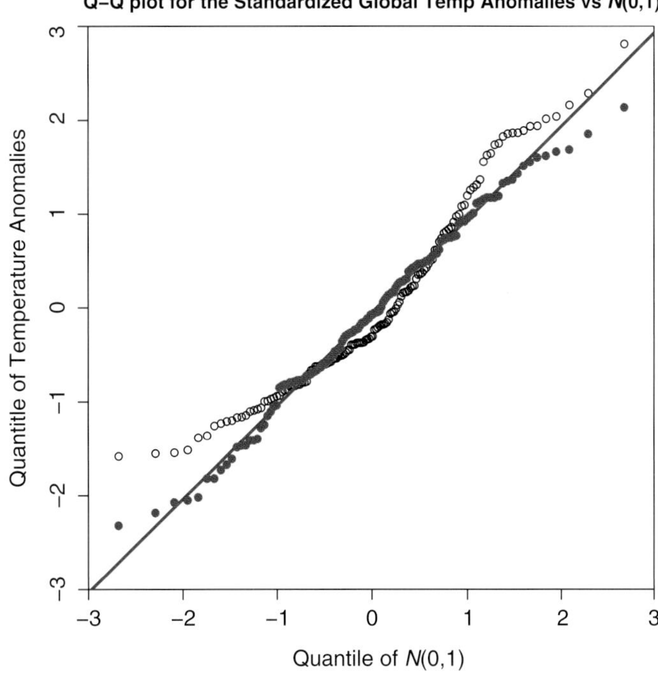

Fig. 3.5 Black empty-circle points are the Q–Q plot of the standardized global average annual mean temperature anomalies vs. standard normal distribution. The purple points are the Q–Q plot for the data simulated by `rnorm(136)`. The red is the distribution reference line of $N(0,1)$.

The black empty circles in Fig. 3.5 compare the quantiles of the standardized global average annual mean temperature anomalies marked on the vertical axis with those of the standard normal distribution marked on the horizontal axis. The purple dots shows a Q–Q plot of a set of 136 random numbers simulated by the standard normal distribution. As expected, the simulated points are located close to the red diagonal line, which is the distribution reference line of $N(0,1)$. On the other hand, the temperature Q–Q plot shows a considerable degree of scattering of the points away from the reference line. We may intuitively conclude that the global average annual temperature anomalies from 1880 to 2015 are not exactly distributed according to a normal (or Gaussian) distribution. However, we may also conclude that the distribution of these temperatures is not very far away from the normal distribution either, because the points on the Q–Q plot are not very far away from the distribution reference line, and also because even the simulated $N(0,1)$ points are noticeably off the reference line for the extremes.

Figure 3.5 can be generated by the following R code.

```
#Q-Q plot for the standardized temperature anomalies
tstand = (tmean15-mean(tmean15))/sd(tmean15)
set.seed(101)
qn=rnorm(136) #Simulate 136 points by N(0,1)
qns=sort(qn)#Sort the points
qq2=qqnorm(qns,col="blue",lwd=2)

setEPS()
postscript("fig0305.eps", height=7, width=7)
par(mar=c(4.5,5,2.5,1), xaxs = "i", yaxs = "i")
qt=qqnorm(tstand,
main = "Q-Q plot for the Standardized Global Temp Anomalies vs N(0,1)",
ylab="Quantile of Temperature Anomalies",
xlab="Quantile of N(0,1)", xlim=c(-3,3),ylim=c(-3,3),
          cex.lab=1.3, cex.axis=1.3)
qqline(tstand, col = "red", lwd=3)
points(qq2$x, qq2$y, pch=19, col="purple")
dev.off()
```

In the R code, we first standardize, also called normalize, the global average annual mean temperature data by subtracting the data mean and dividing by the data's standard deviation. Then, we use these 136 years of standardized global average annual mean temperature anomalies to generate a Q–Q plot, which is shown in Fig. 3.5.

3.3 Probability Distributions

This section describes a few basic probabilistic distributions in addition to the "bell-shaped" normal or Gaussian distribution we often have in mind.

3.3.1 What Is a Probability Distribution?

A probability distribution represents chance or likelihood of occurrence of an event at a certain value or an interval of values. For example, if the daily weather at a location is

Table 3.1 Probability distribution of sky cover

Location	Clear Sky	Cloudy or Precipitation Sky
Las Vegas	0.58	0.42
San Diego	0.40	0.60
Seattle (SEA-TAC)	0.16	0.84

Data source: NOAA National Centers for Environmental Information, June 2018
www1.ncdc.noaa.gov/pub/data/ccd-data/clpcdy15.dat

classified as being in one of two categories: clear weather days, defined as from 0 to 3/10 average sky cover by clouds, and cloudy weather days, defined as from 4/10 to 10/10 average sky cover, then the resulting probability distribution is the probability value of clear and cloudy days. Table 3.1 shows the probabilities of clear weather for three United States cities based on historical data: Seattle 0.16, San Diego, 0.40, and Las Vegas 0.58. The probability distribution table obviously reflects the very different climates of the three cities. Seattle is a Pacific Northwest US city characterized by weather that is often cloudy or rainy, particularly in the winter. San Diego is a Pacific Southwest US city where it rarely rains, but where cloud cover may be relatively large in May and June, the so-called "May Gray and June Gloom." Las Vegas is a US Southwest inland desert city, which often experiences a clear sky during the daytime.

The data of Table 3.1 can also be displayed by the bar chart in Fig. 3.6. This figure visually displays the different cloudiness climates of the three cities. Thus, either the table or the figure demonstrates that a probability distribution can be a good description of important properties of a random variable, such as cloud cover. Here, a random variable means a variable that can take on a value in a random way, such as weather conditions (sunny, rainy, snowy, cloudy, stormy, windy, etc.). Many things that we deal with in our daily lives may involve a random variable, that is to say, a variable which has a random nature, in contrast to a deterministic variable. We describe a random variable by probability and explore what is the probability of the variable having a certain value or a certain interval of values. This description is the probability distribution.

Figure 3.6 can be generated by the following R code.

```
plot.new()
layout(matrix(c(1,2,3), 1, 3, byrow = TRUE),
       widths=c(3,3,3), heights=c(1,1,1))
lasvegas=c(0.58,0.42)
sandiego=c(0.4,0.6)
seattle=c(0.16,0.84)
names(lasvegas)=c("Clear","Cloudy")
names(sandiego)=c("Clear","Cloudy")
names(seattle)=c("Clear","Cloudy")
barplot(lasvegas,col=c("skyblue","gray"), ylim=c(0,1),
        ylab="Probability", cex.lab=1.2)
mtext("Las Vegas", side=3, line=1)
barplot(sandiego,col=c("skyblue","gray"), ylim=c(0,1))
mtext("San Diego", side=3, line=1)
```

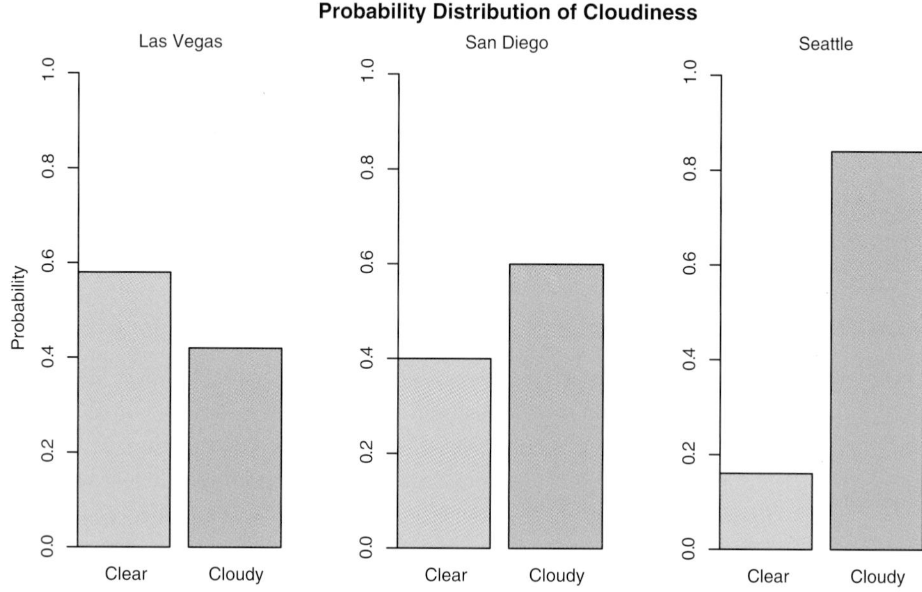

Fig. 3.6 Probability distributions of different climate conditions defined by cloudiness for three cities in the United States.

```
barplot(seattle,col=c("skyblue","gray"), ylim=c(0,1))
mtext("Seattle", side=3, line=1)
mtext("Probability Distribution of Cloudiness",
      cex=1.2,side = 3, line = -1.5, outer = TRUE)
```

A probability distribution can be expressed not only by a table as shown above, but also by bar chart, a curve, or a function $y = f(x)$. Bar charts are used for the random variables which can take on discrete values, such as clear sky or cloudy sky, or intervals of continuous values, such as the temperature in the intervals (0–5, 6–10, 11–15, 16–20, 21–25, 25–30, 31–35)°C for San Diego. A smooth curve or a function $y = f(x)$ is often used to describe a continuous distribution, of which a random variable x can take on any real value in a given range, such as San Diego temperature in the range of $(-50, 50)$°C. In the case of a continuous curve, the curve's vertical coordinate value $f(x)$ is probability density, and an integration of $f(x)$ yields probability. The probability density itself is not probability, but the probability density value times the width of a small interval of the random variable x is probability. For example, $f(x)\Delta x$ is the probability for the random variable to be in the interval $(x, x+\Delta x)$. We call the curve $f(x)$ the probability density function (pdf or PDF). The domain of the pdf $f(x)$ is the entire range of all the possible values of the random variable x. Thus, the probability for x to have a value somewhere in the entire range is one, i.e., the sum of $f(x)\Delta x$ for the entire range is one. Following the method of calculus, when Δx approaches zero and is denoted by dx, the probability one can be expressed as an integral of the pdf $f(x)$:

$$\int_D f(x)dx = 1, \tag{3.11}$$

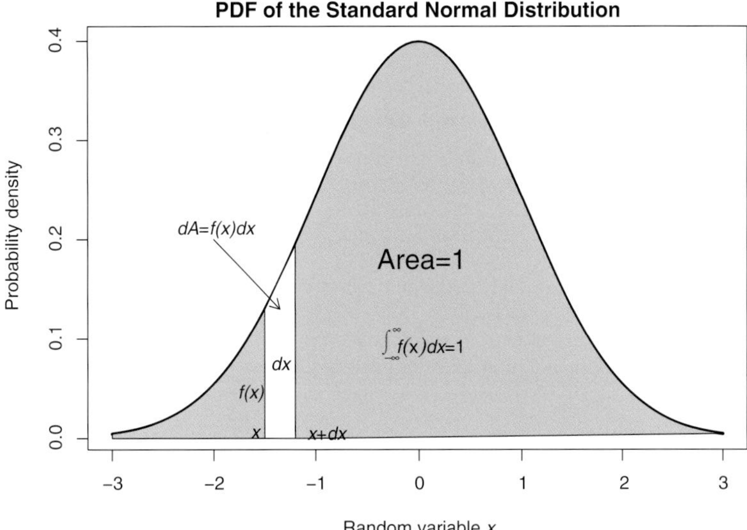

Fig. 3.7 Schematic diagram of a pdf, probability, and the pdf normalization condition.

where D is the domain of the pdf, the entire range of the possible x values, e.g., $D = (-50, 50)^\circ$C in the case of temperature for the US. This formula is called the probability normalization condition, as shown in Fig. 3.7.

Figure 3.7 can be generated by the following R code

```
# Density function, probability, and normalization of N(0,1)
par(mar=c(4.0,4.0,1.5,0.5))
cord.x <- c(-3,seq(-3,3,0.01),-1)
cord.y <- c(0,dnorm(seq(-3,3,0.01)),0)
# Make a curve
curve(dnorm(x,0,1), xlim=c(-3,3), lwd=3,
      main='PDF of the Standard Normal Distribution',
      xlab="Random variable x",
      ylab='Probability density')
# Add the shaded area using many lines
polygon(cord.x,cord.y,col='skyblue')
polygon(c(-1.5,-1.5, -1.2, -1.2),c(0, dnorm(-1.5),
                                    dnorm(-1.2), 0.0),col='white')
text(0,0.18, "Area=1", cex=1.5)
text(-1.65,0.045,"f(x)")
text(-1.35,0.075,"dx")
text(-1.6,0.005,"x")
text(-0.9,0.005,"x+dx")
arrows(-2,0.2,-1.35,0.13, length=0.1)
text(-2,0.21,"dA=f(x)dx")
text(0,0.09,
     expression(paste(integral(f(x)*dx,- infinity,infinity),"=1")))
```

Of course, the normalization condition for a discrete random value, such as clear and cloudy skies, is a summation, rather than the above integral. Consider the San Diego case in Table 3.1. The normalization condition is $0.40 + 0.60 = 1.0$.

3.3.2 Normal Distribution

Figure 3.8 shows five different normal distributions, each of which is a bell-shaped curve with the highest density when the random variable x takes the mean value, and approaches zero as x goes to plus or minus infinity. The figure can be generated by the following R code.

```
#Normal distribution plot
x <- seq(-8, 8, length=200)
plot(x,dnorm(x, mean=0, sd=1), type="l", lwd=4, col="red",
     ylim = c(0,1),
     xlab="Random variable x",
     ylab ="Probability density",
     main=expression(Normal~Distribution ~ N(mu,sigma^2)))
lines(x,dnorm(x, mean=0, sd=2), type="l", lwd=2, col="blue")
lines(x,dnorm(x, mean=0, sd=0.5), type="l", lwd=2, col="black")
lines(x,dnorm(x, mean=3, sd=1), type="l", lwd=2, col="purple")
lines(x,dnorm(x, mean=-4, sd=1), type="l", lwd=2, col="green")
#ex.cs1 <- expression(plain(sin) * phi,  paste("cos", phi))
ex.cs1 <- expression(paste(mu, "=0",~",", "~ sigma, "=1"),
                     paste(mu, "=0",~",", "~ sigma, "=2"),
                     paste(mu, "=0",~",", "~ sigma, "=1/2"),
                     paste(mu, "=3",~",", "~ sigma, "=1"),
                     paste(mu, "=-4",~",", "~ sigma, "=1"))
legend("topleft",legend = ex.cs1, lty=1,
       col=c('red','blue','black','purple','green'), cex=1, bty='n')
```

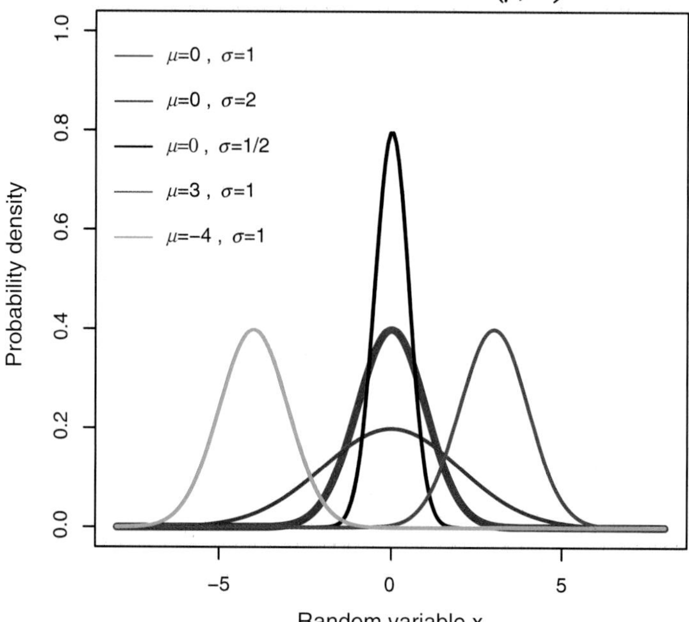

Fig. 3.8 Probability density function for five normal distributions.

The bell-shaped normal distribution curve can be expressed by a mathematical formula

$$f(x|\mu,\sigma^2) = \frac{1}{\sigma\sqrt{2\pi}} \exp\left(-\frac{(x-\mu)^2}{2\sigma^2}\right), \tag{3.12}$$

where μ is the mean and σ is the standard deviation of the normal distribution, and σ^2 is called the variance. The mean is the value one would expect to occur with the highest probability, and is called the expected value. The standard deviation measures how much the actual values deviate away from the mean. The pdf's peak is at the mean. The pdf is flatter for a large standard deviation, and more peaked for a smaller standard deviation. Figure 3.8 clearly shows these properties. Notice how the bell-shaped curve changes due to different values of μ and σ. The mean reflects the mean state of the random variable and hence determines the position of the bell-shaped curve; and the standard deviation reflects the diversity of the random variable and determines the shape of the curve.

Here, x, μ, and σ have the same unit, and, of course, the same dimension.

The probability, or the area, under the entire bell-shaped curve is one. The probability in the interval $(\mu - 1.96\sigma, \mu + 1.96\sigma)$ is 0.95, and that in $(\mu - \sigma, \mu + \sigma)$ is 0.68. These are commonly used properties of a normal distribution. Sometimes we regard 1.96 approximately as 2, and $(\mu - 1.96\sigma, \mu + 1.96\sigma)$ as two standard deviations away from the mean. A corresponding mathematical expression is

$$\int_{\mu-2\sigma}^{\mu+2\sigma} \frac{1}{\sigma\sqrt{2\pi}} \exp\left(-\frac{(x-\mu)^2}{2\sigma^2}\right) dx \approx 0.95. \tag{3.13}$$

One can use an R code to verify this formula:

```
mu=0
sig=1
intg <- function(x){(1/(sig*sqrt(2*pi)))*exp(-(x-mu)^2/(2*sig^2))}
integrate(intg,-2,2)
#0.9544997 with absolute error < 1.8e-11
#Or using the R built-in function dnorm to get the same result
integrate(dnorm,-2,2)
#0.9544997 with absolute error < 1.8e-11
integrate(dnorm,-1.96,1.96)
#0.9500042 with absolute error < 1e-11
```

3.3.3 Student's t-distribution

Figure 3.9 shows Student's t-distribution, or simply the t-distribution. It is used when estimating the mean of a normally distributed variable with a small number of data points and an unknown standard deviation. William Gosset (1876–1937) published the t-distribution under the pseudonym "Student" while working at the Guinness Brewery in Dublin, Ireland. Gosset worked as a brewer, because Guinness hired scientists who could apply their skills to brewing. In 1904, Gosset wrote a report called *The Application of the Law of Error* to the work of the Brewery. In his report, Gosset advocated using statistical methods in the brewing industry. Gosset published under a pseudonym, because the brewery did not allow its scientists to publish their research using their real names, perhaps because the information contained in the research might give a competitive advantage to the

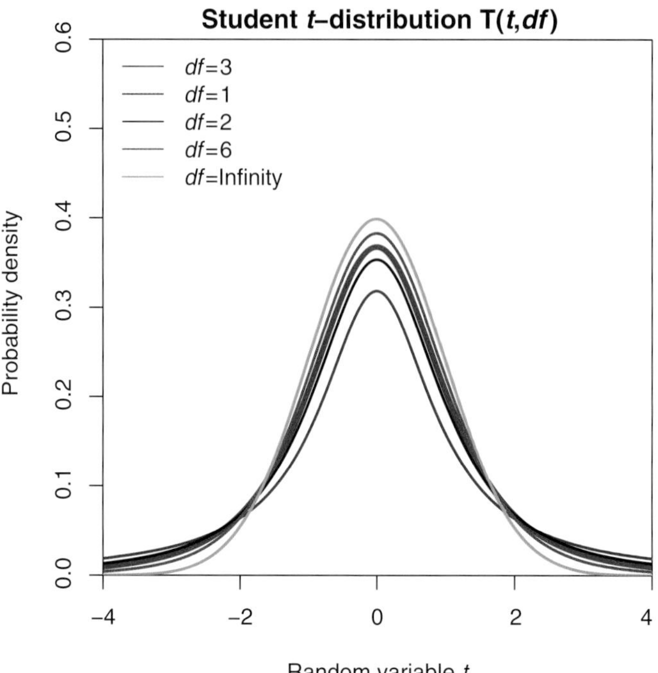

Fig. 3.9 Probability density function for five t-distributions with different degrees of freedom.

brewery. Gosset corresponded with leading statisticians of the time, however, and gained their respect because of his research.

If $x_1, x_2, \cdots, x_n$ are normally distributed data with a given mean μ, an unknown standard deviation, and a small sample n, say, $n < 30$, then

$$t = \frac{\bar{x} - \mu}{S/\sqrt{n}} \tag{3.14}$$

follows a t-distribution with $n-1$ degrees of freedom (df), where

$$\bar{x} = \frac{1}{n}\sum_{i=1}^{n} x_i \tag{3.15}$$

is the estimated sample mean, and

$$S^2 = \frac{1}{n-1}\sum_{i=1}^{n}(x_i - \bar{x})^2 \tag{3.16}$$

is the estimated sample variance.

The random variable t is essentially a measure of the deviation of the sample mean from the given mean value normalized by the estimated standard deviation scaled down by $\sqrt{n}$. The pdf of the random variable t can be plotted by the following R code

```
#Plot t-distribution by R
x <- seq(-4, 4, length=200)
plot(x,dt(x, df=3), type="l", lwd=4, col="red",
```

```
     ylim = c(0,0.6),
     xlab="Random variable t",
     ylab ="Probability density",
     main="Student t-distribution T(t,df)")
lines(x,dt(x, df=1), type="l", lwd=2, col="blue")
lines(x,dt(x, df=2), type="l", lwd=2, col="black")
lines(x,dt(x, df=6), type="l", lwd=2, col="purple")
lines(x,dt(x, df=Inf), type="l", lwd=2, col="green")
#ex.cs1 <- expression(plain(sin) * phi,  paste("cos", phi))
ex.cs1 <- c("df=3", "df=1","df=2","df=6","df=Infinity")
legend("topleft",legend = ex.cs1, lty=1,
       col=c('red','blue','black','purple','green'), cex=1, bty="n")
```

When df, the number of degrees of freedom ($df = n - 1$) is infinity, the t-distribution is exactly the same as the standard normal distribution $N(0,1)$. Even when $df = 6$, the t-distribution is already very close to the standard normal distribution. Thus, t-distribution is meaningfully different from the standard normal distribution only when the sample size is small, say, $n = 5$ (i.e., $df = 4$).

The exact mathematical expression of the pdf for the t-distribution is quite complicated and uses a Gamma function, which is a special function beyond the scope of this book.

3.4 Estimate and Its Error

Assume that the data $(x_1, x_2, \ldots, x_n)$ are drawn from a normal distribution with mean μ and standard deviation σ, but the values of μ and σ are unknown. Then, the sample mean, i.e., the mean of the data, is

$$\bar{x} = \frac{1}{n}\sum_{i=1}^{n} x_i. \tag{3.17}$$

This mean $\bar{x}$ is also normally distributed with mean equal to μ and standard deviation equal to $\sigma/\sqrt{n}$. Here, $\bar{x}$ is an estimate of the unknown mean μ, and $\sigma/\sqrt{n}$ is called the standard error of the estimate. Basic climate data analyses often involve finding an estimate and its error for a climate parameter from the given climate data.

3.4.1 Probability of a Sample inside a Confidence Interval

Given the sample size n, mean μ, and standard deviation σ for a set of normal data, what is the interval $[a,b]$ such that 95% of the sample means will occur within the interval $[a,b]$? Intuitively, the sample mean should be close to the true mean μ most of the time. However, because the sample data are random, the sample means are also random and may be very far away from the true mean. For the example of the global temperature, we might assume that the "true" mean is 14°C and the "true" standard deviation is 0.3°C. Here, "true" is an assumption, however, since no one knows the truth. The sample means are close to 14 most of the time, but climate variations may lead to a sample mean being equal to 16°C or

12 °C, thus far away from the "true" mean 14 °C. We can use the interval $[a,b]$ to quantify the probability of the sample mean being inside this interval. We wish to say that with 95% probability, the sample mean is inside this interval $[a,b]$. This leads to the following confidence interval formula.

For a normally distributed population $(x_1, x_2, \ldots, x_n)$ with the same mean μ and standard deviation σ, the confidence interval of the sample mean $\bar{x} = \sum_{i=1}^n x_i/n$, at the 95% confidence level is

$$(\mu - 1.96\sigma/\sqrt{n}, \mu + 1.96\sigma/\sqrt{n}). \tag{3.18}$$

Namely, with 95% probability, the sample mean $\bar{x}$ is

$$\mu - 1.96\sigma/\sqrt{n} < \bar{x} < \mu + 1.96\sigma/\sqrt{n}. \tag{3.19}$$

Usually, $a = \mu - 1.96\sigma/\sqrt{n}$ is called the lower limit of the confidence interval, and $b = \mu + 1.96\sigma/\sqrt{n}$ the upper limit.

One can easily simulate this confidence interval formula by the following R code.

```
#Confidence interval simulation
mu=14 #true mean
sig=0.3 #true sd
n=50 #sample size
d=1.96*sig/sqrt(n)
lowerlim=mu-d
upperlim=mu+d
ksim=10000 #number of simulations
k=0 #k is the simulation counter
xbar=1:ksim
for (i in 1:ksim)
{
  xbar[i]=mean(rnorm(n, mean=mu, sd=sig))
  if (xbar[i] >= lowerlim & xbar[i] <= upperlim)
    k=k+1
}
print(c(k,ksim))
#[1]  9496 10000

#plot the histogram
hist(xbar,breaks=51,
     main="Histogram of the Simulated
     Sample Mean Temperatures",xaxt="n",
     xlab=expression(paste("Temperature [", degree, "C]")),
     ylim=c(0,600))
axis(1,pos = -20, at = c(13.92, mu, 14.08))
text(14,530,"95% Confidence Interval (13.92, 14.08)")
```

This simulation shows that 9,496 of the 10,000 simulations have the sample means inside the confidence interval. The probability is thus 0.9496, or approximately 0.95. Figure 3.10 displays the histogram of the simulation results. It shows that 9,496 sample means from among 10,000 are in the confidence interval $(13.41, 14.59)$. Only 504 sample means are outside the interval with 254 in $(-\infty, 13.41)$ and 250 in $(14.59, \infty)$. Thus, the confidence level is the probability of the sample mean falling into the confidence interval. Intuitively,

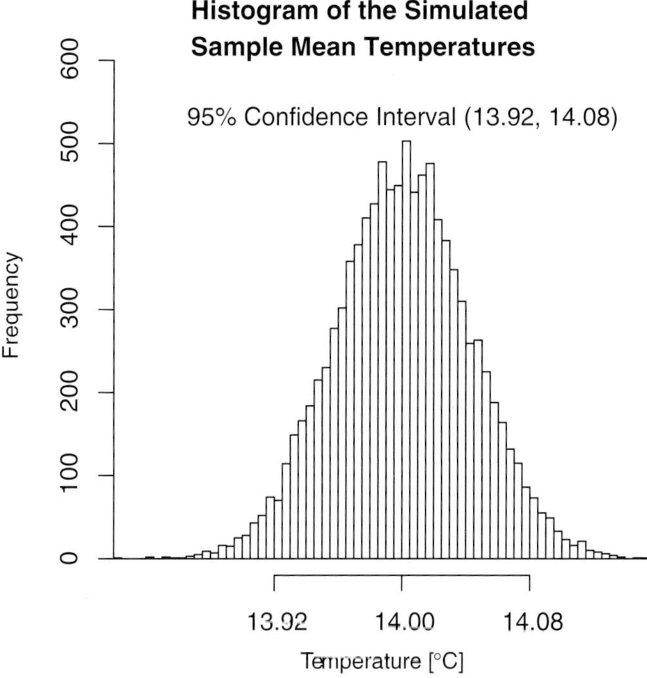

Fig. 3.10 Histogram of 10,000 simulated sample mean temperatures based on the assumption of normal distribution with the "true" mean equal to 14 °C and "true" standard deviation 0.3 °C. Approximately, 95% of the sample means are within the confidence interval (13.92, 14.08), 2.5% in $(14.08, \infty)$, and 2.5% in $(-\infty, 13.92)$.

when the confidence interval is small, the confidence level is low since there is a smaller chance for the sample mean to fall into a smaller interval.

3.4.2 Mean of a Large Sample Size: Approximately Normal Distribution

The purpose of computing the sample mean is to use it as an estimate for the real true mean that we do not know in practice. This estimation is more accurate when the confidence interval is small. The extreme case is that the confidence interval has zero length, which means that with 95% chance, the sample mean is exactly equal to the true mean. The chance to be wrong is only 5%. To be more accurate, our intuition suggests that we need to have a small standard deviation, and have a large sample. The above confidence interval formula (3.18) quantifies this intuition $(\mu - 1.96\sigma/\sqrt{n}, \mu + 1.96\sigma/\sqrt{n})$. A small σ and a large n enable us to have a small confidence interval, and hence an accurate estimation of the mean. Thus, to obtain an accurate result in a survey, one should use a large sample.

3.4.2.1 Confidence Interval of the Sample Mean

This subsection shows a method to find out how large a sample should be, for the case when the confidence probability is given. We also want to deal with the practical situation where the true mean and standard deviation are almost never known. Furthermore, it is usually not known whether the random variable is in fact normally distributed. These two problems can be solved by a very important theoretical result of mathematical statistics, called the Central Limit Theorem (CLT), which says that when the sample size n is sufficiently large, the sample mean $\bar{x} = \sum_{i=1}^{n} x_i/n$ is approximately normally distributed, regardless of the distributions of $x_i (i = 1,2,\ldots,n)$. The approximation becomes better when n becomes larger. Some textbooks suggest that $n = 30$ is good enough to be considered a "large" sample; others use $n = 50$. In climate science, we often use $n = 30$.

When the number of samples is large in this sense, the normal distribution assumption for the sample mean is taken care of. We then compute the sample mean and sample standard deviation by the following formulas

$$\bar{x} = \frac{1}{n}\sum_{i=1}^{n} x_i, \tag{3.20}$$

$$S = \sqrt{\frac{1}{n-1}\sum_{i=1}^{n} (x_i, -\bar{x})^2}. \tag{3.21}$$

The standard error of the sample mean is defined as

$$SE(\bar{x}) = \frac{S}{\sqrt{n}}. \tag{3.22}$$

This gives the size of the "error bar" $\bar{x} \pm SE$ when approximating the true mean using the sample mean.

The error margin at a 95% confidence level is

$$EM = 1.96\frac{S}{\sqrt{n}}, \tag{3.23}$$

where 1.96 comes from the 95% probability in $(\mu - 1.96\sigma, \mu + 1.96\sigma)$ for a normal distribution. When the confidence level α is raised from 0.95 to a larger value, the number 1.96 will be increased to a larger number accordingly.

The confidence interval for a true mean μ is then defined as

$$(\bar{x} - EM, \bar{x} + EM) \text{ or } \left(\bar{x} - 1.96\frac{S}{\sqrt{n}}, \bar{x} + 1.96\frac{S}{\sqrt{n}}\right). \tag{3.24}$$

This means that the given samples imply that the probability for the true mean to be inside the confidence interval $(\bar{x} - EM, \bar{x} + EM)$ is 0.95, or α in general. Similarly, the probability for the true mean to be inside the error bar $\bar{x} \pm SE$ is 0.68. See Fig. 3.11 for the confidence intervals at 95% and 68% confidence levels.

When the sample size n goes to infinity, the error margin EM goes to zero, and accordingly, the sample mean is equal to the true mean. This is correct with 95% probability, and wrong with 5% probability.

One can also understand the sample confidence interval for a new variable

$$z = \frac{\bar{x} - \mu}{S/n}, \tag{3.25}$$

which is a normally distributed variable with mean equal to zero and standard deviation equal to one, i.e., it has standard normal distribution. The variable $y = -z$ also satisfies the standard normal distribution. So, the probability of $-1 < z < 1$ is 0.68, and $-1.96 < z < 1.96$ is 0.95. The set $-1.96 < z < 1.96$ is equivalent to $\bar{x} - 1.96S/\sqrt{n} < \mu < \bar{x} + 1.96S/\sqrt{n}$. Thus, the probability of the true mean being in the confidence interval of the sample mean $\bar{x} - 1.96S/\sqrt{n} < \mu < \bar{x} + 1.96S/\sqrt{n}$ is 1.96. This explanation is visually displayed in Fig. 3.11.

In addition, the formulation $\bar{x} = \mu + zS/\sqrt{n}$ corresponds to a standard statistics problem for an instrument with observational errors:

$$y = x \pm \epsilon, \tag{3.26}$$

where ϵ stands for errors, x is the true but never-known value to be observed, and y is the observational data. Thus, data are equal to the truth plus errors. The expected value of the error is zero and the standard deviation of the error is $S/\sqrt{n}$, also called standard error.

The confidence level 95% comes into the equation when we require that the observed value must lie in the interval $(\mu - EM, \mu + EM)$ with a probability equal to 0.95. This corresponds to the requirement that the standard normal random variable z is found in the interval (z_-, z_+) with a probability equal to 0.95, which implies that $z_- = -1.96$ and $z_+ = 1.96$. Thus, the confidence interval of the sample mean at the 95% confidence level is

$$(\bar{x} - 1.96S/\sqrt{n}, \bar{x} + 1.96S/\sqrt{n}), \tag{3.27}$$

or

$$(\bar{x} - z_{\alpha/2}S/\sqrt{n}, \bar{x} + z_{\alpha/2}S/\sqrt{n}), \tag{3.28}$$

where $z_{\alpha/2} = z_{0.05/2} = 1.96$. So, $1 - \alpha = 0.95$ is used to represent the probability inside the confidence interval, while $\alpha = 0.05$ is the "tail probability" outside of the confidence interval. Outside of the confidence interval means occurring on either the left side or the right side of the distribution. Each side represents $\alpha/2 = 0.025$ tail probability. The red area of Fig. 3.11 indicates the tail probability.

Figure 3.11 can be plotted by the following R code.

```
#Plot confidence intervals and tail probabilities
par(mar=c(2.5,3.5,2.0,0.5))
rm(list=ls())
par(mgp=c(1.4,0.5,0))
curve(dnorm(x,0,1), xlim=c(-3,3), lwd=3,
      main='Confidence Intervals and Confidence Levels',
      xlab="True mean as a random variable", ylab="",
      xaxt="n", cex.lab=1.2)
title(ylab='Probability density', line=2, cex.lab=1.2)
polygon(c(-1.96, seq(-1.96,1.96,len=100), 1.96),
        c(0,dnorm(seq(-1.96,1.96,len=100)),0),col='skyblue')
polygon(c(-1.0,seq(-1.0, 1, length=100), 1),
        c(0, dnorm(seq(-1.0, 1, length=100)), 0.0),col='white')
polygon(c(-3.0,seq(-3.0, -1.96, length=100), -1.96),
        c(0, dnorm(seq(-3.0, -1.96, length=100)), 0.0),col='red')
polygon(c(1.96,seq(1.96, 3.0, length=100), 3.0),
        c(0, dnorm(seq(1.96, 3.0, length=100)), 0.0),col='red')
points(c(-1,1), c(0,0), pch=19, col="blue")
```

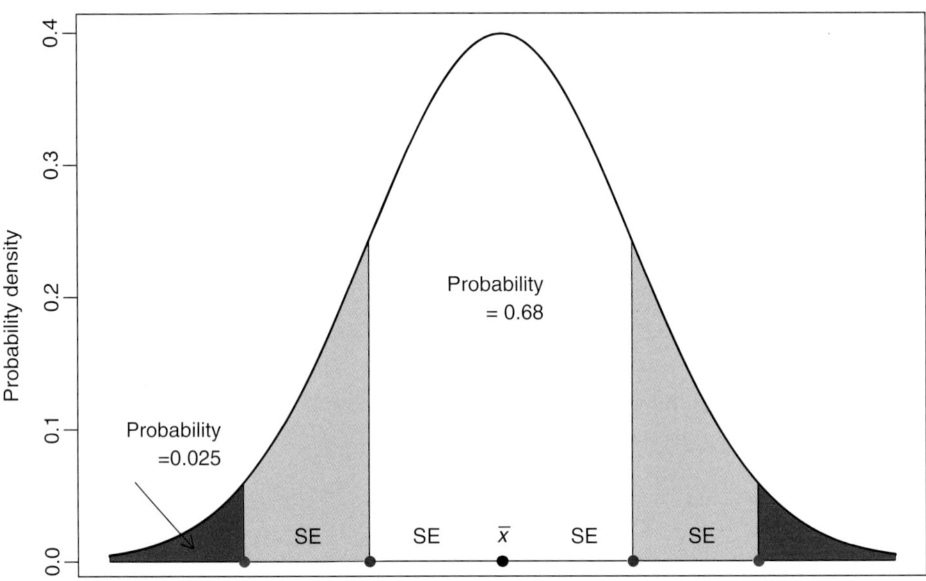

Fig. 3.11 Schematic illustration of confidence intervals and confidence levels of a sample mean for a large sample size. The confidence interval at 95% confidence level is between the two red points, and that at 68% is between the two blue points. SE stands for the standard error, and 1.96 SE is approximately regarded as 2 SE in this figure.

```
points(0,0, pch=19)
points(c(-1.96,1.96),c(0,0),pch=19, col="red")
text(0,0.02, expression(bar(x)), cex=1.0)
text(-1.50,0.02, "SE", cex=1.0)
text(-0.60,0.02, "SE", cex=1.0)
text(1.50,0.02, "SE", cex=1.0)
text(0.60,0.02, "SE", cex=1.0)
text(0,0.2, "Probability
     = 0.68")
arrows(-2.8,0.06,-2.35,0.01, length=0.1)
text(-2.5,0.09, "Probability
     =0.025")
```

In practice, we often regard 1.96 as 2.0, and the 2σ-error bar as the 95% confidence interval.

Example 3.1 Estimate (a) the mean of the 1880–2015 global average annual mean temperatures of the Earth, and (b) the confidence interval of the sample mean at the 95% confidence level.

The answer is that the mean is $-0.2034\,^\circ$C and the confidence interval is $(-0.2545, -0.1524)^\circ$C. These values may be obtained by the following R code.

```
#Estimate the mean and error bar for a large sample
#Confidence interval for NOAAGlobalTemp 1880-2015
setwd("/Users/sshen/climmath")
dat1 <- read.table("data/aravg.ann.land_ocean.90S.90N.v4.0.0.2015.txt")
dim(dat1)
tmean15=dat1[,2]
MeanEst=mean(tmean15)
sd1 =sd(tmean15)
StandErr=sd1/sqrt(length(tmean15))
ErrorMar = 1.96*StandErr
MeanEst
#[1] -0.2034367
print(c(MeanEst-ErrorMar, MeanEst+ErrorMar))
#[1] -0.2545055 -0.1523680
```

3.4.2.2 Estimate the Required Sample Size

The standard error $SE = \sigma/\sqrt{n}$ measures the accuracy of using a sample mean as an estimate of the true mean when the standard deviation of the population is given as σ. A practical problem is to determine the sample size when the accuracy level SE is given. The formula is then

$$n = \left(\frac{\sigma}{SE}\right)^2. \tag{3.29}$$

Example 3.2 The standard deviation of the global average annual mean temperature is given to be 0.3 °C. The standard error is required to be less than or equal to 0.05 °C. Find the minimal sample size required.

The solution is $(0.3/0.05)^2 = 36$. The sample size must be greater than or equal to 36.

3.4.2.3 Statistical Inference for $\bar{x}$ Using a z-score

Figure 3.1 seems to suggest that the average of the global average annual mean temperature anomalies from 1880 to 1939 is significantly below zero. We wish to know whether we can statistically justify that this inference is true, with the probability of being wrong less than or equal to 0.025, or 2.5%. This probability is called the significance level. Figure 3.12 shows the significance level as the tail probability in $(-\infty, z_{0.025})$.

Figure 3.12 can be generated by the following R code.

```
setEPS()
postscript("fig0312.eps", height=7, width=10)
par(mar=c(2.3,3.0,2.0,0.5))
rm(list=ls())
par(mgp=c(1.0,0.5,0))
curve(dnorm(x,0,1), xlim=c(-3,3), lwd=3,
      main='Z-score, p-value, and significance level',
      xlab="z: Standard normal random variable",
      ylab='Probability density',xaxt="n",yaxt="n",
      cex.lab=1.2, ylim=c(-0.1,0.4))
```

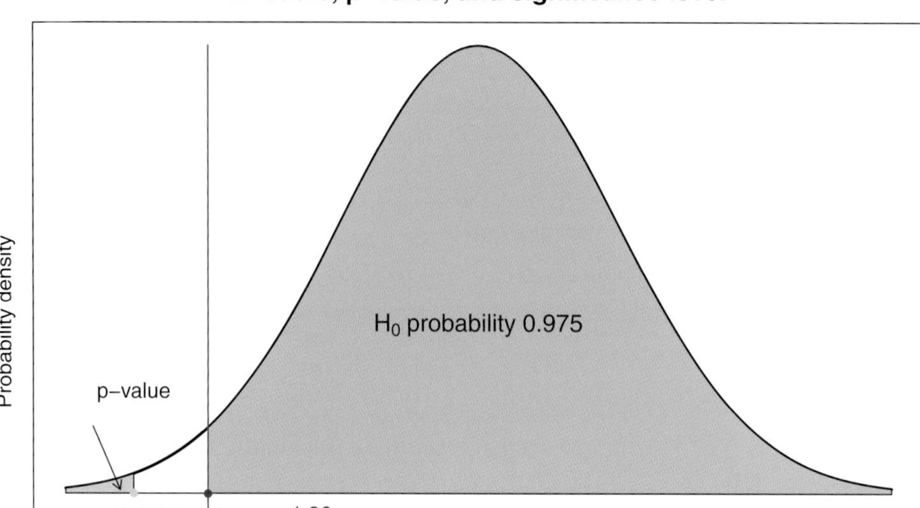

Fig. 3.12 The standard normal distribution chart for statistical inference: z-score, p-value for $\bar{x} < \mu$, and significance level 2.5%. The value $z_{0.025} = -1.96$ is called the critical z-score for this hypothesis test.

```
lines(c(-3,3),c(0,0))
arrows(-3,-0.1,-2.02,-0.1, lwd=12,col='skyblue', length=0.2, code=3)
arrows(3,-0.1,-1.90,-0.1, lwd=12,col='green', length=0.2, code=3)
polygon(c(-3.0,seq(-3.0, -2.5, length=100), -2.5),
        c(0, dnorm(seq(-3.0, -2.5, length=100)), 0.0),col='skyblue')
polygon(c(-1.96,seq(-1.96, 3, length=100), 3),
        c(0, dnorm(seq(-1.96, 3, length=100)), 0.0),col='lightgreen')
points(-1.96,0, pch=19, col="red")
points(-2.5,0,pch=19, col="skyblue")
text(-1.5,-0.02, expression(z[0.025]~'=-1.96'), cex=1.3, col='red')
text(-2.40,-0.02, "z-score", cex=1.3, col='skyblue')
arrows(-2.8,0.06,-2.6,0.003, length=0.1)
lines(c(-1.96,-1.96),c(-0.1, .4),lwd=1.5, col='red')
text(-2.5,0.09, "p-value", cex=1.3)
text(1.0,-0.06, expression(H[0] ~'region'), cex=1.5)
text(-2.5,-0.06, expression(H[1] ~'region'), cex=1.5)
text(0,0.15, expression(H[0] ~'probability 0.975'), cex=1.5)
dev.off()
```

To make the justification, we compute a parameter

$$z = \frac{\bar{x} - \mu}{S/\sqrt{n}}, \tag{3.30}$$

where $\bar{x}$ is the sample mean, S is the sample standard deviation, and n is the sample size. This z value is called the z-statistic, or simply the z-score, which follows the standard normal distribution, because the sample size $n = 60$ is large. From the z-score, we can determine the probability of the random variable z being in a certain interval, such as $(-\infty, z_s)$. This significance level 2.5% corresponds to $z_s = -1.96$ according to Fig. 3.11. Thus, the z-score can quantify how significantly is z different from zero, which is equivalent to the sample mean being significantly different from the assumed or given value. The associated probability, e.g., the probability in $(-\infty, z)$, is called the p-value that measures the chance of a wrong inference. We want this p-value to be small in order to be able to claim significance. The typical significance levels used in practice are 5%, 2.5%, and 1%. Choosing which level to use depends on the nature of the problem. For drought conditions, one may use 5%, while for flood control and dam design, one may choose 1%. A statistical inference is significant when the p-value is less than the given significance level.

For our problem of 60 years of data from 1880–1939, the sample size is $n = 60$. The sample mean can be computed by an R command `xbar=mean(tmean15[1:60])`, and the sample standard deviation can be computed by `S=sd(tmean15[1:60])`. The results are $\bar{x} = -0.4500$ and $S = 0.1109$.

When $\mu = 0$, the z-score computed using formula (3.30) is -31.43. The probability in the interval $(-\infty, z)$ is tiny, namely 4.4×10^{-217}, which can be regarded as zero. We can thus conclude that the sample mean from 1880–1939 is significantly less than zero at a p-value equal to 4.4×10^{-217}, which means that our conclusion is correct at a significance level of 2.5%.

A formal statistical terminology for the above inference is called a hypothesis test, which tests a null hypothesis

$$H_0 : \bar{x} \geq 0, \quad \text{(Null hypothesis: the mean is not smaller than zero)} \tag{3.31}$$

and an alternative hypothesis

$$H_1 : \bar{x} < 0, \quad \text{(Alternative hypothesis: the mean is smaller than zero).} \tag{3.32}$$

Our question of the average temperature from 1880–1939 is whether to reject the null hypothesis and confirm the alternative hypothesis. The method is to examine where the z-score point is on a standard normal distribution chart and what is the corresponding p-value. Thus, the statistical inference becomes a problem of z-score and p-value using the standard normal distribution chart (see Fig. 3.12). Our z-score is -31.43 in the H_1 region, and our p-value is 4.4×10^{-217}, much less than 0.025. We thus accept the alternative hypothesis, i.e., we reject the null hypothesis with a tiny p-value 4.4×10^{-217}. We conclude that the 1880–1939 mean temperature is significantly less than zero.

One can similarly formulate a hypothesis test for a warming period from 1981–2015 and ask whether the average temperature during this period is significantly greater than zero. The two hypotheses are

$$H_0 : \bar{x} \leq 0, \quad \text{(Null hypothesis: the mean is not greater than zero)} \tag{3.33}$$

and an alternative hypothesis

$$H_1 : \bar{x} > 0, \quad \text{(Alternative hypothesis: the mean is greater than zero).} \tag{3.34}$$

One can follow the same procedure to compute the z-score, see whether it is in the H_0 region or H_1 region, and compute the p-value. Finally an inference can be made based on the z-score and the p-value.

3.4.3 Mean of a Small Sample Size t-Test

The hypothesis test in the above subsection is based on the standard normal distribution for the cases of a large sample size, say, at least 30. When the sample size is small, the sample mean satisfies a t-distribution, not a normal distribution.

3.4.3.1 $H_1 : \bar{T} > 0$ Test for the 2006–2016 Global Average Annual Temperature

Thus, when the sample size n is small, say less than 10, and the variance is to be estimated, then we should use a t-distribution, because

$$t = \frac{\bar{x} - \mu}{S/\sqrt{n}} \tag{3.35}$$

follows a t-distribution of the degrees-of-freedom (df or dof) equal to $n - 1$. Figure 3.9 shows that the t-distribution is flatter than the corresponding normal distribution (of the same sample mean and sample variance) and has fatter tails. When the dof increases to infinity, the t-distribution approaches the normal distribution of the sample mean and sample variance.

The hypothesis test procedure is the same as before, except the standard normal distribution is now replaced by the t-distribution with dof equal to $n - 1$.

Example 3.3 Test whether the global average annual mean temperature from 2006–2015 is significantly greater than the 1961–1990 climatology, i.e., whether the sample mean is greater than zero.

The two hypotheses are

$$H_0 : \bar{T} \leq 0, \quad \text{(Null hypothesis: The 2006–2015 mean is not greater than zero)} \tag{3.36}$$

and

$$H_1 : \bar{T} > 0, \quad \text{(Alternative hypothesis: The mean is greater than zero).} \tag{3.37}$$

One can follow the same procedure as in the last section to compute the t-score, see whether it is in the H_0 region or H_1 region, and to compute the p-value. We have 10 years of data from 2006–2015, which is a small sample with a sample size $n = 10$. The following R code computes the sample mean 0.4107, standard deviation 0.1023, t-score 12.6931, p-value 2.383058×10^{-7}, and the critical t-value: $t_{0.975} = 2.2622$. The t-score is in the alternative hypothesis region with a very small p-value. Therefore, we conclude that the average temperature from 2006–2015 is significantly greater than zero.

```
 #Hypothesis test for NOAAGlobalTemp 2006-2015
setwd("/Users/sshen/climmath")
dat1 <- read.table("data/aravg.ann.land_ocean.90S.90N.v4.0.0.2015.txt")
tm0615=dat1[127:136,2]
MeanEst=mean(tm0615)
MeanEst
#[1] 0.4107391
```

```
sd1 =sd(tm0615)
sd1
#[1] 0.1023293
n=10
t_score=(MeanEst -0)/(sd1/sqrt(n))
t_score
#[1] 12.69306
1-pt(t_score, df=n-1)
#[1] 2.383058e-07 #p-value
qt(1-0.025, df=n-1)
#[1] 2.262157 #critical t-score
```

For the standard normal distribution, $z_{0.975} = 1.96 < t_{0.975} = 2.2622$, because the t-distribution is flatter than the corresponding normal distribution and has fatter tails. Thus, the critical t-scores are larger.

Clearly, one should use the t-test to make the inference when the sample size is very very small, say, $n = 7$. However, it is unclear whether one should use the t-test or the z-test if the sample size is, say, 27. The recommendation is to always use the t-test if you are not sure whether the z-test is applicable, because the t-test has been mathematically proven to be accurate, while the z-test is an approximation. Since the t-distribution approaches the normal distribution when dof approaches infinity, the t-test will yield the same result as the z-test when the z-test is applicable.

3.4.3.2 Compare Temperatures of Two Short Periods

A common question in climate science is whether the temperature in one decade is significantly greater than the temperature in another. The task is thus to compare the temperatures of two decades.

The general problem is whether the sample mean of the data $\{T_{11}, T_{12}, \ldots, T_{1n_1}\}$ and the sample mean of another set of data $\{T_{21}, T_{22}, \ldots, T_{2n_2}\}$ are significantly different from each other. The t-statistic for this problem can be computed using the following formula:

$$t = \frac{\bar{T}_2 - \bar{T}_1}{S_{pooled}\sqrt{\frac{1}{n_1} + \frac{1}{n_2}}}, \tag{3.38}$$

where $\bar{T}_1$ and $\bar{T}_2$ are the two sample means

$$\bar{T}_1 = \frac{T_{11} + T_{12} + \cdots + T_{1n_1}}{n_1}, \tag{3.39}$$

$$\bar{T}_2 = \frac{T_{21} + T_{22} + \cdots + T_{2n_2}}{n_2}, \tag{3.40}$$

S_{pooled} is the pooled sample standard deviation

$$S_{pooled} = \sqrt{\frac{(n_1 - 1)S_1^2 + (n_2 - 1)S_2^2}{n_1 + n_2 - 2}}, \tag{3.41}$$

and S_1 and S_2 are the two sample standard deviations

$$S_1 = \sqrt{\frac{(T_{11} - \bar{T}_1)^2 + (T_{12} - \bar{T}_1)^2 + \cdots + (T_{1n_1} - \bar{T}_1)^2}{n_1 - 1}}, \tag{3.42}$$

$$S_2 = \sqrt{\frac{(T_{21} - \bar{T}_2)^2 + (T_{22} - \bar{T}_2)^2 + \cdots + (T_{2n_2} - \bar{T}_2)^2}{n_2 - 1}}. \tag{3.43}$$

This t-statistic follows a t-distribution of dof equal to $n_1 + n_2 - 2$.

Example 3.4 Investigate whether the global average annual mean temperature in the decade of 1991–2000 is significantly different from the previous decade.

The two statistical hypotheses are

$$H_0 : \bar{T}_1 = \bar{T}_2 \quad \text{(Null hypothesis: The temperatures of the two decades are the same)} \tag{3.44}$$

and

$$H_1 : \bar{T}_1 \neq \bar{T}_2 \quad \text{(Alternative hypothesis: The two decades are different).} \tag{3.45}$$

This is a two-sided test. The alternative region is a union of both sides $(-\infty, t_{0.025})$ and $(t_{0.975}, \infty)$ if the significance level is set to be 5%. We will compute the t-score using formula (3.38). The result is below:

a. the t-score is 2.5784,
b. the H_0 region is $(-2.1009, 2.1009)$,
c. the p-value is 0.009470,
d. the mean of the temperature anomalies in 1981–1990 is 0.036862 °C, and
e. the mean of the temperature anomalies in 1991–2000 is 0.161255 °C.

The t-score is outside the H_0 region. Thus, the H_0 hypothesis is rejected. The 1991–2000 mean temperature anomaly 0.161255 °C is significantly different from the 1981–1990 mean 0.036862 °C with a p-value equal to 1%. The temperature difference of the two decades is $0.124392 = 0.161255 - 0.036862$ °C which is significantly different from zero.

The above results were obtained by the following R code.

```
#Hypothesis test for global temp for 1981-1990 and 1991-2000
setwd("/Users/sshen/climmath")
dat1 <- read.table("data/aravg.ann.land_ocean.90S.90N.v4.0.0.2015.txt")
tm8190=dat1[102:111,2]
tm9100=dat1[112:121,2]
barT1=mean(tm8190)
barT2=mean(tm9100)
S1sd=sd(tm8190)
S2sd=sd(tm9100)
n1=n2=10
Spool=sqrt(((n1 - 1)*S1sd^2 + (n2 - 1)*S2sd^2)/(n1 + n2 -2))
t = (barT2 - barT1)/(Spool*sqrt(1/n1 + 1/n2))
tlow = qt(0.025, df= n1 + n2 -2)
tup = qt(0.975, df= n1 + n2 -2)
```

```
paste("t-score=", round(t,digits=5),
      "tlow=", round(tlow,digits=5),
      "tup=", round(tup,digits=5))
#[1] "t-score= 2.57836 tlow= -2.10092 tup= 2.10092"
pvalue = 1-pt(t,  df= n1 + n2 -2)
paste( "p-value=", pvalue)
#[1] "p-value= 0.00947040009284539"
paste("1981-90 temp=", barT1, "1991-00 temp=",barT2)
#[1] "1981-90 temp= 0.0368621 1991-00 temp= 0.1612545"
barT2 - barT1
#[1] 0.1243924
```

The above is a two-sided test to determine if a sample mean is different from zero. However, the time series of the global temperature in Fig. 3.1 had already indicated that the 1991–2000 decade is warmer than 1981–1990. If we take this as a given prior knowledge, then we should use the one-sided test with the following two hypotheses

$$H_0 : \bar{T}_1 > \bar{T}_2 \quad \text{(Null hypothesis: The temperatures of the two decades are the same)} \tag{3.46}$$

and

$$H_1 : \bar{T}_1 \leq \bar{T}_2 \quad \text{(Alternative hypothesis: The two decades are different).} \tag{3.47}$$

The t-score is the same as the above, but the critical t-score is now $t_{0.95} = 1.734$. Again, the t-score 2.57836 is in the H_1 region.

3.5 Statistical Inference of a Linear Trend

When studying climate change, one often makes a linear regression and asks if a linear trend is significantly positive, or negative, and different from zero. For example, is the linear trend of the global average annual mean temperature from 1880–2015 shown in Fig. 3.1 significantly greater than zero? This is again a t-test problem. The estimated trend $\hat{b}$ from a linear regression follows a t-distribution.

With the given data pairs $\{(x_i, y_i), i = 1, 2, \cdots, n\}$ and their regression line

$$\hat{y} = \hat{b}_0 + \hat{b}_1 x, \tag{3.48}$$

the t-score for the trend $\hat{b}_1$ is defined by the following formula

$$t = \frac{\hat{b}_1}{S_n/\sqrt{S_{xx}}}, \qquad dof = n-2. \tag{3.49}$$

Here,

$$S_n = \sqrt{\frac{SSE}{n-2}} \tag{3.50}$$

with the sum of squared errors SSE defined as

$$SSE = \sum_{i=1}^{n} \left[y_i - (\hat{b}_0 + \hat{b}_1 x_i)\right]^2, \tag{3.51}$$

and

$$S_{xx} = \sum_{i=1}^{n} (x_i - \bar{x})^2 \tag{3.52}$$

with the sample means of x-data and y-data defined as

$$\bar{x} = \frac{\sum_{i=1}^{n} x_i}{n} \tag{3.53}$$

$$\bar{y} = \frac{\sum_{i=1}^{n} y_i}{n}, \tag{3.54}$$

the regression slope estimated by

$$\hat{b}_1 = \frac{\sum_{i=1}^{n} (x_i - \bar{x})(y_i - \bar{y})}{S_{xx}}, \tag{3.55}$$

and the regression intercept estimated by

$$\hat{b}_0 = \bar{y} - \hat{b}_1 \bar{x}. \tag{3.56}$$

The dof of this t-score is $n-2$. With this dof and a specified significance level, one can then find the critical t-values and determine whether the t-score is in the H_0 region or H_1 region.

We wish to use the t-inference procedure to check if the 1880–2015 temperature anomalies trend is significantly greater than zero. The statistical hypotheses are

$$H_0 : \bar{b}_1 < 0 \text{ (Null hypothesis: The trend is not greater than zero)} \tag{3.57}$$

and

$$H_1 : \bar{b}_1 \geq 0 \text{ (Alternative hypothesis: The trend is greater than zero).} \tag{3.58}$$

This is a one-sided test. The critical t-score is now $t_{0.95} = 1.734$. The summary of the linear regression command gives the needed statistical values:

a. the trend is $\hat{b}_1 = 0.667791\,^\circ$C per century,
b. the t-score for $\hat{b}_1$ is 20.05,
c. the p-value is 1×10^{-16}, and
d. the critical t value is 1.6563 by an R command `qt(0.95, 134)`.

Clearly, the t-score 20.05 is in the H_1 region. We conclude that the trend is significantly greater than zero with a p-value equal to 1×10^{-16}.

The above results were computed by the following R code.

```
setwd("/Users/sshen/climmath")
dat1 <- read.table("data/aravg.ann.land_ocean.90S.90N.v4.0.0.2015.txt")
tm=dat1[,2]
x = 1880:2015
summary(lm(tm ~ x))
#Coefficients
#  Estimate Std. Error t value Pr(>|t|)
#(Intercept) -1.321e+01  6.489e-01  -20.36   <2e-16 ***
#  x            6.678e-03  3.331e-04   20.05   <2e-16 ***
```

Sometimes one may need to check if the trend is greater than a specified value β_1. Then, the t-score is defined by the following formula

$$t = \frac{\hat{b}_1 - \beta_1}{S_n/\sqrt{S_{xx}}}, \qquad \text{dof} = n - 2. \tag{3.59}$$

In this case, the t-score must be computed from the formulas, not from the summary of a linear regression by R.

3.6 Free Online Statistics Tutorials

This statistics chapter has presented a very brief course in statistics, but it provides a sufficient statistics basis and R codes for doing simple statistical analyses of climate data. This chapter also provides the foundation for expanding a reader's statistics knowledge and skills by studying more comprehensive or advanced materials on climate statistics. A few free statistics tutorials available online are introduced below.

The manuscript by David Stephenson (2005) of the University of Reading, United Kingdom, provides the basics of statistics with climate data as examples:

http://empslocal.ex.ac.uk/people/staff/dbs202/cag/courses/MT37C/course-d.pdf

This online manuscript is appropriate for readers who have virtually no statistics background.

Eric Gilleland (2009) of NCAR authored a slide series of lecture notes for using R to do climate statistics, in particular the analysis of extreme values:

www.maths.lth.se/seamocs/meetings/Malta_Posters_and_Talks/
MaltaShortCourseSlides4.pdf

This set of lecture notes provides many R codes for analyzing climate data, such as risk estimation. The material is very useful for climate data users, and does not require much mathematical background.

The report *Statistical methods for the analysis of simulated and observed climate data applied in projects and institutions dealing with climate change impact and adaptation* by the Climate Service Center, Hamburg, Germany, is particularly useful for weather and climate data.

www.climate-service-center.de/imperia/md/content/csc/projekte/csc-report13_englisch_final-mit_umschlag.pdf

This online report provides a "user's manual" for a large number of statistical methods used for climate data analysis with real climate data examples. The material is an excellent reference for users of the statistics for climate data.

3.7 Chapter Summary

This chapter has provided a brief introduction to useful statistical concepts and methods for climate science and has included the following materials.

(i) Formulas to compute the most commonly used statistical indices:
- Mean as an average, median as a datum that is larger than 50% of the data being analyzed,
- Standard deviation as a measure of the width of the probability distribution,
- Variance as the square of the standard deviation,
- Skewness as a measure of the degree of asymmetry in distribution, and
- Kurtosis as a measure of the peakedness of the data distribution compared to that of the normal distribution.

(ii) The commonly used statistics plots:
- Histogram for displaying the probability distribution of the data,
- Scatter plot for showing the coherent structure of the data,
- Linear regression line for providing a linear model for the data,
- Box plot for quantifying the probability distribution of the data, and
- Q–Q plot for checking whether the data are normally distributed.

(iii) Normal distribution determined by mean μ and standard deviation σ, displayed as a bell-curve, and mathematically expressed by an exponential function

$$f(x|\mu,\sigma^2) = \frac{1}{\sigma\sqrt{2\pi}} \exp\left(-\frac{(x-\mu)^2}{2\sigma^2}\right). \tag{3.60}$$

(iv) The concept of confidence interval (CI) within which an estimate lies with a given probability expressed as a percentage, called the confidence level (CL).

(v) Student's t-distribution when the standard deviation is unknown and has to be computed from the data.

(vi) Using the hypothesis test approach to make objective conclusions based on data with a given significance level, t-score, and p-value.

R codes and climate data examples are given to demonstrate these concepts and the use of the relevant formulas. This chapter is a kind of "user's manual" covering the most commonly used statistical methods in climate data analysis, so that users can arrive at credible conclusions based on the data, together with a given error probability. With this background plus some R programming skill, you will have sufficient knowledge to meet the needs of routine statistical analysis of climate data.

To this end, you may ask what is statistics? The answer is apparently not unique; but the purpose of statistics is clear: to draw credible and objective conclusions from the data. The word "statistics" comes from the Latin "status" meaning "state." We use the term "statistics" to mean a suite of scientific methods for analyzing climate data and for drawing conclusions about climate phenomena. Statistical methods are routinely used for analyzing and drawing conclusions from climate data, such as for calculating the climate "normal" of precipitation at a weather station and for quantifying the reliability of the calculation. Statistical methods have often been used for demonstrating that global warming is occurring, based on a significant upward trend of data such as surface air temperature (SAT) anomalies, and on establishing a given significance level for this trend.

You may want to keep in mind that although the statistical calculations and methods themselves are "objective," the interpretations and consequences of the conclusions are highly context-dependent. Think of the differences between the odds of rejecting a meal in a good restaurant and the odds of rejecting a flight on an unsafe airplane. Always remember that meaningful interpretation of the statistical results of analyzing climate data requires substantial domain knowledge of climate science.

References and Further Readings

[1] Hennemuth, B., S. Bender, K. Bulow, N. Dreier, E. Keup-Thiel, O. Kruger, C. Mudersbach, C. Radermacher, and R. Schoetter, 2013: *Statistical methods for the analysis of simulated and observed climate data, applied in projects and institutions dealing with climate change impact and adaptation.* CSC Report 13, Climate Service Center, Germany URL:

www.climate-service-center.de/imperia/md/content/csc/projekte/csc-report13_englisch_final-mit_umschlag.pdf

> This free statistics recipe book outlines numerous methods for climate data analysis, which are collected and edited by climate science professionals and have examples of real climate data.

[2] Gilleland, E., 2009: *Statistical Software for Weather and Climate: The R Programming Language*. URL:

www.maths.lth.se/seamocs/meetings/Malta_Posters_and_Talks/MaltaShortCourseSlides4.pdf

> This is a set of slides developed by the author for an interdisciplinary workshop. The slides cover R codes from basic statistics to advanced methods for climate data analysis, such as extreme value analysis (EVA).

[3] IPCC, 2013: *Climate Change 2013: The Physical Science Basis. Contribution of Working Group I to the Fifth Assessment Report of the Intergovernmental Panel on Climate Change* [Stocker, T. F., D. Qin, G.-K. Plattner, M. Tignor, S. K. Allen, J. Boschung, A. Nauels, Y. Xia, V. Bex and P. M. Midgley (eds.)], Cambridge University Press, New York.

This is an IPCC report, famous around the world. The IPCC assessment reports appear about every six years, and they are the definitive summary of our scientific understanding of climate change. This particular volume, the Working Group One report, deals with the physical science of the climate system. At some 1500 pages, with lots of charts, graphs, and technical language, it is not easy reading. It and the many other IPCC reports are available for free at the IPCC website, `www.ipcc.ch`.

[4] Stephenson, D.B., 2005: *Data Analysis Methods in Weather and Climate Research.* Lecture Notes, University of Reading, UK URL:

http://empslocal.ex.ac.uk/people/staff/dbs202/cag/courses/MT37C/course-d.pdf

This short book gives a very clear description of statistical concepts relevant to environmental science and is based on the author's professional experience in the correct use of proper statistical methods for the analysis of climate data.

Exercises

All exercises in this chapter require the use of a computer.

3.1 Assume that the average bank balance of US residents is \$5,000. Assume that the bank balances are normally distributed. A group of 25 samples was taken. The sample data have a mean equal to \$5,000 and standard deviation of \$1,000. Find the confidence interval of this group of samples at 95% confidence level.

3.2 Random variable x satisfies a normal distribution with mean equal to 15, and standard deviation equal to 0.4. Use R and numerical integration to find the probability for x to be in the interval $[-14.8, 15.2]$.

3.3 The two most commonly used datasets of global ocean and land average annual mean surface air temperature (SAT) anomalies are those credited to the research groups originally led by Dr. James E. Hansen of NASA (relative to the 1951–1980 climatology period) and Professor Phil Jones, of the University of East Anglia (relative to the 1961–1990 climatology period):

http://cdiac.ornl.gov/trends/temp/hansen/hansen.html

http://cdiac.ornl.gov/trends/temp/jonescru/jones.html

(a) Find the average anomalies for each period of 15 years, starting at 1880.

(b) Use the t-distribution to find the confidence interval of each 15-year period SAT average at the 95% confidence level using the t-distribution. You can use either Hansen's data or Jones' data. Figure SPM.1(a) of IPCC 2013 (AR5) is a helpful reference.

(c) Find the hottest and the coldest 15-year periods from 1880–2014, which is divided into nine disjoint 15-year periods. Use the t-distribution to check whether the temperature difference in the hottest 15-year period minus that in the coldest 15-year period is significantly greater than zero. Do this problem for either Hansen's data or Jones' data.

(d) Discuss the differences between the Hansen and Jones datasets.

3.4 To test if the average of temperature in Period 1 is significantly different from that in Period 2, one can use the t-statistic

$$t^* = \frac{\bar{x}_1 - \bar{x}_2}{\sqrt{\frac{s_1^2}{n_1} + \frac{s_2^2}{n_2}}}, \tag{3.61}$$

where $\bar{x}_i$ and s_i^2 are the sample mean and variance of the Period i ($i = 1,2$). The degree of freedom (i.e., df) of the relevant t-distribution is equal to the smaller of $n_1 - 1$ and $n_2 - 1$. The null hypothesis is that the two averages do not have significant differences, i.e., their difference is zero (in a statistical sense with a confidence interval). The alternative hypothesis is that the difference is significantly different from zero. Now you can choose to use a one-sided test when the difference is positive. Use a significance level of 5% or 1%, or another level of your own choosing.

(a) Choose two 15-year periods which have very different average anomalies. Use the higher one minus the lower one. Use the t-test method for a one-sided test to check if the difference is significantly greater than zero. Do this for the global average annual mean temperature data from either Hansen's dataset or Jones' dataset.

(b) Choose two 15-year periods which have very similar average anomalies. Use the higher one minus the lower one. Use the t-test method for a two-sided test to check if the difference is not significantly different from zero. Do this for the global average annual mean temperature data from either Hansen's dataset or Jones' dataset.

3.5 (a) Plot the histogram of the Hansen dataset in Problem 3.2.

(b) Plot the histogram of the Jones dataset in Problem 3.2.

(c) Discuss the similarities and differences between the two histograms.

3.6 (a) Plot the two box plots of the Hansen and Jones datasets in Problem 3.2 on the same figure.

(b) Discuss the box plot results.

3.7 Generate a scatter plot based on the Hansen and Jones datasets in Problem 3.2, and discuss the results.

3.8 (a) Generate a Q–Q plot relative to the standard normal distribution for the Jones monthly dataset in Problem 3.2 for every month from January to December, and place all the 12 Q–Q plots on the same figure.

(b) Explore which month's temperature anomalies are the closest to the normal distribution.

3.9 (a) Use R to evaluate the linear trend of January monthly mean global average surface air temperature anomalies based on the Jones dataset from 1900 to 1999. Output your result in the unit: [°C/century].

(b) Repeat (a) for each of the other eleven months.

(c) Generate a table to present the twelve trends, and their corresponding p-values based on the t-distribution.

(d) Following the method of Section 3.5, use both formulas and text to discuss the tabular results.

3.10 (a) Download the monthly surface air temperature data for a station of your interest using the Global Historical Climatology Network (GHCN) dataset produced by the National Centers for Environmental Information, NOAA, United States.

(b) Plot the time series and their linear trend lines for January and July.

(c) Discuss your results.

3.11 Perform all the same procedures as Problem 3.10, but for the monthly precipitation data.

4 Climate Data Matrices and Linear Algebra

Mathematicians have described some aspects of pure mathematics as "beautiful," using terms such as "elegant," "deep," and "general." This book, however, is about topics in mathematics that have proven to be *useful* in the application of mathematics to climate science. In deciding what to include in this book, our criteria have been practical rather than aesthetic. To illustrate the usefulness of linear algebra and matrices in climate science, one instructive example is the analysis of sea level pressure data using a singular value decomposition (SVD) method. The term El Niño originally meant the occasional appearance, every few years, of unusually warm surface water in the eastern tropical Pacific. Thanks to research in climate science, we now realize that El Niño is part of a large-scale complex of changes in the atmosphere and ocean with far-reaching effects. SVD and related mathematical methods have played an important role in this major scientific advance.

Emanuel Lasker (1868–1941), one of the greatest chess players of all time, was chess champion of the world for 27 years. He also had earned a doctorate in mathematics, but he gave up mathematics for chess. Lasker once said, "In mathematics, if I find a new approach to a problem, another mathematician might claim that he has a better, more elegant solution. In chess, if anybody claims he is better than I, I can checkmate him." The practical experience of playing a number of games of chess with each other is the best way to determine which of two chess players is stronger. The practical experience of carrying out a large amount of climate research with various mathematical methods is the best way to determine which methods are most *useful* in climate science.

4.1 Matrix as a Data Array

In general, a matrix is a rectangular array of numbers or symbols or expressions, which are called elements, arranged in rows or columns. A table such as that shown in Fig. 4.1 is a matrix, consisting of N rows and Y columns of numbers. Figure 4.1 shows the 10 annual precipitation anomalies from the year 1900 to 1909 for the 15 five-degree latitude–longitude boxes centered at 2.5°E longitude for different latitudes over the Northern Hemisphere ranging from latitudes 2.5°N to 72.5°N. The matrix shown in Fig. 4.1 thus has 15 rows ($N = 15$), and 10 columns ($Y = 10$). An anomaly of a climate parameter is defined as its actual value minus its normal value that is an average of 30 or more years.

Many other types of climate data can also be represented as matrices (which is the plural of matrix). Precipitation data [units: mm/day] at multiple stations and multiple days can also form a matrix, normally with stations [each marked by a station identifier, or station ID]

Lat	Lon	1900	1901	1902	1903	1904	1905	1906	1907	1908	1909
2.5	2.5	0.283240	-0.131860	-0.190500	0.160040	-0.878110	0.080356	0.059193	-0.136900	0.200420	0.822600
7.5	2.5	0.172670	0.830550	-0.180350	-0.203630	-0.238590	0.425310	0.002805	0.102780	0.254050	0.516200
12.5	2.5	0.024392	0.152030	-0.034115	-0.062696	-0.192070	0.074360	0.201970	-0.011311	0.035259	0.272010
17.5	2.5	0.006780	0.066783	-0.084581	-0.008636	-0.038109	-0.001092	0.088250	0.011047	0.029358	0.082329
22.5	2.5	0.021162	0.079977	0.020016	-0.022142	-0.027032	0.065704	0.012937	-0.003823	0.032545	0.028636
27.5	2.5	0.049846	0.057413	0.026621	0.019914	-0.002651	0.071242	0.012837	0.001567	0.051857	0.099650
32.5	2.5	0.107740	0.143510	0.061613	0.076137	0.147760	0.137890	-0.074612	0.110300	0.087752	0.126920
37.5	2.5	0.128250	0.211940	0.113010	0.027472	0.183710	0.125550	-0.267500	0.215980	0.007609	0.055573
42.5	2.5	0.158490	0.800950	0.292690	0.172930	0.272010	0.126370	-0.017183	0.184880	0.118980	0.200520
47.5	2.5	-0.112800	0.243130	-0.121630	-0.076247	-0.047231	0.110160	0.080978	-0.091371	0.016172	-0.060487
52.5	2.5	-0.199840	-0.381070	-0.217570	-0.107760	-0.124700	-0.117470	-0.062448	-0.171070	-0.277650	-0.132690
57.5	2.5	-0.076619	-0.515070	0.005342	0.016647	0.137820	0.038041	0.131370	-0.196490	-0.132480	0.014887
62.5	2.5	-0.261760	-0.402600	0.137200	-0.214960	0.249210	0.147550	0.866120	-0.453910	-0.026134	0.053409
67.5	2.5	0.034079	0.223610	0.314090	-0.044832	0.130470	0.201260	0.554170	-0.054434	0.185870	0.308950
72.5	2.5	-0.119680	0.022949	0.004324	-0.050248	0.251330	-0.233080	-1.043800	0.363850	-0.315400	-0.113080

Fig. 4.1 Annual precipitation anomalies data of the Northern Hemisphere at longitude 2.5°E [units: mm/day]. The annual total of the anomalies should be multiplied by 365.

represented in rows, and time [units: day] represented in columns. The daily minimum surface air temperature (Tmin) data for the same stations and the same days form another matrix. In general, a space–time climate data table always forms a matrix. Conventionally, the spatial locations correspond to the rows, and the time coordinate corresponds to the columns.

Another matrix example, taken from everyday life, is that the ages of members of an audience, sitting in a movie theater in seats arranged in rows and columns, also form a matrix. The weights of these audience members form another matrix. Their bank account balances form still another matrix, and so on. Thus, a matrix is a data table, and extensive mathematical methods have been developed in the twentieth century to study matrices. Computer software systems, such as R, have also been developed in recent years that greatly facilitate working with matrices.

This chapter will discuss the following topics:

(i) matrix algebra of addition, subtraction, multiplication, and division (i.e., inverse matrix);
(ii) linear equations;
(iii) space–time decomposition, eigenvalues, eigenvectors, and the climate dynamics interpretation of a space–time climate data matrix;
(iv) a matrix application example in balancing the mass in a chemical reaction equation by solving a set of linear equations; and
(v) a matrix application in multivariate linear regression.

4.2 Matrix Algebra

Matrix algebra is quite different from the algebra for a few scalars of x, y, z as we learned in high school. For example, matrix multiplication does not have the commutative property, i.e., matrix A times matrix B is not always the same as matrix B times matrix A. This section describes a set of rules for doing matrix algebra.

4.2.1 Matrix Equality, Addition, and Subtraction

An $m \times n$ matrix A has m rows and n columns and can be written as

$$A = \begin{bmatrix} a_{11} & a_{12} & \cdots & a_{1n} \\ a_{21} & a_{22} & \cdots & a_{2n} \\ \vdots & \vdots & & \vdots \\ a_{m1} & a_{m2} & \cdots & a_{mn} \end{bmatrix}, \tag{4.1}$$

or

$$A_{m \times n} = [a_{ij}], \tag{4.2}$$

or simply

$$A = [a_{ij}], \tag{4.3}$$

where $a_{ij}, i = 1, \ldots, m, j = 1, \ldots, n$ are the mn elements of the rectangular matrix A, and $m \times n$ is often called the size or order of a matrix. We say that A is an m times n matrix, or an m-by-n matrix. The elements of the matrix shown in Fig. 4.1 are the precipitation anomaly data, $m = 15$, and $n = 10$. Therefore, Fig. 4.1 is a 15 times 10 matrix.

Matrix $A = [a_{ij}]$ is equal to matrix $B = [b_{ij}]$ if and only if $a_{ij} = b_{ij}$ for all the elements. That is, A is identical to B. We can understand this by considering the two identical photos. A black and white photo is a matrix of pixel brightness values. Two photos are identical only when the corresponding brightness values of the two photos are the same. Thus, when considering that a matrix is equal to another, we may regard the equality as two same-size photos, maps, or climate charts being identical to one another.

Matrix addition is simply the addition of the corresponding elements:

$$A + B = [a_{ij} + b_{ij}]. \tag{4.4}$$

Of course, two matrices can be added together only when they have the same size $m \times n$. If the two matrices represent dimensional data, such as precipitation, then their units must be the same in order to add corresponding elements. However, each element in a matrix can represent a different climate parameter. For example, a climate data matrix for San Diego for 24 hours may have its first row representing temperature, the second precipitation, the third atmospheric pressure, the fourth relative humidity, etc., while the first columns represent time from 1:00 (i.e., 1:00 a.m.) to 24:00 (i.e., 12:00 a.m.).

Matrix subtraction is defined in a similar way:

$$A - B = [a_{ij} - b_{ij}]. \tag{4.5}$$

Example 4.1 Matrix subtraction:

$$\begin{bmatrix} 1 & 1 \\ 1 & -1 \end{bmatrix} - \begin{bmatrix} 1 & 2 \\ 3 & 0 \end{bmatrix} = \begin{bmatrix} 0 & -1 \\ -2 & -1 \end{bmatrix}. \tag{4.6}$$

The R code for the above matrix subtraction can be written as follows:

```
matrix(c(1,1,1,-1), nrow=2) - matrix(c(1,3,2,0), nrow=2)
#      [,1] [,2]
#[1,]    0   -1
#[2,]   -2   -1
```

One can use matrix subtraction to quantify the differences of climate between two space–time domains when the climate data for each domain are in a matrix form.

4.2.2 Matrix Multiplication

While matrix addition and subtraction are similar to those of scalars, matrix multiplication has several distinct properties.

4.2.2.1 A Row Vector Times a Column Vector

The single-column n-row matrix is often called an n-dimensional vector, or an n-dimensional column vector. Similarly, one can define an n-dimensional row vector as a single-row n-column matrix.

A row vector of n elements times a column vector of the same number of elements is equal to a scalar, which is the sum of the products of each pair of corresponding elements:

$$\begin{bmatrix} a_1 & a_2 & \cdots & a_n \end{bmatrix} \begin{bmatrix} b_1 \\ b_2 \\ \vdots \\ b_n \end{bmatrix} = a_1b_1 + a_2b_2 \cdots + a_nb_n. \tag{4.7}$$

This is also called the dot product of two vectors $\mathbf{a} = (a_1, a_2, \ldots, a_n)$ and $\mathbf{b} = (b_1, b_2, \ldots, b_n)$, denoted by

$$\mathbf{a} \cdot \mathbf{b} = a_1b_1 + a_2b_2 \cdots + a_nb_n. \tag{4.8}$$

In two- or three-dimensional space, the dot product of two vectors has a simple geometric interpretation:

$$\mathbf{a} \cdot \mathbf{b} = ||\mathbf{a}|| \, ||\mathbf{b}|| \cos\gamma, \tag{4.9}$$

where $||\mathbf{a}||$ stands for the length of a vector $\mathbf{a}$, and γ is the angle between $\mathbf{a}$ and $\mathbf{b}$. The proof of the equivalence of the above two expressions (4.8) and (4.9) is given in Appendix A.

When $\mathbf{a}$ is force and $\mathbf{b}$ is the displacement of an object moved by the force, then $\mathbf{a} \cdot \mathbf{b}$ is the work done by the force on the object.

Example 4.2 Dot product of two vectors in a two-dimensional space:

$$\begin{bmatrix} 1 & 1 \end{bmatrix} \begin{bmatrix} 1 \\ 0 \end{bmatrix} = 1. \tag{4.10}$$

The same result can be obtained from the geometric formula (4.9). The length of the first vector is $\sqrt{2}$ since it is the diagonal vector of a unit square in the first quadrant, and the length of the second vector is 1 since it is the unit square's side on the x-axis. The angle between the two vectors is 45°. Thus, the dot product of the two vectors is $\sqrt{2} \times 1 \times \cos(45°) = 1$.

4.2.2.2 A Scalar Times a Matrix

A scalar c times a matrix $A = [a_{ij}]$ is formed by multiplying the scalar into every element of the matrix.

$$c \times A = [c \times a_{ij}]. \tag{4.11}$$

The product is again a matrix of the same size. A physically meaningful product requires that the dimension of the scalar and the dimensions of the matrix elements are compatible. For example, if the matrix is the price data of many products, and if the scalar is the number of sales of each product, then the scalar times the matrix gives the revenue matrix of all the products.

Example 4.3 A scalar 3 times a 2-by-2 matrix

$$A = \begin{bmatrix} 1 & 1 \\ 1 & -1 \end{bmatrix}$$

is

$$3 \times A = 3 \begin{bmatrix} 1 & 1 \\ 1 & -1 \end{bmatrix} = \begin{bmatrix} 3 & 3 \\ 3 & -3 \end{bmatrix}. \tag{4.12}$$

4.2.2.3 An *m*-by-*n* Matrix Times an *n*-by-*k* Matrix

The multiplication of two matrices is defined as a set of dot products between the row vectors of the first matrix and the column vectors of the second matrix. Because of the requirement to form dot products, the column number of the first matrix must be the same as the row number of the second matrix. $A_{m\times n}B_{n\times k}$ is defined as a new matrix $C_{m\times k} = [c_{ij}]$, where the element c_{ij} is the dot product of the ith row vector of $A_{m\times n}$ and jth column vector of $B_{n\times k}$. Therefore, we may write,

$$A_{m\times n}B_{n\times k} = \left[\sum_{l=1}^{n} a_{il}b_{lj}\right]_{m\times k}. \tag{4.13}$$

Thus, $A_{m\times n}B_{n\times k}$ is equal to a matrix of $m \times k$ dot products, and this can be tedious to compute.

Example 4.4 Matrix multiplications:

$$\begin{bmatrix} 1 & 1 \\ 1 & -1 \end{bmatrix} \begin{bmatrix} 1 & 3 \\ 2 & 4 \end{bmatrix} = \begin{bmatrix} 3 & 7 \\ -1 & -1 \end{bmatrix}, \tag{4.14}$$

and

$$\begin{bmatrix} 1 & 3 \\ 2 & 4 \end{bmatrix} \begin{bmatrix} 1 & 1 \\ 1 & -1 \end{bmatrix} = \begin{bmatrix} 4 & -2 \\ 6 & -2 \end{bmatrix}. \tag{4.15}$$

In the above formula (4.14), the first element of the right-hand-side matrix is 3, which is the product of the first row vector of A: $(1,1)$, and the first column vector of B: $(1,2)$. Their dot product is

$$(1,1)\cdot(1,2) = 1\times 1 + 1\times 2 = 3. \tag{4.16}$$

In the same way, one can verify every element of the above multiplication results.

An R code for the above products is below.

```
matrix(c(1,1,1,-1), nrow=2) %*% matrix(c(1,2,3,4), nrow=2)
#       [,1] [,2]
#[1,]     3    7
#[2,]    -1   -1
matrix(c(1,2,3,4), nrow=2) %*% matrix(c(1,1,1,-1), nrow=2)
#        [,1] [,2]
#[1,]      4   -2
#[2,]      6   -2
```

In the above example, the second matrix multiplication (4.15) involves the same matrices as the first one (4.14), but in a different order: If Eq. (4.14) is denoted by AB, then Eq. (4.15) is BA. Clearly, the results are different. In general,

$$AB \neq BA \tag{4.17}$$

for a matrix multiplication. Thus, matrix multiplication does not have the commutative property which the multiplication of two scalars x and y does have: $xy = yx$.

4.2.2.4 Transpose Matrix, and Diagonal, Identity, and Zero Matrices

Although a space–time climate data matrix often has its rows to mark spatial locations and columns to mark time, the row and column roles may need to exchange for some applications, which use rows to mark time and columns to mark spatial locations. This operation of exchanging rows and columns is called matrix transpose. The transposed matrix of A is denoted by A^t. The columns of A^t are the rows of A.

Example 4.5

$$A = \begin{bmatrix} 1 & 2 \\ 3 & 4 \end{bmatrix}, \quad A^t = \begin{bmatrix} 1 & 3 \\ 2 & 4 \end{bmatrix} \tag{4.18}$$

It is obvious that

$$(A+B)^t = A^t + B^t. \tag{4.19}$$

However, a true but less obvious formula is the transpose of a matrix multiplication:

$$(AB)^t = B^t A^t. \tag{4.20}$$

When a matrix whose only non-zero elements are on the diagonal of elements, this matrix is called a diagonal matrix:

$$D = \begin{bmatrix} d_1 & & & \\ & d_2 & & \\ & & \ddots & \\ & & & d_n \end{bmatrix}. \tag{4.21}$$

An identity matrix is a special type of diagonal matrix, whose diagonal elements are all one and whose off-diagonal elements are all zero, and is denoted by I:

$$I = \begin{bmatrix} 1 & & & \\ & 1 & & \\ & & \ddots & \\ & & & 1 \end{bmatrix}, \tag{4.22}$$

which plays a role in matrix operations similar to the role of the value 1.0 in the familiar real number system.

A zero matrix is a matrix whose elements are all zero. If two matrices are the same, then their difference is a zero matrix.

4.2.2.5 Matrix Division and Inverse

The division of a scalar y by another non-zero scalar x can be written as y times the inverse of x:

$$y/x = y \times x^{-1}. \tag{4.23}$$

Thus, the division problem becomes a multiplication problem when the inverse is found. The inverse of x is defined as $x^{-1} \times x = 1$.

Matrix division is defined in the same way:

$$A/B = A \times B^{-1}, \tag{4.24}$$

where B^{-1} is the inverse matrix of B defined as

$$B^{-1}B = I, \tag{4.25}$$

where I is the identity matrix.

Example 4.6 The R command to find the inverse of matrix B is `solve(B)` as shown by a numerical example below.

```
solve(matrix(c(1,1,1,-1), nrow=2))
#      [,1] [,2]
#[1,]   0.5  0.5
#[2,]   0.5 -0.5
```

That is

$$\begin{bmatrix} 1 & 1 \\ 1 & -1 \end{bmatrix}^{-1} = \begin{bmatrix} 0.5 & 0.5 \\ 0.5 & -0.5 \end{bmatrix}. \tag{4.26}$$

One can verify this "by hand," or by the following R code

```
matrix(c(0.5,0.5,0.5,-0.5), nrow=2) %*% matrix(c(1,1,1,-1), nrow=2)
#      [,1] [,2]
#[1,]     1    0
#[2,]     0    1
```

Finding the inverse matrix of a matrix B "by hand" is usually very difficult and involves a long procedure for a large matrix, say, a 4×4 matrix. Modern climate models can involve multiple $n \times n$ matrices with n from several hundred to several million, or even billion. In this book, we use R to find inverse matrices and do not attempt to explain how to find the inverse of a matrix by hand. Mastering the material in this book will not require you to have this skill. A typical linear algebra textbook will devote a large portion of its material to finding an inverse of a matrix. A commonly used scheme is called echelon reduction through row operations. For detailed information, see the excellent text *Introduction to Linear Algebra* by Gilbert Strang (2016).

4.3 A Set of Linear Equations

A somewhat different matrix example is the coefficient matrix of a system of linear equations. Solving a system of linear equations is very common in science and engineering. Finding numerical solutions for a climate model based on partial differential equations will usually involve solving large systems of linear equations.

As an introduction to the coefficient matrix of linear equations, let us look at a simple elementary school mathematics problem: The sum of the ages of two brothers is 20 years and the difference of the ages is 4 years. What are the ages of the two brothers? One can easily guess that the older brother is 12 years old, and the younger one is 8.

If we form a set of equations, this would be

$$\begin{aligned} x_1 + x_2 &= 20 \\ x_1 - x_2 &= 4 \end{aligned} \tag{4.27}$$

where x_1 and x_2 stand for the brothers' ages.

The matrix form of these equations would be

$$Ax = b \tag{4.28}$$

which involves three matrices:

$$A = \begin{bmatrix} 1 & 1 \\ 1 & -1 \end{bmatrix}, \quad x = \begin{bmatrix} x_1 \\ x_2 \end{bmatrix}, \quad b = \begin{bmatrix} 20 \\ 4 \end{bmatrix}. \tag{4.29}$$

Here, Ax means a matrix multiplication: $A_{2\times 2}x_{2\times 1}$.

The matrix notation of a system of two linear equations can be extended to systems of many linear equations, hundreds or millions of equations in climate modeling and climate data analysis. Typical linear algebra textbooks introduce matrices in this way by describing linear equations in a matrix form. However, this approach may be less intuitive for climate science, which emphasizes data. Thus, our book uses data to introduce matrices as shown at the beginning of this chapter.

Although one can easily guess that the solution to the above simple matrix equation (4.28) is $x_1 = 12$ and $x_2 = 8$, a more general method for computing the solution may be using the R code shown below:

```
solve(matrix(c(1,1,1,-1),nrow=2),c(20,4))
#[1] 12  8  #This is the result x1=12, and x2=8.
```

In this R command, `matrix(c(1,1,1,-1),nrow=2)` yields the A matrix, and `c(20,4)` the b vector.

This type of R program can solve more complicated systems of linear equations, such as a system with 1,000 unknowns rather than two unknowns, as in this example.

The solution may be represented as

$$x = A^{-1}b, \tag{4.30}$$

where A^{-1} is the inverse matrix of A and was found earlier in Eq. (4.26). One can verify that

$$A^{-1}b = \begin{bmatrix} 0.5 & 0.5 \\ 0.5 & -0.5 \end{bmatrix} \begin{bmatrix} 20 \\ 4 \end{bmatrix} = \begin{bmatrix} 12 \\ 8 \end{bmatrix} \tag{4.31}$$

is indeed the solution of the system of two linear equations.

4.4 Eigenvalues and Eigenvectors of a Square Space Matrix

"Eigenvalue" is a partial translation of the German word "eigenwert," meaning "self-value" or "intrinsic value." In German, "eigen" can mean "self" or "own," as in "one's own," and "wert" means "value."

A square matrix is a matrix for which the number of rows is equal to the number of columns. The matrix thus has the shape of a square. Similarly, other matrices may be called rectangular matrices, or tall matrices.

4.4.1 Matrices of Data Anomalies, Standardized Anomalies, Covariance, and Correlation

In climate science, one often considers the covariance or correlation among N stations, N grid boxes, or N grid points. The covariance between station i and station j is denoted by c_{ij}, and the corresponding correlation is denoted by r_{ij}. Then $C = [c_{ij}]_{N\times N}$ is the covariance matrix, and $Corr = [r_{ij}]_{N\times N}$ is the correlation matrix. They can be estimated by the observational or model data.

Let us use an $N \times Y$ matrix

$$X = [x_{it}]_{N\times Y} \tag{4.32}$$

to represent the climate data of these N stations with Y time steps. A time step can be a month, or a year, or a week, depending on the application needs and the data availability. The covariance and correlation matrices of the N stations for a climate parameter are square matrices. Both covariance and correlation matrices can be estimated using the anomaly data, which are defined as data minus their temporal mean of each station, i.e.,

$$A_{N\times Y} = [a_{it}]_{N\times Y} = [x_{it} - \bar{x}_i]_{N\times Y}, \quad \text{(Anomaly data matrix)}, \tag{4.33}$$

where

$$\bar{x}_i = \frac{1}{Y}\sum_{t=1}^{Y} x_{it} \tag{4.34}$$

is the climate data mean of station i $(i = 1,2,\ldots,N)$. The mean is also called climatology or climate normal of the climate parameter. Thus, the temporal mean of the anomaly data for each station should be zero. However, in many climate applications, the temporal data are not continuous and were missing in the earlier period or in the middle. It is impossible to compute the temporal mean for the entire period of length Y. Instead, the climatology is computed using a sub-interval of data period, whose length is smaller than Y, for example, using 30 years from 1971 to 2000. The 30-year climatology has almost become a recent community standard because of the period's best space–time coverage of observations. Consequently, the temporal mean of the anomaly data based on the 30-year climatology is often non-zero.

Another case is that some applications require us to compute climatologies using 10 years or shorter because of climate change. In any event, climatology and anomalies are commonly used terms in climate data analysis. Their exact definitions vary depending on practical applications. Chapter 11 of this book has more detailed discussions on climatology, anomalies, statistics, and visualization for the historical climate data which are incomplete with missing values.

When the anomaly matrix is normalized by the standard deviation of each station, then the result is called the standardized anomaly matrix:

$$A_{sd} = \left[\frac{x_{it} - \bar{x}_i}{s_i}\right]_{N\times Y}, \quad \text{(Standardized anomaly matrix)}, \tag{4.35}$$

where

$$s_i = \left[\frac{1}{Y} \sum_{t=1}^{Y} (x_{it} - \bar{x}_i)^2 \right]^{1/2} \tag{4.36}$$

is the estimated standard deviation of station i ($i = 1,2,\ldots,N$). Strictly speaking from statistical theory, this standard deviation formula should use $Y - 1$ as the denominator, rather than simply Y, according to formula (3.2) in Chapter 3. However, climate scientists often use Y, not $Y - 1$, perhaps because formula (4.36) has a clear meaning as the mean of square errors. In practice, the difference between the two formulas is small since Y is usually greater or equal to 30. Climate data analysis applications often use 30 years of data, say from 1971 to 2000, to compute climatology and standard deviation.

An advantage of using the standardized anomaly matrix is that it is dimensionless. A climate parameter, such as temperature, often has a larger variance over the land than ocean. The standardized anomaly data show better homogeneity between ocean and land.

Then, the covariance and correlation matrices can be estimated by the following formulas

$$C = \frac{1}{Y} A(A^t) \quad \text{(Covariance matrix)}, \tag{4.37}$$

$$Corr = \frac{1}{Y} A_{sd}(A_{sd})^t \quad \text{(Correlation matrix)}. \tag{4.38}$$

Once more, applications often use a 30-year mean, not the mean of the entire time period of length Y, to compute the covariance and correlation matrices.

Example 4.7 This example's data matrix A has $N = 2$ and $Y = 3$. The data's covariance matrix is covm.

```
A=matrix(c(1,-1,2,0,3,1),nrow=2)
A
#      [,1] [,2] [,3]
#[1,]     1    2    3
#[2,]    -1    0    1
covm=(1/(dim(A)[2]))*A%*%t(A)
covm #is the covariance matrix.
#          [,1]      [,2]
#[1,] 4.6666667 0.6666667
#[2,] 0.6666667 0.6666667
```

4.4.2 Eigenvectors and Their Corresponding Eigenvalues

The covariance matrix times a vector u yields a new vector in a different direction.

```
u=c(1,-1)
v=covm%*%u
v
#      [,1]
#[1,]     4
#[2,]     0
#u and v are in different directions.
```

In general, when we consider a given square matrix C and a given vector u, the product Cu is usually not in the same direction as u, as shown in the above example. However, there is always a "self-vector" vector w for each covariance matrix C such that Cw is in the same direction as w, i.e., Cw and w are parallel, and this fact is expressed as

$$Cw = \lambda w, \tag{4.39}$$

where λ is a scalar which has the property that it simply scales w so that the above equation holds. This scalar λ is called an eigenvalue (i.e., a "self-value," "own-value," or "characteristic value" of the matrix C), and w is called an eigenvector.

R can calculate the eigenvalues and eigenvectors of a covariance matrix `covm` with a command `eigen(covm)`. The output is in an R data frame, which has two storages: `ew$values` for eigenvalues and `ew$vector` for eigenvectors, as shown below.

```
ew=eigen(covm)
eigen(covm)$values
#[1] 4.7748518 0.5584816
eigen(covm)$vectors
#[,1]        [,2]
#[1,] -0.9870875  0.1601822
#[2,] -0.1601822 -0.9870875
#Verify the eigenvectors and eigenvalues
covm%*%ew$vectors[,1]/ew$values[1]
#[,1]
#[1,] -0.9870875
#[2,] -0.1601822
#This is the first eigenvector
```

A 2×2 covariance matrix has two eigenvalues, and two eigenvectors (λ_1, w_1) and (λ_2, w_2), which are shown below from the R computation above for our example covariance matrix C:

$$\lambda_1 = 4.7748518, \quad \mathbf{w_1} = \begin{bmatrix} -0.9870875 \\ -0.1601822 \end{bmatrix}, \tag{4.40}$$

$$\lambda_2 = 0.5584816, \quad \mathbf{w_2} = \begin{bmatrix} 0.1601822 \\ -0.9870875 \end{bmatrix}. \tag{4.41}$$

The eigenvectors are frequently called modes, or empirical orthogonal functions (EOFs) in climate science. The term EOF was coined by Edward Lorenz (1917–2008) in his 1956 paper on statistical weather prediction (Lorenz 1956). The first few eigenvectors of a large climate covariance matrix of climate data often represent some typical patterns of climate variability. Examples which are of great importance in climate science include the El Niño–Southern Oscillation (ENSO), North American Oscillation (NAO), and Pacific Decadal Oscillation (PDO). Several calculation and visualization examples of using R to compute EOFs by SVD for the gridded space–time climate data are included in Chapter 10.

It is often the case that the components of the first mode, i.e., the elements of the first eigenvector, have the same sign, either all positive or all negative. The components of the second mode will then have half negative and half positive signs. Exceptions can occur, however.

Each eigenvalue is equal to the variance of the data projection on the corresponding eigenvector, and is thus positive. The sum of all the eigenvalues of such an $N \times N$ matrix represents the total variance of the climate system observed at these N stations. The first eigenvalue λ_1 is the largest, corresponding to the largest spatial variability of the climate field under study. The eigenvalues sizes follow the order $\lambda_1 \geq \lambda_2 \geq \cdots$.

However, in carrying out an analysis of climate data, one can often find the important patterns as eigenvectors more directly from the anomaly data matrix A without computing the covariance matrix C explicitly. This is known as the singular value decomposition (SVD) approach (Golub and Reinsch 1970), which we discuss next. It separates the space–time anomaly data into a space part, a time part, and what we may think of as a variation in energy part. This mathematical method of space–time decomposition is universally applicable to any data that we sample in space and time, and it can often help to develop physical insight and scientific understanding of the phenomena and their properties as contained in the observational data. Efficient computing methods of SVD have been extensively researched and developed since the 1960s. Gene H. Golub (1932–2007) was a leading researcher in this effort and is remembered as "Professor SVD" by his Stanford colleagues and the world mathematics community.

4.5 An SVD Representation Model for Space–Time Data

We encounter space–time data every day, a simple example being the air temperature at different locations at different times. If you take a plane to travel from San Diego to New York, you may experience the temperature at San Diego in the morning when you depart and that at New York in the evening after your arrival. Such data have many important applications. We may need to examine the precipitation conditions around the world at different days in order to monitor agricultural yields. A cellphone company may need to monitor its market share and the temporal variations of that quantity in different countries. A physician may need to monitor a patient's symptoms in different areas of the body at different times. The observed data in all these examples can form a space–time data matrix with the row position corresponding to the spatial location and the column position corresponding to time, as in Table 4.1.

Table 4.1 Space–time data table

	Time 1	Time 2	Time 3	Time 4
Space 1	*D*11	*D*12	*D*13	*D*14
Space 2	*D*21	*D*22	*D*23	*D*24
Space 3	*D*31	*D*32	*D*33	*D*34
Space 4	*D*41	*D*42	*D*43	*D*44
Space 5	*D*51	*D*52	*D*53	*D*54

Graphically, the space–time data may typically be plotted as a time series at each given spatial position, or as a spatial map at each given time. Although these straightforward graphical representations can sometimes provide very useful information as input for signal detection, the signals are often buried in the data and may need to be detected by different linear combinations in space and time. Sometimes the data matrices are extremely large, with millions of data points in either space or time. Then the question arises as to how can we extract the essential information in such a big data matrix? Can we somehow manage to represent the data in a more simple and yet more useful way? A very useful approach to such a task involves a space–time separation. Singular value decomposition (SVD) is a method designed for this purpose. SVD decomposes a space–time data matrix into a spatial pattern matrix U, a diagonal energy level matrix D, and a temporal matrix V^t, i.e., the data matrix A is decomposed into

$$A_{N\times Y} = U_{N\times m} D_{m\times m} (V^t)_{m\times Y}, \tag{4.42}$$

where N is the spatial dimension, Y is the temporal length, $m = min(N,Y)$, and V^t is the transpose of V. The columns of U are a set of spatial eigenvectors. They are orthonormal vectors, each of which is orthogonal to another and has its length equal to one. Thus, the word "orthonormal" comes from two words "orthogonal" and "normal." The orthonormal property can be expressed by the following formula:

$$\mathbf{u}_l \cdot \mathbf{u}_k = \delta_{lk}, \tag{4.43}$$

where $\mathbf{u}_l$ and $\mathbf{u}_k$ are column vectors of U, and δ_{lk} is the Kronecker delta equal to zero when $k \neq l$ and one when $k = l$. The columns of V are also a set of orthonormal vectors known as temporal eigenvectors.

Usually, the elements of the U and V matrices are unitless (i.e., dimensionless), and the unit of the D elements is the same as the elements of the data matrix. For example, if A is a space–time precipitation data matrix with a unit [mm/day], then the unit of the D elements is also [mm/day].

Example 4.8 SVD for the 2×3 data matrix A in Example 4.7 of Section 4.4.1.

```
#Develop a 2-by-3 space-time data matrix for SVD
A=matrix(c(1,-1,2,0,3,1),nrow=2)
A
#     [,1] [,2] [,3]
#[1,]    1    2    3
#[2,]   -1    0    1
#Perform SVD calculation
msvd=svd(A)
msvd
msvd$d
#[1] 3.784779 1.294390
msvd$u
#           [,1]       [,2]
#[1,] -0.9870875 -0.1601822
#[2,] -0.1601822  0.9870875
msvd$v
```

```
#             [,1]       [,2]
#[1,] -0.2184817 -0.8863403
#[2,] -0.5216090 -0.2475023
#[3,] -0.8247362  0.3913356
#One can verify that A=UDV', where V' is transpose of V.
verim=msvd$u%*%diag(msvd$d)%*%t(msvd$v)
verim
#      [,1]         [,2] [,3]
#[1,]     1 2.000000e+00    3
#[2,]    -1 1.665335e-16    1
round(verim)
#      [,1] [,2] [,3]
#[1,]     1    2    3
#[2,]    -1    0    1
#This is the original data matrix A
```

The covariance of the space–time matrix A is a spatial matrix:

$$C = \frac{1}{Y} A A^t, \tag{4.44}$$

where Y is the number of columns of A and is equal to the length of time being considered.

```
covm = (1/(dim(A)[2]))*A%*%t(A)
eigcov = eigen(covm)
eigcov$values
#[1] 4.7748518 0.5584816
eigcov$vectors
#             [,1]       [,2]
#[1,] -0.9870875  0.1601822
#[2,] -0.1601822 -0.9870875
```

Thus, the eigenvectors of a covariance matrix are the same as the SVD eigenvectors of the anomaly data matrix. The eigenvalues of the covariance matrix and the SVD have the following relationship

```
((msvd$d)^2)/(dim(A)[2])
#[1] 4.7748518 0.5584816
eigcov$values
#[1] 4.7748518 0.5584816
```

Therefore, the EOFs from a given space–time dataset can be calculated directly by using an SVD command and do not need the step of calculating the covariance matrix. With efficient SVD algorithms, this shortcut can save significant amounts of time for an EOF analysis, also known as the principal component analysis (PCA) in the statistics community, compared to the traditional covariance matrix approach. Therefore, the result is extremely helpful for the EOF analysis, which is an indispensable modern tool of climate data analysis. The result is formally described as a theorem whose proof is also provided as follows.

Theorem 4.1 *The eigenvectors* $\mathbf{u}_k$ *of the covariance matrix* $C = (1/Y)AA^t$

$$\mathbf{C}\mathbf{u}_k = \lambda_k \mathbf{u}_k, \quad (k = 1, 2, \ldots, N), \tag{4.45}$$

are the same as the SVD spatial modes of $A = UDV^t$. *The eigenvalues* λ_k *of* $C_{N\times N}$ *and the SVD eigenvalues* d_k *of* $A_{N\times Y}$ *have the following relationship*

$$\lambda_k = d_k^2/Y \ \ (k = 1,2,\ldots,Y), \quad \textit{when } Y \le N, \tag{4.46}$$

where Y *is the total time length (i.e., time dimension) of the anomaly data matrix* A, *and* N *is the total number of stations for* A *(i.e., space dimension).*

Proof The SVD of the space–time data matrix is

$$A = UDV^t. \tag{4.47}$$

The data matrix A's corresponding covariance matrix is thus

$$\begin{aligned} C &= \frac{1}{Y}AA^t \\ &= \frac{1}{Y}UDV^t(UDV^t)^t \\ &= \frac{1}{Y}UDV^t(VDU^t) \\ &= \frac{1}{Y}UD(V^tV)DU^t \\ &= \frac{1}{Y}UDIDU^t \\ &= \frac{1}{Y}UD^2U^t. \end{aligned} \tag{4.48}$$

In the above, we have used $V^tV = I_Y$ is a $Y \times Y$ identity matrix according to the SVD definition. The identity matrix's dimension is Y if $Y \le N$. Otherwise, $V^tV = I_N$.

The expression $C = \frac{1}{Y}UD^2U^t$ is the EOF expansion of the covariance matrix and means that a covariance matrix consists of EOFs and its associated variance, or "energy."

The covariance matrix's eigenvalue problem is

$$CU = \frac{1}{Y}UD^2U^tU = \frac{1}{Y}UD^2 = U\Lambda, \tag{4.49}$$

where

$$\Lambda = \frac{1}{Y}D^2 \tag{4.50}$$

is the diagonal eigenvalue matrix with its elements as

$$\lambda_k = d_k^2/Y \ \ (k = 1,2,\ldots,Y). \tag{4.51}$$

In Eq. (4.49), we used $U^tU = I_Y$ based on the SVD definition. Equation (4.49) becomes an eigenvalues problem because Λ is a diagonal matrix: $C\mathbf{u}_k = \lambda_k\mathbf{u}_k$ $(k = 1,2,\ldots,Y)$, where $\mathbf{u}_k$ is the kth column vector of the space matrix U. Thus, Eq. (4.49) implies the first part of the theorem: The space eigenvectors U from the SVD of the space–time data matrix A are the same as the eigenvectors of the corresponding covariance matrix C.

Equation (4.51) is exactly the second part of the theorem. The proof is thus complete.

In the above proof, we implicitly assumed that the columns of data matrix A are independent when $Y \leq N$. Independence of a set of vectors means that no one vector can be expressed in terms of the others, so no vectors can be replaced. Real climate data always satisfy this independence assumption.

4.6 SVD Analysis of Southern Oscillation Index

This section is an application example of the SVD method for constructing an optimally weighted Southern Oscillation Index (WSOI).

SOI is an index that monitors ENSO and is the standardized measure of an important atmospheric pressure gradient. Specifically, it is the atmospheric sea level pressure (SLP) at Tahiti (18°S, 149°W) minus that at Darwin (12°S, 131°E). It thus measures the SLP difference between the eastern tropical Pacific and the western tropical Pacific. During a "normal" (non-El Niño) year, Darwin's anomaly pressure is lower than that of Tahiti, and it is this pressure difference which maintains the easterly trade winds and is dynamically consistent with the existence of the western Pacific warm pool, an area of higher ocean surface temperatures compared to the eastern tropical Pacific. When the wind direction reverses, the pressure anomalies have the opposite sign, leading to westerlies in the region where trade winds are normally found, and to the accumulation of anomalously warm surface water in the central or eastern tropical Pacific. This latter ocean temperature pattern is the characteristic signal of El Niño.

The SLP data of these two stations between January 1951 and December 2015 can be downloaded from the NOAA Climate Prediction Center (CPC)'s website www.cpc.ncep.noaa.gov/data/indices/ We will first examine SOI from the standardized Tahiti and Darwin SLP data, and then use the same data but the SVD approach to develop a new southern oscillation index: WSOI, which can more accurately quantify the El Niño signal.

4.6.1 Standardized SLP Data and SOI

Figure 4.2 shows the standardized Tahiti and Darwin SLP data, SOI, Cumulative SOI index (CSOI), and Atlantic Multi-decadal Oscillation (AMO) index. The AMO index is defined as the standardized average sea surface temperature (SST) over the North Atlantic region (80°W–0°E, 0°N–60°N). The AMO data from January 1951 to December 2015 can be downloaded from the NOAA Earth System Research Laboratory website www.esrl.noaa.gov/psd/data/correlation/amon.us.data

It appears that the CSOI and AMO indexes follow a similar nonlinear trend. Both CSOI and AMO decrease from 1950s to a bottom in the 1970s, then increase to around year 1998 followed by a plateau lasting about a decade, and then start to decrease around 2006. CSOI has a much smaller variance than the AMO index and has a more clear signal.

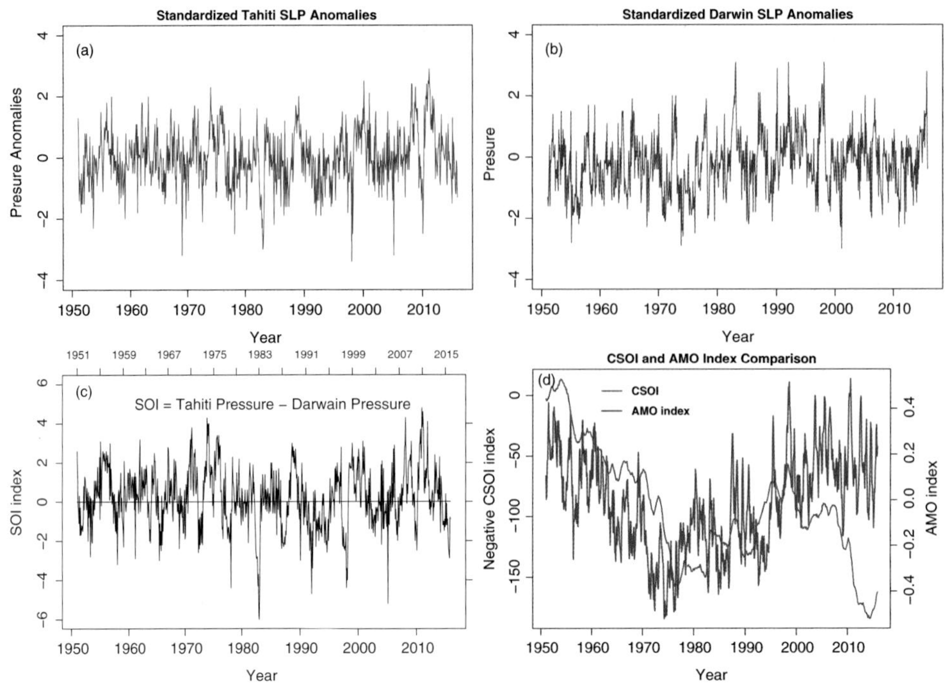

Fig. 4.2 Standardized sea level pressure anomalies of Tahiti (upper left panel), that of Darwin (upper right). SOI time series (lower left), and the negative CSOI time series and AMO time series (lower right).

The first three panels of Fig. 4.2 can be generated from the following R code.

```
#SOI and the Standardized SLP data at Darwin and Tahiti
setwd("/Users/sshen/climmath")
setEPS()
postscript("fig0402a.eps", height=4.9, width=7)
par(mar=c(4.0,4.2,1.5,0.5))
Pta<-read.table("data/PSTANDtahiti.txt", header=F)
# Remove the first column that is the year
ptamon<-Pta[,seq(2,13)]
#Convert the matrix into a vector according to mon:
#Jan 1951, Feb 1951, ..., Dec 2015
ptamonv<-c(t(ptamon))
#Generate time ticks from Jan 1951 to Dec 2015
xtime<-seq(1951, 2016-1/12, 1/12)
# Plot the Tahiti standardized SLP anomalies
plot(xtime, ptamonv,type="l",xlab="Year",
     ylab="Presure Anomalies", cex.lab=1.3, cex.axis=1.3,
     main="Standardized Tahiti SLP Anomalies", col="red",
     xlim=range(xtime), ylim=c(-4,4))
text(1952,3.5, "(a)", cex=1.3)
dev.off()

# Do the same for Darwin SLP
setEPS()
postscript("fig0402b.eps", height=4.9, width=7)
par(mar=c(4.0,4.2,1.5,0.5))
```

```
Pda<-read.table("data/PSTANDdarwin.txt", header=F)
pdamon<-Pda[,seq(2,13)]
pdamonv<-c(t(pdamon))
plot(xtime, pdamonv,type="l",cex.lab=1.3, cex.axis=1.3,
     xlab="Year",ylab="Presure",
     main="Standardized Darwin SLP Anomalies", col="blue",
     xlim=range(xtime), ylim=c(-4,4))
text(1952,3.5, "(b)", cex=1.3)
dev.off()

#Plot the SOI index
setEPS()
postscript("fig0402c.eps", height=4.9, width=7)
par(mar=c(4.0,4.2,2.0,0.5))
SOI <- ptamonv-pdamonv
plot(xtime, SOI ,type="l", cex.lab=1.3, cex.axis=1.3,
     xlab="Year",ylab="SOI index",
     col="black",xlim=range(xtime), ylim=c(-6,6), lwd=1)
#Add ticks on top edge of the plot box
axis (3, at=seq(1951,2015,4), labels=seq(1951,2015,4))
#Add ticks on the right edge of the plot box
axis (4, at=seq(-4,4,2), labels=seq(-4,4,2))
lines(xtime, rep(0,length(xtime)))
text(1985, 5, "SOI - Tahiti Pressure   Darwain Pressure", cex=1.3)
text(1952,5.6, "(c)", cex=1.3)
#abline(lm(SOI ~ xtime), col="red", lwd-2)
dev.off()
```

The following R code can generate the fourth panel of Fig. 4.2.

```
#CSOI and AMO time series comparison
setEPS()
postscript("fig0402d.eps", height=4.9, width=7)
#par(mar=c(4.0,4.2,1.5,0.5))
par(mar=c(4,4.2,1.6,4))
cnegsoi<--cumsum(ptamonv-pdamonv)
#par(mgp=c(2,2,4,0))
plot(xtime, cnegsoi,type="l", cex.lab=1.3, cex.axis=1.3,
     xlab="Year",ylab="Negative CSOI index",
     main="CSOI and AMO Index Comparison",
     col="purple",xlim=range(xtime), ylim=range(cnegsoi), lwd=1.5)
legend(1960,15, col=c("purple"),lty=1,lwd=2.0,
       legend=c("CSOI"),bty="n",text.font=2,cex=1.0)
text(1951,12, "(d)", cex=1.3)
#AMO data and plot
amodat=read.table("data/AMO1951-2015.txt", header=FALSE)
amots=as.vector(t(amodat[,2:13]))
par(new=TRUE)
plot(xtime, amots,type="l",col="darkgreen",
     cex.lab=1.3, cex.axis=1.3,
     lwd=1.5,axes=FALSE,xlab="",ylab="")
legend(1960,0.45, col=c("darkgreen"),lty=1,lwd=2.0,
       legend=c("AMO index"),bty="n",text.font=2,cex=1.0)
#Suppress the axes and assign the y-axis to side 4
axis(4, cex.axis=1.3)
mtext("AMO index",side=4,line=3, cex=1.3)
dev.off()
```

4.6.2 Weighted SOI Computed by the SVD Method

The space–time data matrix of the SLP at Tahiti and Darwin from January 1951–December 2015 can be obtained from

```
ptada = cbind(ptamonv,pdamonv)
```

This is a matrix of two columns: the first column is the Tahiti SLP and the second column is the Darwin SLP. Because normally the spatial position is indicated by row and the time is indicated by column, we transpose the matrix `ptada = t(ptada)` This is the 1951–2015 standardized SLP data for Tahiti and Darwin: 2 rows and 780 columns.

```
dim(ptada)
[1]   2 780
```

Make the SVD space–time separation: `svdptd = svd(ptada)`

Verify this separation by reconstructing the original space–time data matrix using the SVD results

```
recontd=svdptd$u%*%diag(svdptd$d[1:2])%*%t(svdptd$v)
```

One can verify that `recontd=ptada`.

The spatial matrix U is a 2×2 orthogonal matrix since there are only two points. Each column is an eigenvector of the covariance matrix $C = (1/t)AA^t$, where $A_{n\times t}$ is the original data matrix of n spatial dimension and t temporal dimension. The SVD decomposition of the matrix A becomes

$$A_{2\times 780} = U_{2\times 2}D_{2\times 2}V_{2\times 780}. \tag{4.52}$$

Our EOF matrix U is

```
U=svdptd$u
U
#             [,1]      [,2]
#[1,] -0.6146784 0.7887779
#[2,]  0.7887779 0.6146784
```

The two column vectors of U are the eigenvectors of the covariance matrix, i.e., the EOFs. The EOFs represent spatial patterns or modes. The first column is the first spatial mode which is $\mathbf{u}_1 = (-0.61, 0.79)$, meaning opposite signs of Tahiti and Darwin, which justifies the SOI index as one location's SLP minus that of another location. The SOI thus assigns equal weights to Tahiti and Darwin, but opposite signs, with one being positive and the other negative. The above SVD result suggests a pair of unequal and optimal weights, which leads to the following weighted SOI:

$$WSOI1 = -0.6147P_{Tahiti} + 0.7888P_{Darwin}. \tag{4.53}$$

The WSOI1 is optimal in the sense that it has the largest variance among all the linear combinations of Tahiti and Darwin SLP time series:

$$\hat{P}(t) = a_1 P_{Tahiti}(t) + a_2 P_{Darwin}(t) \tag{4.54}$$

under a normalization constraint

$$a_1^2 + a_2^2 = 1. \tag{4.55}$$

This mode's energy level, i.e., the temporal variance, is $d_1 = 31.35$ given by

```
svdptd$d
#[1] 31.34582 22.25421
D=diag(svdptd$d)
D
#          [,1]     [,2]
#[1,] 31.34582  0.00000
#[2,]  0.00000 22.25421
```

which forms the diagonal matrix D in the SVD formula. One can verify that the variance of WSOI1 is 31.35.

In nature, the second eigenvalue is often much smaller than the first, but that is not true in this example. Here the second mode's energy level is $d_2 = 22.25$, which is equal to 71% of the first energy level 31.35.

The second weighted SOI mode, i.e. the second column $\mathbf{u}_2$ of U, is thus

$$WSOI2 = 0.7888 P_{Tahiti} + 0.6147 P_{Darwin}. \tag{4.56}$$

From the SVD formula $A = UDV^t$, the above two weighted SOIs are U^tA:

$$U^tA = DV^t, \tag{4.57}$$

because U is an orthogonal matrix and $U^{-1} = U^t$.

The V matrix is given by

```
V=svdptd$v
V
#            [,1]          [,2]
#  [1,] -5.820531e-02  1.017018e-02
#  [2,] -4.026198e-02 -4.419324e-02
#  [3,] -2.743069e-03 -8.276652e-02
#  ......
```

The first temporal mode $\mathbf{v}_1$ is the first row of V^t and is called the first principal component (PC1). The above formulas imply that

$$\mathbf{v}_1 = WSOI1/d_1 \tag{4.58}$$

$$\mathbf{v}_2 = WSOI2/d_2. \tag{4.59}$$

The two PCs are orthonormal vectors, meaning the dot product of two different PC vectors is zero, and that of the same two PC vectors is one. The two EOFs are also orthonormal vectors. Thus, the SLP data at Tahiti and Darwin have been decomposed into a set of spatially and temporally orthonormal vectors: EOFs and PCs, together with energy levels.

The WSOIs' standard deviations are d_1 and d_2, reflecting each WSOI's oscillation magnitude and frequency.

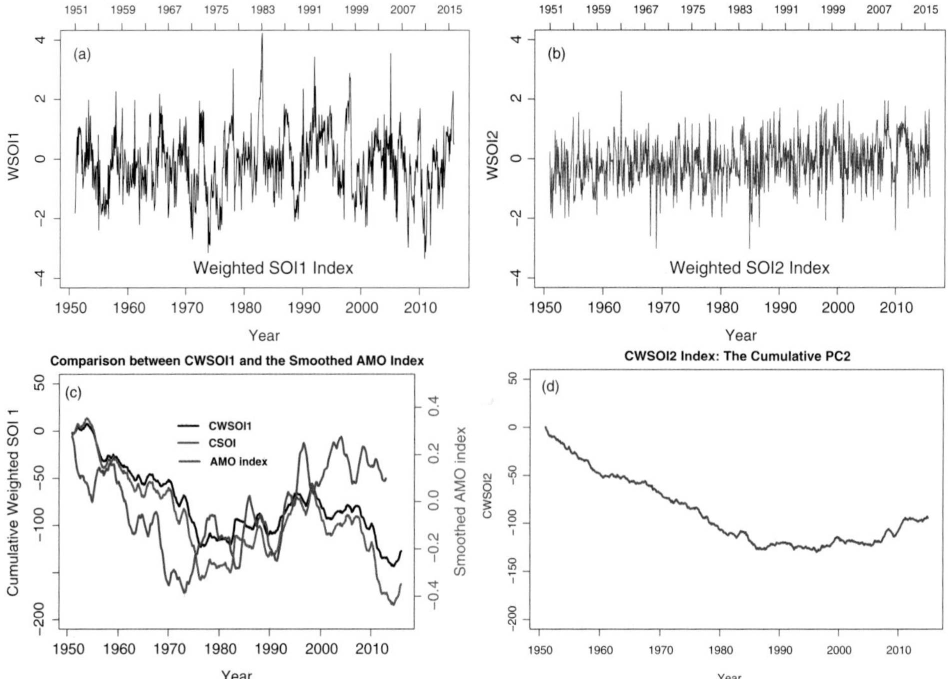

Fig. 4.3 Weighted SOI1 (upper left panel), weighted SOI2 (upper right), cumulative WSOI1 (lower left), and cumulative WSOI2 (lower right).

We also have the relations

$$d_k PC_k = WSOI_k \ (k = 1, 2). \tag{4.60}$$

The two WSOIs are shown in Fig. 4.3.

```
U=svdptd$u #The matrix of the EOF spatial patterns
V=svdptd$v #The matrix of the PC tempral patterns
D=diag(svdptd$d) #The diagonal matrix of standard deviations
recontada=U%*%D%*%t(V) #Verify the SVD decomposition
svd_error = ptada - recontada
max(svd_error)
#[1] 3.996803e-15
min(svd_error)
#[1] -4.218847e-15

#Plot WSOI1
setEPS()
postscript("fig0403a.eps", height=4.9, width=7)
par(mar=c(4,4.2,2.0,0.5))
xtime<-seq(1951, 2016-1/12, 1/12)
wsoi1=D[1,1]*t(V)[1,]
plot(xtime, wsoi1,type="l", cex.lab=1.3, cex.axis=1.3,
     xlab="Year",ylab="WSOI1",
     col="black",xlim=range(xtime), ylim=c(-4,4))
axis (3, at=seq(1951,2015,4), labels=seq(1951,2015,4))
```

```
text(1985,-3.7,"Weighted SOI1 Index", cex=1.5)
text(1952,3.5, "(a)", cex=1.3)
dev.off()

#Plot WSOI2
setEPS()
postscript("fig0403b.eps", height=4.9, width=7)
par(mar=c(4,4.2,2.0,0.5))
xtime<-seq(1951, 2016-1/12, 1/12)
wsoi2=D[2,2]*t(V)[2,]
plot(xtime, wsoi2,type="l", cex.lab=1.3, cex.axis=1.3,
     xlab="Year",ylab="WSOI2",
     col="blue",xlim=range(xtime), ylim=c(-4,4))
axis (3, at=seq(1951,2015,4), labels=seq(1951,2015,4))
text(1985,-3.7,"Weighted SOI2 Index", cex=1.5)
text(1952,3.5, "(b)", cex=1.3)
dev.off()
```

The cumulative WSOIs can be plotted by the following R commands

```
setEPS()
postscript("fig0403c.eps", height=4.9, width=7)
par(mar=c(4,4.2,2.0,4.2))
cwsoi1=cumsum(wsoi1)
plot(xtime, cwsoi1,type="l",cex.lab=1.3, cex.axis=1.3,
     xlab="Year",ylab="Weighted SOI 1",
     col="black",lwd=3, ylim=c(-200,50),
     main="Comparison between CWSOI1 and the Smoothed AMO Index")
text(1951,40, "(c)", cex=1.3)
#axis (3, at=seq(1951,2015,4), labels=seq(1951,2015,4))
legend(1970,20, col=c("black"),lty=1,lwd=3.0,
       legend=c("CWSOI1"),bty="n",text.font=2,cex=1.0)
#Superimpose CSOI time series on this CWSOI1
cnegsoi<--cumsum(ptamonv-pdamonv)
lines(xtime, cnegsoi,type="l",col="purple", lwd=3.0)
legend(1970,2, col=c("purple"),lty=1,lwd=3.0,
       legend=c("CSOI"),bty="n",text.font=2,cex=1.0)
#24-month ahead moving average of the monthly AMO index
amodat=read.table("data/AMO1951-2015.txt", header=FALSE)
amots=as.vector(t(amodat[,2:13]))
#install.packages("TTR")
library("TTR")
amomv=SMA(amots,n=24, fill=NA)
#Average of the previous n points
par(new=TRUE)
xtime=seq(1951,2015,len=780)
plot(xtime, amomv[37:816],type="l",col="darkgreen",
     lwd=3,axes=FALSE,xlab="",ylab="",
     ylim=c(-0.5, 0.5))
legend(1970,0.23, col=c("darkgreen"),lty=1,lwd=3.0,
       legend=c("AMO index"),bty="n",text.font=2,cex=1.0)
#Suppress the axes and assign the y-axis to side 4
axis(4, cex.axis=1.3, col.axis="darkgreen")
mtext("Smoothed AMO index",
      side=4,line=3, cex=1.3, col="darkgreen")
dev.off()
```

```
#Plot cumulative WSOI2: CWSOI2
setEPS()
postscript("fig0403d.eps", height=4.9, width=7)
par(mar=c(4,4.2,2.0,0.5))
cwsoi1=cumsum(wsoi1)
wsoi2=D[2,2]*t(V)[2,]
cwsoi2=cumsum(wsoi2)
plot.new()
#par(mar=c(4,4,3,4))
plot(xtime, cwsoi2,type="l",xlab="Year",ylab="CWSOI2",
     col="blue",lwd=3, ylim=c(-200,50),
     main="CWSOI2 Index: The Cumulative PC2")
dev.off()
```

When the cumulative WSOI1 decreases, so does the Southern Hemisphere surface air temperature from 1951 to 1980. When the cumulative WSOI1 increases, so does the temperature from the 1980s to the peak in 1998. Later, the cumulative WSOI1 decreases to a plateau from 1998 to 2002, then remains on the plateau until 2007, then decreases again. This also agrees with the nonlinear trend of the Southern Hemisphere surface air temperature anomalies before 1998.

The CWSOI2 decreases from 1951 to the 1980s and remains in a flat valley until its further increase from around 2007. This increase coincides with the persistent global surface air temperature increase in the last decade.

This example illustrates clearly that SVD results may lead to physical interpretations and help provide physical insight. SVD is thus a valuable and convenient tool to use.

4.6.3 Visualization of the ENSO Mode Computed from the SVD Method

The space–time data matrix `ptada` of the SLP at Tahiti and Darwin from January 1951 to December 2015 has 2 rows for space and 780 columns for time. The U matrix from the SVD is a 2×2 matrix. Its first column represents the El Niño mode. Note that the eigenvectors are determined except for a positive or negative sign. Because Tahiti has a positive SST anomaly during an El Niño, we choose Tahiti 0.61 and hence make Darwin -0.79. This is the negative first eigenvector from the SVD. The second mode is Tahiti 0.79 and Darwin 0.61. These two modes are orthogonal because $(-0.79, 0.61) \cdot (0.61, 0.79) = 0$. They are displayed in Fig. 4.4, which may be generated by the following R code.

```
#Display the two ENSO modes on a world map
library(maps)
library(mapdata)

plot.new()
par(mfrow=c(2,1))

par(mar=c(0,0,0,0)) #Zero space between (a) and (b)
map(database="world2Hires",ylim=c(-70,70), mar = c(0,0,0,0))
grid(nx=12,ny=6)
points(231, -18,pch=16,cex=2, col="red")
text(231, -30, "Tahiti 0.61", col="red")
```

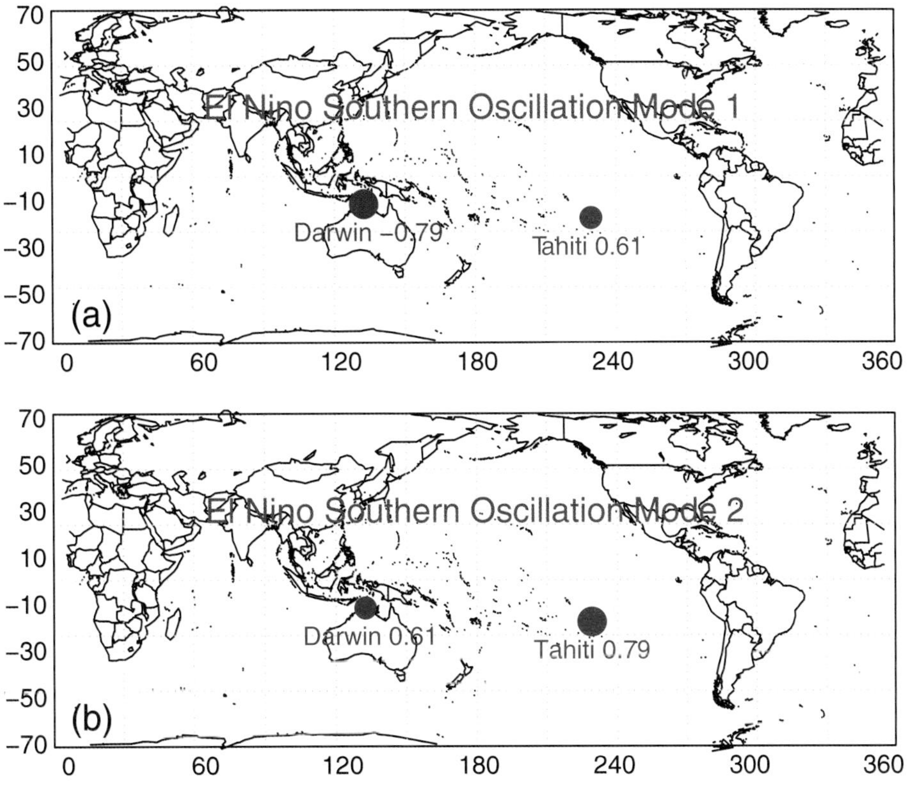

Fig. 4.4 The two orthogonal EOF modes from the Tahiti and Darwin standardized SLP data. The four round dots represent the four components in the 2-by-2 *U* matrix resulted from the SVD calculation. The size of each dot represents the value of the corresponding element of the *U* matrix. Red color indicates positive values, and blue indicates negative values. ENSO Mode 1 is the vector EOF1 which is the first column of the *U* matrix, and ENSO Mode 2 is EOF2 that is the second column vector of *U*.

```
points(131, -12,pch=16,cex=2.6, col="blue")
text(131, -24, "Darwin -0.79", col="blue")
axis(2, at=seq(-70,70,20),
     col.axis="black", tck = -0.05, las=2, line=-0.9,lwd=0)
axis(1, at=seq(0,360,60),
     col.axis="black",tck = -0.05, las=1, line=-0.9,lwd=0)
text(180,30, "El Nino Southern Oscillation Mode 1",col="purple",cex=1.3)
text(10,-60,"(a)", cex=1.4)
box()

par(mar=c(0,0,0,0)) #Plot mode 2
map(database="world2Hires",ylim=c(-70,70),  mar = c(0,0,0,0))
grid(nx=12,ny=6)
points(231, -18,pch=16,cex=2.6, col="red")
text(231, -30, "Tahiti 0.79", col="red")
points(131, -12,pch=16,cex=2, col="red")
text(131, -24, "Darwin 0.61", col="red")
text(180,30, "El Nino Southern Oscillation Mode 2",col="purple",cex=1.3)
```

```
axis(2, at=seq(-70,70,20),
     col.axis="black", tck = -0.05, las=2, line=-0.9,lwd=0)
axis(1, at=seq(0,360,60),
     col.axis="black",tck = -0.05, las=1, line=-0.9,lwd=0)
text(10,-60,"(b)", cex=1.4)
box()
```

More advanced visualization of EOFs and PCs by R is described in Chapter 10 of this book for more complex data, such as those from a climate model output.

4.7 Mass Balance for Chemical Equations in Marine Chemistry

This section describes another application of linear algebra in climate science. Marine chemistry or atmospheric chemistry involves various kinds of chemical reaction equations which describe the conservation of mass during the chemical reactions. Then, the problem is how to systematically achieve the mass balance, i.e., to determine the numbers of molecules on each side of an equation that depicts a chemical reaction. Here we use photosynthesis as an example to illustrate a linear algebra approach to this problem.

In the process of photosynthesis, plants convert the solar radiant energy carried by photons, plus carbon dioxide (CO_2) and water (H_2O), into glucose ($C_6H_{12}O_6$) and oxygen (O_2). The chemical equation for this conversion could be written schematically as

$$CO_2 + H_2O \longrightarrow C_6H_{12}O_6 + O_2 \tag{4.61}$$

However, the conservation of mass requires that the atomic weights on both sides of the equation be equal. The photons have no mass. Thus, the chemical equation as written above is incorrect. The correct equation should specify precisely how many CO_2 molecules react with how many H_2O molecules to generate how many $C_6H_{12}O_6$ and O_2 molecules. Suppose that these coefficients are x_1, x_2, x_3, x_4. We then have

$$x_1CO_2 + x_2H_2O \longrightarrow x_3C_6H_{12}O_6 + x_4O_2 \tag{4.62}$$

Making the number of atoms of carbon on the left and right sides of the equation equal yields

$$x_1 = 6x_3 \tag{4.63}$$

because water and oxygen contain no carbon. Doing the same for hydrogen atoms leads to

$$2x_2 = 12x_3. \tag{4.64}$$

Similarly, the balance of oxygen atoms is

$$2x_1 + x_2 = 6x_3 + 2x_4. \tag{4.65}$$

We thus have three equations in four variables. Therefore, these equations have infinitely many solutions. We can set any variable fixed, and express the other three using this fixed variable. Since the largest molecule is glucose, we set its coefficient x_3 fixed. Then we have

$$x_1 = 6x_3, \quad x_2 = 6x_3, \quad x_4 = 6x_3. \tag{4.66}$$

Thus, the chemical equation is

$$6x_3CO_2 + 6x_3H_2O \longrightarrow x_3C_6H_{12}O_6 + 6x_3O_2. \tag{4.67}$$

If we want to produce one glucose molecule, i.e., $x_3 = 1$, then we need 6 carbon dioxide and 6 water molecules:

$$6CO_2 + 6H_2O \longrightarrow C_6H_{12}O_6 + 6O_2. \tag{4.68}$$

Similarly, one can write chemical equations for many common reactions, such as iron oxidation

$$3Fe + 4H_2O \longrightarrow 4H_2 + Fe_3O_4, \tag{4.69}$$

and the redox reaction in a human body which consumes glucose and converts the glucose into energy, water, and carbon dioxide

$$C_6H_{12}O_6 + 6O_2 \longrightarrow 6CO_2 + 6H_2O. \tag{4.70}$$

4.8 Multivariate Linear Regression Using Matrix Notations

This section is an application of matrix algebra in statistical data analysis, particularly on a linear regression for more than one variable. The linear regression in Section 3.5 discussed the fitting of a straight line on the xy-plane $y = b_1x + b_0$ to a pair of data vectors of length N: $[x_d]_{N\times 1}, [y_d]_{N\times 1}$. It resulted in correlation, trend, and other regression quantities. An example is the correlation between the January SOI as x and US temperature as y. The non-trivial correlation can suggest that there may be a physical mechanism to explain how the January SOI influences the US January temperature.

Here, we use $y = b_1x + b_0$ as the representation of a deterministic fitting function. This expression does not involve random variables. Most statistics books would consider random linear models, which distinguish between random variables and their deterministic estimators by data. Our simple linear mathematical model formulation here is equivalent to using only the deterministic estimators.

The US January temperature can be influenced by multiple factors in addition to the SOI. These factors may include the North Atlantic sea surface temperature (SST), the North Pacific SST, Arctic pressure conditions, etc. Then, the linear model becomes

$$y = b_0 + b_1x_1 + b_2x_2 + \cdots + b_nx_n. \tag{4.71}$$

When $n = 1$, this degenerates into the single variable regression.

In the following we will present a few R examples of multivariate regression and its applications. The mathematical theory behind the R code for a multivariate regression deals mostly with the matrix operations, which can be found from any standard textbooks covering multivariate regression.

Example 4.9 This example shows a two-variable regression.

$$y = b_0 + b_1 x_1 + b_2 x_2. \tag{4.72}$$

Geometrically, this is an equation of a plane in a 3D space (x_1, x_2, y). Given three points not on a straight line, a plane can be determined uniquely. This means specifying three x_1 coordinate values, three x_2 coordinate values and three y coordinate values, which can be done by means of the following R code:

```
x1=c(1,2,3) #Given the coordinates of the 3 points
x2=c(2,1,3)
y=c(-1,2,1)
df=data.frame(x1,x2,y) #Put data into the data.frame format
fit <- lm(y ~ x1 + x2, data=df)
fit#Show the regression results
#Call:
#  lm(formula = y ~ x1 + x2, data = df)
#Coefficients:
#  (Intercept)          x1          x2
#-5.128e-16    1.667e+00    -1.333e+00

1.667*x1-1.333*x2  #Verify that 3 points determining a plane
#[1] -0.999  2.001  1.002
```

Example 4.10 This example will show that four arbitrarily specified points cannot all be on a plane. The fitted plane has the shortest distance squares, i.e., the least squares (LS), or minimal mean square error (MMSE). Thus, the residuals are non-zero, in contrast to the zero residuals in the previous example.

```
u=c(1,2,3,1)
v=c(2,4,3,-1)
w=c(1,-2,3,4)
mydata=data.frame(u,v,w)
myfit <- lm(w ~ u + v, data=mydata)
summary(myfit)#Show the result
#Coefficients:
#  Estimate Std. Error t value Pr(>|t|)
#(Intercept)   1.0000      1.8708    0.535     0.687
#u             2.0000      1.2472    1.604     0.355
#v            -1.5000      0.5528   -2.714     0.225
```

Example 4.11 This example will show a general multivariate linear regression using R. It has three independent variables, one dependent variable, and ten data points. For R program simplicity, the data are generated by an R random number generator. Again, R requires that the data be put into data frame format so that a user can clearly specify which are independent variables, also called explainable variables, and which is the dependent variable, also called response variable.

```
dat=matrix(rnorm(40),nrow=10,
           dimnames=list(c(letters[1:10]), c(LETTERS[23:26])))
fdat=data.frame(dat)
fit=lm(Z~ W + X + Y, data=fdat)
summary(fit)
```

```
#Coefficients
#              Estimate  Std. Error  t value  Pr(>|t|)
#(Intercept)   0.36680    0.16529    2.219    0.0683
#W             0.11977    0.20782    0.576    0.5853
#X            -0.53277    0.19378   -2.749    0.0333
#Y            -0.04389    0.14601   -0.301    0.7739
```

Thus, the linear model is

$$Z = 0.37 + 0.12W - 0.53X - 0.04Y. \tag{4.73}$$

The 95% confidence interval for W's coefficient is $0.12 \pm 2 \times 0.21$, that for X's coefficient is $-0.53 \pm 2 \times 0.19$, and that for Y's coefficient is $-0.04389 \pm 2 \times 0.15$. Each confidence interval includes zero. Thus, there is no significant non-zero trend for the Z data with respect to W, X, Y. This result is to be expected, because the data are randomly generated and thus should not have a trend.

In practical applications, a user can simply convert the data into the same data frame format as shown here. Then, R command

```
lm(formula = Z~ W + X + Y, data = fdat)
```

can do the regression job.

R can also do nonlinear regression with specified functions, such as quadratic functions and exponential functions. See examples from the URLs

https://www.zoology.ubc.ca/~schluter/R/fit-model/
https://stat.ethz.ch/R-manual/R-devel/library/stats/html/nls.html

4.9 Chapter Summary

This chapter introduces matrices and linear algebra from the perspective of space–time climate data, with rows corresponding to spatial locations (e.g., the order of grid boxes, grid point IDs, or climate station IDs) and columns corresponding to time (e.g., the month stamps January 1901, February 1901, ..., December 2017 for the monthly data, or the day stamps 1 June 2017, 2 June 2017, ..., 30 June December 2017 for the daily data of June 2017). This way of introducing matrices as organized collections of data is different from that of most textbooks which describe a matrix as originating from a set of linear equations. This particular way to arrange a space–time climate data matrix follows the similar philosophy of the netCDF data file, which has recently become a very popular format for representing extensive climate datasets.

The most essential methods of linear algebra have to do with matrix properties (e.g., eigenvalues, eigenvectors, and SVD matrices) and matrix operations (e.g., solving linear equations, and making linear transformations). Our chapter has described the following linear algebra methods:

(i) Matrix arithmetic: addition of a matrix plus another matrix, subtraction of a matrix minus another matrix, multiplication of a matrix times another matrix, and multiplication of a matrix by a scalar, and division of the identity matrix by a matrix which is a way to define the inverse matrix of the latter.

(ii) An eigenvector w of a matrix C is such a special vector that Cw is in w's own direction. Thus, there exists a scalar λ such that $Cw = \lambda w$, where λ is called an eigenvalue of the matrix C corresponding to the eigenvector w.

(iii) SVD decomposes the space–time data matrix A into three matrices: a dimensionless spatial orthogonal matrix U as spatial eigenvectors, a dimensionless temporal orthogonal matrix V as temporal eigenvectors, and a diagonal "energy level" matrix D related to the eigenvalues of A's covariance matrix, i.e.,

$$A = UDV^t. \tag{4.74}$$

The column vectors of U may represent the intrinsic spatial patterns of the climate data (e.g., the warming pattern of the eastern tropical Pacific for the El Niño based on the SST data), and those of V may represent the intrinsic temporal patterns (e.g., the El Niño peaks in the time series determined by a column vector of V). See the specific patterns shown in the examples of Section 10.3 on the SVD analysis for the reanalysis data. The diagonal values of D show the "energy level" (or the strength) of the corresponding spatial and temporal patterns. In the last few years, the SVD method has become more widely used in science and engineering because of its universal power in decomposing any space–time matrix. Mathematicians have now begun to modernize the proofs of the theorems of linear algebra by using the SVD approach, instead of the traditional approach known as echelon reduction through row operations.

(iv) Mathematical models based on linear equations have been developed and employed for many climate science applications, such as the mass balance model for chemical equations.

(v) Multivariate linear regression has been formulated here from the perspective of matrix algebra.

(vi) R codes have been written for all these methods, which can be conveniently used for solving climate data analysis problems in research and practical applications.

Although the basic materials of linear algebra appeared as early as the seventeenth century, widespread university classroom education in linear algebra in the United States began only in the 1950s and later (Tucker 1993). At that time, electronic computers began to become available to solve large numbers of linear equations for engineering applications. Today's vastly increased computing power now provides opportunities to modernize educational aspects of this branch of mathematics, such as using SVD to compute various properties of a matrix and to prove numerous theorems of linear algebra. Climate mathematics takes full advantage of this technical progress. Many graduate students in climate science now use EOFs and PCs in their research, whereas the EOF-PC techniques were rarely taught as part of climate science education before the 1990s.

References and Further Readings

[1] Golub, G. H. and C. Reinsch, 1970: Singular value decomposition and least squares solutions. *Numerische mathematik*, 14, 403–420.

This seminal paper established an important method, known as the Golub–Reinsch algorithm, to compute the eigenvalues of a covariance matrix from a space–time matrix A without actually first computing the covariance matrix AA^t. This algorithm makes the SVD computation very efficient, which helps scientists consider SVD as a genuine linear algebra method, not a traditionally regarded statistical method based on a covariance matrix.

[2] Lorenz, E. N., 1956: Empirical orthogonal functions and statistical weather prediction. *Scientific Report No. 1, Statistical Forecasting Project.* Air Force Research Laboratories, Office of Aerospace Research, USAF, Bedford, Massachusetts.

Edward N. Lorenz (1917–2008) was an American meteorologist and applied mathematician, who has been best known for his theories of chaos and for coining the term "butterfly effect." Although he was not the first to invent the mathematical method of empirical orthogonal functions (EOFs), Lorenz, together with other scientists, such as Gerald R. North, made EOF a very popular method for climate data analysis.

[3] Strang, G., 2016: *Introduction to Linear Algebra*. 5th Edition, Wellesley-Cambridge Press, Wellesley, MA 02482.

Gilbert Strang (1934–) is an American mathematician and educator. His textbooks and pedagogy have been internationally influential. This text is one of the very few basic linear algebra books that includes excellent materials on SVD, probability, and statistics.

[4] Tucker, A., 1993: The growing importance of linear algebra in undergraduate mathematics. *College Mathematics Journal*, 24, 3–9.

This paper describes the historical development of linear algebra, such as the term "matrix" being coined by J. J. Sylvester in 1848, and pointed out that "tools of linear algebra find use in almost all academic fields and throughout modern society." The use of linear algebra in the big data era is now even more popular.

Exercises

Exercises 4.2–4.8 require the use of a computer.

4.1 The following are the SVD results

```
mat
     [,1] [,2]
[1,]    1    1
[2,]    1   -1

svd(mat)
$d
[1] 1.414214 1.414214

$u
           [,1]       [,2]
[1,] -0.7071068 -0.7071068
[2,] -0.7071068  0.7071068

$v
     [,1] [,2]
[1,]   -1    0
[2,]    0   -1
```

Use $A = UDV^t$ to recover the first column of

```
mat
     [,1] [,2]
[1,]    1    1
[2,]    1   -1
```

Show detailed calculations of all the relevant matrices and vectors, and use space–time decomposition to describe your results.

For extra credit: Describe the spatial and temporal modes, and their corresponding variances or energies.

4.2 Use R and the updated Darwin and Tahiti standardized SLP data to reproduce the EOFs and PCs and to plot the EOF pattern maps and PC time series.

4.3 Do the same procedures in the previous problem but for original non-standardized data. Comment on the difference of the results from the previous problem.

4.4 (a) Download the monthly precipitation data at five different stations over the United States from the USHCN website:

https://www.ncdc.noaa.gov/ushcn

(b) Use R to organize the January data from 1951 to 2010 into the space–time format.

(c) Compute the climatology of each station as the 1971–2000 mean.

(d) Compute the space–time anomaly data matrix A as the original space–time data matrix minus the climatology.

(e) Use R to make the SVD decomposition of the space–time anomaly data matrix $A = UDV^t$.

(f) Output the U and D matrices.

4.5 In the previous problem, use R and the formula UDV^t to reconstruct the original data matrix A. This is a verification of the SVD decomposition, and is also called EOF-PC reconstruction.

4.6 Use R to plot the maps of the first three EOF modes from the U matrix in the previous problem in a way similar to the two El Niño mode maps shown in Fig. 4.4. Try to explain the climate meaning of the EOF maps.

4.7 Use R to plot the first three PC time series from the V matrix in Problems 4.4 and 4.5. Try to explain the climate meaning of the time series.

4.8 (a) A covariance matrix C can be computed from a space–time observed anomaly data matrix X which has N rows for spatial locations and Y columns for time in years:

$$C = X \cdot X^t / Y. \tag{4.75}$$

This is an $N \times N$ matrix. Choose an X data matrix from the USHCN annual total precipitation data at three California stations from north to south [Berkeley, CA (040693); Santa Barbara, CA (047902); Cuyamaca, CA (042239)] and five years from 2001 to 2005. Consider the anomaly data with respect to the 2001–2005 mean and use R to calculate a covariance matrix for $N = 3$ and $Y = 5$.

(b) Use R to find the inverse matrix of the covariance matrix C.

(c) Use R to find the eigenvalues and eigenvectors of C.

(d) Use R to make SVD decomposition of the data matrix $X = UDV^t$. Explicitly write out the three matrices U, D, and V.

(e) Use R to explore the relationship between the eigenvalues of C and the matrix D.

(f) Compare the eigenvectors of C and the matrix U.

(g) Plot the PC time series and describe their behavior.

4.9 The burning of methane (CH_4) with oxygen (O_2) produces water (H_2O) and carbon dioxide (CO_2). Balance the chemical reaction equation.

4.10 The burning of propane (C_3H_8) with oxygen (O_2) produces water (H_2O) and carbon dioxide (CO_2). Balance the chemical reaction equation.

4.11 The burning of gasoline (C_8H_{18}) with oxygen (O_2) produces water (H_2O) and carbon dioxide (CO_2). Balance the chemical reaction equation.

5 Energy Balance Models for Climate

This chapter is devoted to a class of climate models known as energy balance models (EBMs). These models have been developed as highly idealized and simplified representations of certain key aspects of a climate system. An EBM is developed on the basis of the principle of energy balance: the climate system in question is assumed to be in a balanced or equilibrium state, such that the incoming solar energy entering the system is equal to the outgoing energy leaving the system. The motions of an atmosphere and an ocean are not explicitly considered in an EBM.

Such an extremely simplified model can aid us in understanding a very complex system, such as Earth's climate, and in the case of systems much simpler than the Earth's climate, an EBM may be capable of simulating some climate parameters of the system fairly realistically. In fact, some bodies in the solar system have neither atmosphere nor ocean, and our Moon is one such body. Furthermore, the Moon has no water. As we shall show, a simple EBM can simulate the Moon's surface temperature quite realistically.

This chapter will also describe applications of an EBM to the planet Earth, despite the drastic simplification that atmospheric and oceanic motions are not considered at all. The unknown variable in an EBM applied to the Earth's climate is the Earth's surface air temperature. This chapter considers only a few very simple EBMs, such as a stationary zero-dimensional model with and without albedo feedbacks, and discusses the EBM solutions, their stability, and their climate interpretations and limitations.

5.1 EBM for Modeling the Moon's Surface Temperature

In 1961, the United States set a goal: "landing a man on the Moon and returning him safely to the Earth" within the decade. This ambitious goal was met by the Apollo program, which first landed astronauts on the Moon in 1969. Altogether, six Apollo flights between 1969 and 1972 landed a total of 12 men who walked on the Moon. The Apollo astronauts left clear boot prints on the Moon, and also brought back samples of lunar soils and rocks, which cover the surface of the Moon. The lunar soils are fine particles from disintegrated rocks. Any atmosphere or ocean which the Moon might once have had must have escaped long ago, because the Moon is so small that its gravity, only 1/6 that of the Earth, is too weak to allow the Moon to retain a significant atmosphere or an ocean.

5.1.1 Moon–Earth–Sun Orbit and Lunar Surface

The Moon rotates around its axis very slowly, compared to the Earth, and takes approximately 27 Earth days per revolution, compared to the Earth's rotation of one Earth day per revolution. The Moon's rotation around its own axis is synchronized with its orbiting around the Earth, which means that rotating around its axis once and orbiting around the Earth once use the same time. Thus, the Moon always presents the same side to the Earth. The Moon's axis is tilted at a small angle, only 1.54 degrees, compared to 23.4 degrees for the Earth. The Earth's axial tilt used here is the angle between its rotational axis and a line perpendicular to the plane of Earth's orbit. The Moon's axial tilt is defined in a similar way. Figure 5.1 shows the relative positions and orbits of the Moon, Earth, and Sun. One can also search and watch some Internet videos to see the relative motions of the Moon, Earth and Sun.

The axial tilt is responsible for the seasons. Thus, the Moon has only a weak seasonality compared to the Earth. The sunlit hemisphere of the Moon faces the Sun and receives solar radiation. The solar energy absorbed by the Moon heats the Moon's surface, and some of the heat is also conducted a short distance beneath the surface. The strength of the solar radiation received by the Moon is about the same as that of the Earth, because the Moon and the Earth are both about the same distance from the Sun, with a solar constant approximately equal to 1,368 watts per square meter [W m^{-2}]. A portion of the solar radiation reaching the Moon's surface is reflected back to space, and the magnitude of this portion is determined by the Moon's reflectivity, which is called the albedo and denoted by α. The term albedo may have originated in the mid-nineteenth century and is derived from Latin word *albus*, meaning white. An absolutely white body, or perfect reflector, which reflects all the radiation reaching it, is said to have an albedo equal to one, and an absolutely

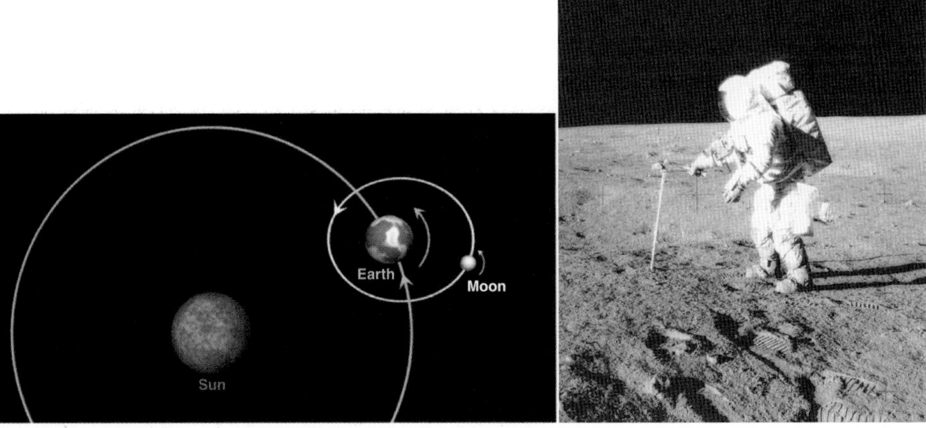

Fig. 5.1 (a) The orbits of the Moon and the Earth around the Sun, not to scale. The Earth–Sun distance is about 150 million km. The Earth–Moon distance varies from about 356 thousand to about 407 thousand km. *Source: The Science: The Earth and Moon* https://moonblink.info/Eclipse/why/solsys (b) Footprints of Apollo 12-astronauts on the Moon in 1969. *Credit: NASA*

blackbody, or perfect absorber, which absorbs all the radiation reaching it, is said to have an albedo equal to zero. The lunar surface is fairly dark and has an average albedo around 0.12, according to NASA Goddard Space Flight Center

https://nssdc.gsfc.nasa.gov/planetary/factsheet/moonfact.html

although the albedo still varies with location. Thus, on average, some 88% of the incident solar radiation is absorbed by the lunar surface. The Moon's surface has no water and almost no atmosphere, only 25,000 [kg] compared to 5.1×10^{18} [kg] of the Earth's atmosphere. The absorbed solar radiation heats the upper part of the Moon's regolith to a depth of about 0.5 to 1.0 meters, and the Moon does not have water or air to be heated. The high surface temperature gives rise to a large vertical temperature gradient that allows heat to be conducted from the Moon's surface downward into the lunar regolith. The regolith, which is found on the Earth, the Moon, and several other bodies, is a layer of loose, heterogeneous deposits covering solid rock. It includes dust, soil, broken rock, etc. The word regolith combines two Greek words, *regos* (blanket) and *lithos* (rock). The lunar regolith is typically several meters thick. The energy thus stored beneath the surface is released during the lunar night. Because the Moon lacks both air and water, the nighttime temperature of the lunar surface depends on the heat released from beneath the surface.

Again because of the lack of any atmosphere or ocean, the lateral energy flow on the Moon's surface is small and is neglected in our EBM modeling.

5.1.2 Moon's Surface Temperature

Figure 5.2 shows a snapshot of the temperature for the entire lunar surface based on the satellite remote sensing of the United States NASA Diviner Lunar Radiometer Experiment (Williams et al. 2017). Diviner is a satellite that orbits the Moon. It was launched in 2009 and has an orbit only about 50 km above the Moon's surface. The Moon's equatorial surface temperature has a large range. At local noon it is about 390 K (or 117°C), and at midnight it is about 100 K (−173°C).

In advance of the Apollo mission, NASA scientists were able to use an energy balance model (EBM) to make fairly accurate predictions of the Moon's surface temperature at a given location and time. The data were important for planning the first manned landing on the Moon in 1969 and several subsequent landings.

Figure 5.2 can be plotted by the following R code.

```
#NASA Diviner Data Source:
#http://pds-geosciences.wustl.edu/lro/lro-l-dlre-4-rdr-v1/lrodlr_1001/data/gcp/
setwd("/Users/sshen/climmath")
d19=read.table("data/tbol_snapshot.pbin4d-19.out-180-0.txt", header=FALSE)
dim(d19)
#[1] 259200      3 #259200 grid points at 0.5 lat-lon resolution
#259200=720*360, starting from (-179.75, -89.75) going north
#then back to south pole then going north
#until the end (179.75, 89.75)
m19=matrix(d19[,3],nrow=360)
dim(m19)
#[1] 360 720
```

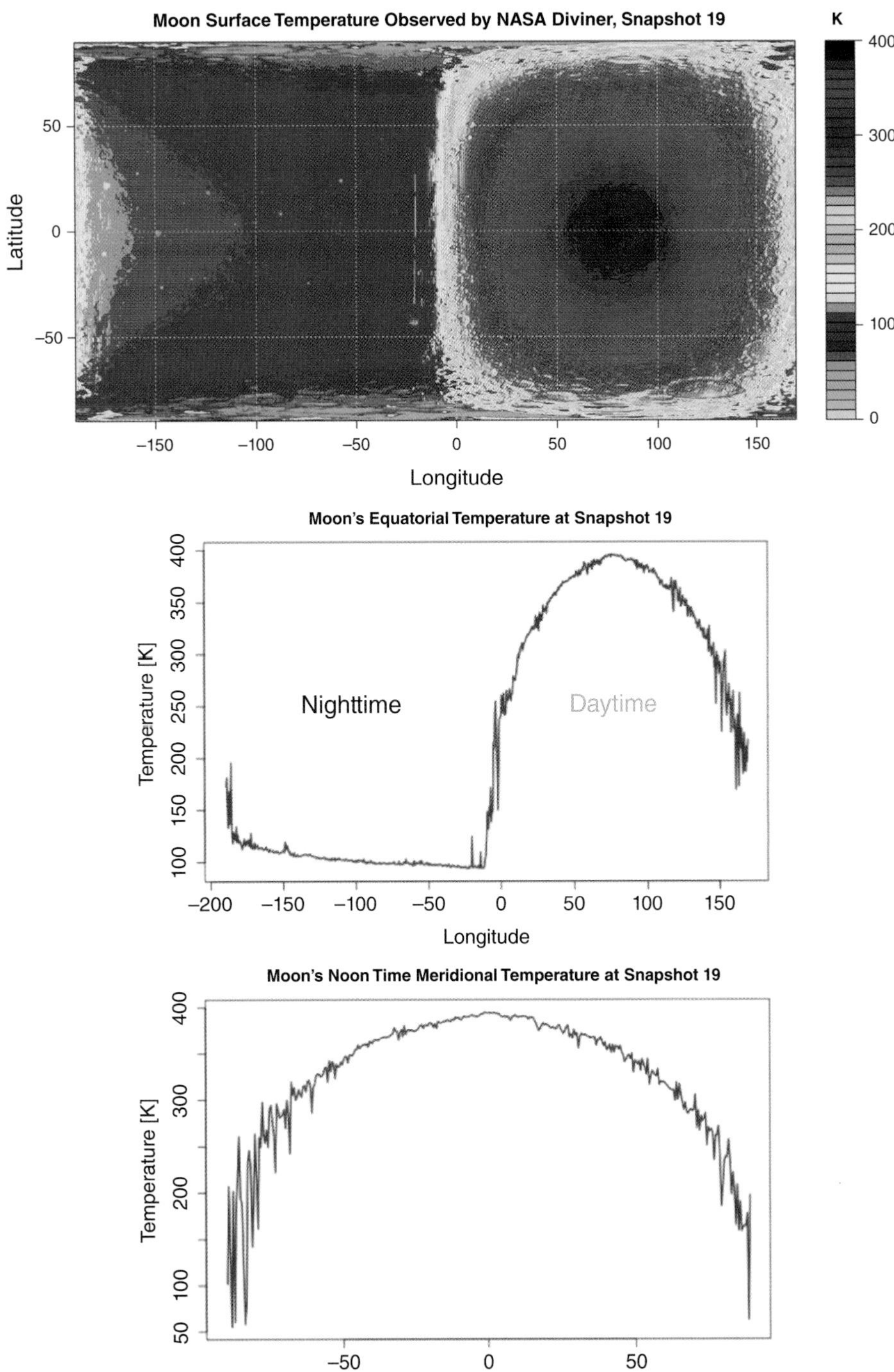

Fig. 5.2 A snapshot of the temperature for the lunar surface based on Diviner satellite data. Top panel: The temperature for the entire lunar surface as a function of latitude and longitude; Middle panel: The equatorial temperature as a function of longitude; Lower panel: Temperature as a function of latitude along the noon-time meridional line.

```
library(maps)
Lat1=seq(-89.75,by=0.5,len=360)
Lon1=seq(-189.75,by=0.5, len=720)
  mapmat=t(m19)
  #mapmat=pmin(mapmat,10)
  #mapmat= mapmat[,seq(length(mapmat[1,]),1)], no flipping
  plot.new()
png(filename=paste("Moon Surface Temperature, Snapshot=", 19,".png"),
      width=800, height=400)
  int=seq(0,400,length.out=40)
  rgb.palette=colorRampPalette(c('skyblue',  'green', 'blue', 'yellow',
      'orange', 'pink','red', 'maroon', 'purple', 'black'),
      interpolate='spline')
  filled.contour(Lon1, Lat1, mapmat,
                 color.palette=rgb.palette, levels=int,
                 plot.title=title("Moon Surface Temperature Observed
    by NASA Diviner, Snapshot 19",
                 xlab="Longitude", ylab="Latitude"),
                 plot.axes={axis(1); axis(2);grid()},
                 key.title=title(main="K"))
  dev.off()

#Plot the equator temperature for a snapshot
  plot.new()
  png(filename=paste("Moon's Equatorial Temperature at Snapshot",
                     19,".png"),
      width=600, height=400)
  plot(Lon1,m19[180,],type="l", col="red",lwd=2,
       xlab="Longitude", ylab="Temperature [K]",
       main="Moon's Equatorial Temperature at Snapshot 19")
  text(-100,250,"Nighttime",cex=2)
  text(80,250,"Daytime",cex=2, col="orange")
  dev.off()

  #Plot the noon time meridional temperature for a snapshot
  plot.new()
  png(filename=paste("Moon's Noon Meridional Temperature at Snapshot",
                     19,".png"), width=600, height=400)
  plot(Lat1,m19[,540],type="l", col="red",lwd=2,
       xlab="Latitude", ylab="Temperature [K]",
       main="Moon's Noon Time Meridional Temperature at Snapshot 19")
  dev.off()
```

The average temperature of the Moon's bright side is 303 K (or 30°C), and that of the dark side is 125 K (or −148°C). These are calculated from the Diviner data by the following R code.

```
  #Compute the bright side average temperature
  bt=d19[129601:259200,]
  aw=cos(bt[,2]*pi/180)
  wbt=bt[,3]*aw
  bta=sum(wbt)/sum(aw)
  bta
  #[1] 302.7653  K

  #Compute the dark side average temperature
  dt=d19[0:12960,]
  aw=cos(dt[,2]*pi/180)
```

```
wdt=dt[,3]*aw
dta=sum(wdt)/sum(aw)
dta
#[1] 124.7387  K
```

5.1.3 EBM Prediction for the Moon Surface Temperature

A simple EBM can approximately simulate the lunar surface temperature except in the polar regions, where the solar radiance energy data have a large uncertainty.

Figure 5.3 shows the balance of energy. In the absence of both an atmosphere and an ocean, the lateral heat transfer is negligible because of the very small lateral temperature gradient. Thus, an approximate energy balance can safely be assumed to be established locally, such as along the noon-time equator. The solar constant of the Moon is the same as that of the Earth: $S = 1,368$ W m^{-2}, because the Moon and the Earth are almost the same distance from the Sun. The solar constant is the power flux of solar radiation through a plane that is perpendicular to the parallel rays of solar radiation at the Earth's mean distance from the Sun. The solar constant, although called "constant," actually varies with

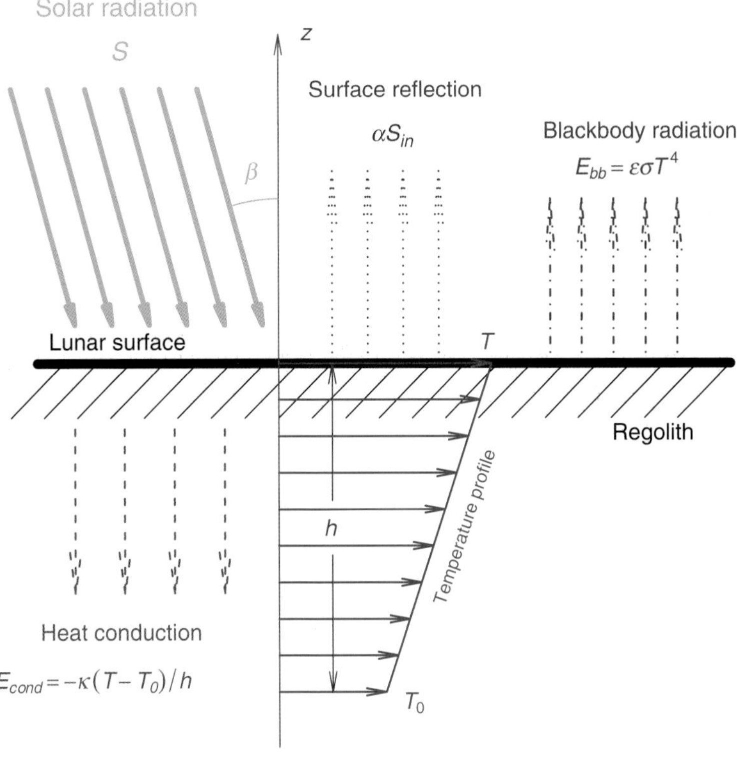

Fig. 5.3 Schematic diagram of fluxes included in an EBM for the lunar surface temperature. The EBM is a balanced power flux model of the Moon's surface at a given location. The incoming energy is from the incident solar radiation minus the surface reflection, and the outgoing energy consists of the blackbody radiation to space and thermal conduction into the lunar regolith.

time around this value both randomly and periodically because of solar activity, such as the 11-year sunspot cycle. According to a NASA article entitled "The Inconstant Sun," the solar constant S during low sunspot activity can be about 1,365 W m^{-2} and can be about 1,368 W m^{-2} during high sunspot activity

http://science.nasa.gov/science-news/science-at-nasa/2003/17jan_solcon/

Keeping in mind this observed natural variability in S, we have sometimes used slightly different values for S in this book, including 1,365 and 1,368 W m^{-2}.

Note that watt is a unit for power, not energy, which is an integration of power with respect to time. Thus, the energy balance here is a balance of "power" per unit area, i.e., power flux, in the units of W m^{-2}. Thus, strictly speaking, an energy balance model should perhaps be called a power balance model, but the term EBM is widely used, and we have chosen to follow the prevailing custom and call these models EBMs.

The power flux reaching the lunar surface at a given location is

$$S_{in} = S\cos\beta \tag{5.1}$$

where β is the solar zenith angle for the location of the EBM. Because the Moon's axial tilt is so small, the solar zenith angle β at a location on the Moon is approximately equal to this location's latitude ϕ.

A portion of the solar radiation arriving at the lunar surface is reflected back to space and thus does not affect the lunar surface temperature. This portion is determined by the surface albedo α multiplied by the solar radiation flux

$$S_r = \alpha S_{in}. \tag{5.2}$$

Thus, the incoming power flux is the solar power flux arriving at the Moon's surface at this point minus the reflected power flux

$$E_{in} = S_{in} - S_r = (1-\alpha)S_{in} = (1-\alpha)S\cos\phi. \tag{5.3}$$

The Moon's surface has an average albedo value of about 0.12. Thus, about 88% of the solar energy reaching the Moon at a given location is absorbed by the Moon's surface, and this energy heats the lunar regolith to a depth of about 0.5 to 1.0 meters. This makes the lunar equatorial surface very hot at noon.

The heating of the upper lunar regolith beneath the surface occurs via a thermal conduction process which can be modeled by Fourier's law:

$$E_{cond} = \kappa\frac{T - T_0}{h} \tag{5.4}$$

where $\kappa = 7.4 \times 10^{-4}$ $[\mathrm{W\ m^{-1}\ K^{-1}}]$ is the thermal conductivity of the Moon's surface regolith (according to the NASA Diviner Lunar Radiometer Experiment result in Hayne et al. (2017)), T is the surface temperature to be predicted, T_0 is the deep crust's temperature, which is assumed to be constant at 260 K, and h [m] is the lunar crust's depth that can be reached by the thermal conduction from the surface and is assumed to be 0.4 meters in our examples in this subsection (see Fig. 7 of Vasavada et al. (2012)). Here we use the word "crust" in the geological sense, meaning the outermost layer of a planet or other body, such as the Moon.

The thermal conductivity of the Moon's deep crust is larger and is estimated to be $\kappa = 34 \times 10^{-4}$ $[\mathrm{W\,m^{-1}\,K^{-1}}]$ at a depth of one meter, according to Hayne et al. (2017). The small thermal conductivity of the Moon's surface regolith implies that the Moon's surface is an excellent insulator, even better than wool, the thermal conductivity of which is about 0.04–0.20 $[\mathrm{W\,m^{-1}\,K^{-1}}]$. Thus, the thermal conduction process at the Moon surface is very weak. The slow rotation of the Moon allows the incoming radiative heat at the surface during the day to be slowly conducted to the deep crust and also allows the deep crust's heat to be conducted slowly to the surface during the night.

In contrast, the thermal conductivity of Earth's surface is much larger with $\kappa = 0.15 - 4$ $[\mathrm{W\,m^{-1}\,K^{-1}}]$ for soil and $\kappa = 0.591$ $[\mathrm{W\,m^{-1}\,K^{-1}}]$ for water. These terrestrial values are thus on the order of one thousand times larger than the lunar conductivity.

We assume here that the energy radiated by a body having a temperature T is governed by the Stefan–Boltzmann blackbody radiation law

$$E_{bb} = \epsilon\sigma T^4, \tag{5.5}$$

where ϵ is the lunar surface's emissivity, and $\sigma = 5.670367 \times 10^{-8}$ $[\mathrm{W\,m^{-2}\,K^{-4}}]$ is called the Stefan–Boltzmann constant. The energy balance equation is then based on the incoming energy being equal to the outgoing energy, as described by the following equation

$$E_{in} = E_{bb} + E_{cond}, \tag{5.6}$$

or

$$(1-\alpha)S\cos\phi = \epsilon\sigma T^4 + \kappa\frac{T - T_0}{h}. \tag{5.7}$$

Given the values of the parameters $\alpha, S, \phi, \epsilon, \sigma, \kappa, T_0$, and h, one can solve this nonlinear equation to predict the lunar surface temperature T at a given location. For example, for the equator at noon, this EBM predicts a temperature 384 K, a value which compares well with the Diviner observation of 389 K. The R code for finding this solution is below.

```
#Equator noon temperature of the moon from an EBM
lat=0*pi/180
sigma=5.670367*10^(-8)
alpha= 0.12
S=1368
ep=0.98
k=7.4*10^(-4)
h=0.4
T0=260
fEBM=function(T){(1-alpha)*S*cos(lat) -(ep*sigma*T^4 + k*(T-T0)/h)}
#Numerically solve this EBM: fEBM=0
uniroot(fEBM,c(100,420))
#[1] 383.6297
```

Using the same R code, we can can predict the noontime temperature at latitude 60°N: `lat=60*pi/180`. The result is 323 K, which compares well with the Diviner observation of 318 K.

During the lunar night, a point on the dark side of the Moon receives no solar radiation: $S = 0$, but the decrease in nighttime lunar surface temperature is mitigated, because thermal conduction allows the heat stored beneath the lunar surface during the daytime to be released to the lunar surface. This thermal conduction process is again governed by Fourier's law, but with a much larger gradient if we assume that the lunar surface is on average an insulation layer with a depth of 2 cm. Thus, the thermal conduction is through this layer. Hence, we choose $h = 0.02$ [m]. With these data: $\phi = 0, S = 0, h = 0.02$, the EBM predicts the midnight temperature of the equator to be 100 K, while the Diviner observation is 101 K.

It is remarkable that the simple EBM can reasonably simulate the observed lunar surface temperature distribution except in the polar regions. The polar regions simulation is more difficult, mainly because our zenith angle approximation for the calculation of the incoming solar radiation is not as accurate there. The Diviner observations also have a large uncertainty in the polar regions. NASA scientists used the EBM approach to predict the lunar surface temperature in advance of the Apollo landings. They also employed a more complex model than we have developed here for the thermal conduction, in order to improve the accuracy of the prediction.

Next, we turn to developing an EBM for the climate of the Earth, which is very different and more difficult, in large part because the lateral energy exchanges on Earth occurring through the motions of air and water are much too important to be neglected.

5.2 EBM for the Global Average Surface Temperature of the Earth: A Zero-Dimensional Climate Model

In a similar way, the assumption of an exact balance between the incoming solar energy to the Earth's surface and the outgoing energy from the Earth to space through radiation and reflection forms an equation, which is an EBM for Earth's climate. This is perhaps the simplest non-trivial climate model, yet an important one, useful in the study of climate science.

The EBM we used to model the lunar surface temperature at a given location considered only the vertical energy balance. Because of the air and water flows on the Earth's surface, an EBM for a single location of the Earth must consider the energy balance, not only in the vertical direction, but also in the lateral directions. Modeling the lateral energy exchange is very difficult. For this reason, we choose to describe an extremely simple EBM for Earth that does not need to consider the lateral energy exchange. This EBM deals with the energy balance for the global average temperature of the entire spherical surface of the Earth. Because the EBM is for the average temperature of the entire Earth surface system, the lateral energy exchange is an internal activity. It redistributes energy in the climate system but has no role in either the energy coming into the system, or in the energy leaving the system. Thus, the complex lateral energy exchange due to the extremely complicated air and water flows does not need to be considered in our global Earth EBM. Consequently, the model does not have spatial location dependence. Of course, our result will only be a single temperature value, representing the global average temperature for the entire surface of the Earth.

In this global Earth EBM, our fundamental assumption is still that the energy that comes into the Earth system is equal to the energy that goes out from the system. Thus, the Earth's surface is in an equilibrium state of a constant temperature without temporal variation. Such a simple EBM without dependence on space or time is thus a stationary zero-dimensional climate model for the Earth.

5.2.1 The Incoming Power from the Solar Radiation to the Earth

At any given time only one side of the Earth faces the Sun and receives solar radiation as parallel rays shown in Fig. 5.4. The Earth's solar constant S is defined as the power flux of solar irradiance rays through a plane that is perpendicular to the rays. Because the Moon and the Earth have about the same distance to the Sun, the Earth's solar constant is approximately the same as that of the Moon: $S = 1,368\ [\mathrm{W\ m^{-2}}]$.

Because the entire Earth receives solar radiance on one side at a given moment, the total solar energy flux to the entire Earth surface is equivalent to that incident on a circular disk whose radius R is the same as that of the Earth, about 6,400 km. The disk's area is πR^2. That energy, however, is distributed over the entire Earth surface whose area is $4\pi R^2$. Thus, the solar irradiance received by the Earth's surface per unit area is

$$S_{solar} \quad S\frac{\pi R^2}{4\pi R^2} \quad S/4 \quad 1,368/4 \quad 342\ [\mathrm{W\ m^{-2}}]. \tag{5.8}$$

Some solar irradiance is reflected back to space. This amount is determined by the Earth's albedo. The albedo of fresh white snow is 0.80–0.90, while that of fresh black asphalt is

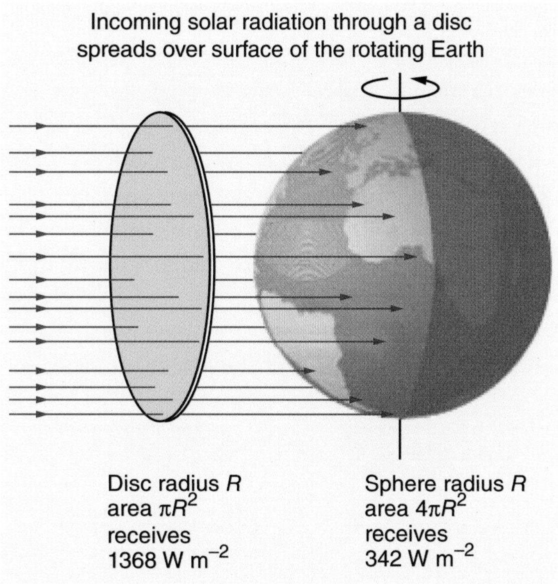

Fig. 5.4 The side of the Earth facing the Sun receives solar radiation as parallel rays that pass through a circular disc of radius R. *Source: Free Course: Climate Change:* www.open.edu/openlearn

0.04. Thus, 80–90% of radiation incident on the fresh white snow is reflected, while only 4% is reflected from the fresh black asphalt. The albedo of green grass is about 0.25, and that of desert sand is about 0.40. According to NASA Earth Observatory https://earthobservatory.nasa.gov/IOTD/view.php?id=84499, the average albedo of the entire Earth surface is approximately 0.30, most of which is due to clouds. Thus, the solar power absorbed by Earth is equal to the total power radiated to Earth $S/4$ minus that reflected by the Earth $\alpha(S/4)$:

$$E_{in} = S/4 - \alpha(S/4) = (1 - \alpha)S/4. \tag{5.9}$$

If $S = 1,368$ $[\mathrm{W\ m^{-2}}]$ and $\alpha = 0.3$, then $E_{in} = 239$ $[\mathrm{W\ m^{-2}}]$ is the solar energy absorbed by the atmosphere and Earth's surface. This value $E_{in} = 239$ $[\mathrm{W\ m^{-2}}]$ may be interpreted as the total incoming power 341 $[\mathrm{W\ m^{-2}}]$ minus the total reflected power 102 $[\mathrm{W\ m^{-2}}]$.

The actual solar energy absorbed by the Earth can have significant variations due to the variable conditions of the Earth's surface and clouds (through variations of the albedo α value) and due to changes in the Sun (through variations of the solar constant S value). These variations can be as large as a few percent.

5.2.2 The Outgoing Power from Long-Wave Radiation Emitted by the Earth

The other part of the EBM is a model for the radiation emitted by the Earth. Every body in the Universe having a temperature above absolute zero radiates energy; the Earth is such a body. People also emit radiation. For example, one can feel the heat radiation from a person or a group of people nearby. A human body radiates energy via infrared waves, which are invisible to our eyes but can be detected by night vision goggles or by cameras which are sensitive to the infrared portion of the electromagnetic spectrum. The Earth also radiates mainly via infrared waves, which are long compared to the wavelengths of the incoming solar irradiance. In general, we know from basic physics that the energy emitted by a hotter body has a higher energy and shorter wavelengths than the energy emitted by a cooler body. The Earth and its atmosphere are much colder than the Sun. Solar radiation is peaked in short wavelengths in the visible portion of the electromagnetic spectrum, while the Earth's surface and atmosphere radiate mainly in invisible long waves peaked in the infrared portion of the spectrum.

The long-wave radiation power emitted per unit area from the Earth's surface can be quantified by the Stefan–Boltzmann blackbody radiation law

$$E_{bb} = \epsilon\sigma T^4, \tag{5.10}$$

where

$$\sigma = 5.670367 \times 10^{-8}\ [\mathrm{W\ m^{-2}\ K^{-4}}] \tag{5.11}$$

is called the Stefan–Boltzmann constant, and $0 < \epsilon \leq 1$ is the dimensionless emissivity of the Earth's surface. In the Stefan–Boltzmann law, temperature has the unit of kelvin, which is 273.15 plus Celsius degrees. We first take the emissivity ϵ to be one, i.e., the Earth's surface is assumed to be a perfect radiating body. Under this assumption, the radiation is at its maximum, depends only on the Earth's temperature, and is not reduced or blocked in any way.

5.2.3 EBM as a Power Balance

If the Earth's average surface temperature does not change and the incoming energy is equal to the outgoing energy,[1] then we have an equation of energy balance $E_{in} = E_{bb}$, i.e.,

$$\epsilon\sigma T^4 = (1-\alpha)(S/4). \tag{5.12}$$

We can easily solve this algebraic equation

$$T^4 = \frac{(1-\alpha)(S/4)}{\epsilon\sigma} = \frac{(1-0.30)(1{,}368/4)}{1.0\times 5.670373\times 10^{-8}} = 42.22\times 10^8\ [\mathrm{K}]^4 \tag{5.13}$$

Then, the temperature is

$$T = (42.22\times 10^8)^{1/4} = 255\ [\mathrm{K}], \tag{5.14}$$

or

$$T = 255 - 273.15 \approx -18\ [^\circ\mathrm{C}]. \tag{5.15}$$

As an estimate of the Earth's surface air temperature, this T value is about 33°C too low compared to observations. The observed average surface temperature of the Earth is about +15°C. A uniform surface air temperature of $T = -18$°C is far below the freezing point of water. Such a frozen planet is referred to as "snowball Earth."

To alter our EBM so that it remains very simple but yields a uniform Earth with a more realistic average surface temperature T, we might decide to "tune" the model by adjusting the parameters α and ϵ in the above EBM solution:

$$T = \left[\frac{(1-\alpha)(S/4)}{\epsilon\sigma}\right]^{1/4} - 273.15\quad [^\circ\mathrm{C}]. \tag{5.16}$$

One way to do this is to change the Earth's ϵ from 1.0 to 0.6. This would reduce the planet's radiation to space and hence would produce a warmer Earth. We might claim that such an adjustment in emissivity is justified by the greenhouse effect, caused by water vapor, carbon dioxide, methane, ozone, etc. Then

$$T = \left[\frac{(1-0.30)(1{,}368/4)}{0.6\times 5.670373\times 10^{-8}}\right]^{1/4} - 273.15 = 16\quad [^\circ\mathrm{C}]. \tag{5.17}$$

This number seems reasonable and is close to the recently observed global average surface air temperature (SAT) of the Earth: 15°C (59°F).

Nonetheless, if we are honest with ourselves, we would have to admit that we have chosen the value $\epsilon = 0.6$, because it yields a result that we know agrees with observations. Nonetheless, this ad hoc procedure provides a "knob to turn" for us to tune our very simple climate model. However, it also illustrates a danger that we must always be alert to in trying to model the climate, or anything else: we must always resist the temptation to alter

[1] Strictly speaking, it is the balance of power: incoming power is equal to the outgoing power.

a model in an arbitrary way, motivated by knowing in advance what model result will be "the right answer." Therefore, quantitatively plausible physical mechanisms and credible observational evidence ideally should be cited to support any tuning of a mathematical model for climate.

5.3 EBM for the Global Average Surface Temperature of an Earth with a Nonlinear Albedo Feedback

To make our model a bit closer to reality, we might next try to make reflectivity dependent on the surface temperature T. A sufficiently low temperature would then be assumed to result in snow or ice covering the Earth's surface, producing a larger albedo (i.e., a whiter surface, or a larger reflectivity value), and conversely, a higher temperature would imply an Earth's surface with a smaller albedo (i.e., not as white). A simple model for this situation is

$$\alpha(T) = \begin{cases} 0.7 & T \leq 250 \\ -0.0133T + 4.0333 & 250 < T < 280 \\ 0.3 & T > 280 \end{cases} \tag{5.18}$$

Thus, when the uniform temperature is below 250 K or -23.15°C, the Earth's surface is assumed to be ice and snow with albedo equal to 0.7; and when the temperature is above 280 K or 6.85°C, the Earth's surface is assumed to be water and/or land with albedo equal to 0.3. The temperature interval between 250 and 280 K is a transitional zone[2] with the albedo gradually decreasing from 0.7 to 0.3. Figure 5.5 shows the albedo variation with temperature, based on the above-defined piecewise function. This figure can be generated by the following R code:

```
#Define a piecewise albedo function
a1=0.7
a2=0.3
T1=250
T2=280
ab= function(T) {ifelse(T < T1, a1,
            ifelse(T < T2,((a1-a2)/(T1-T2))*(T-T2)+ a2, a2))}
#Define the range of temperature
T=seq(200,350,len=200)
#Plot the albedo as a nonlinear function of T
setwd("/Users/sshen/climmath")
png(file="fig05-05.png", width=400, heigh=300)
plot(T, ab(T), type="l", lwd=2.0,
     ylim=c(0,1), xlab="Surface Temperature [K]",
     ylab="Albedo", main="Nonlinear Albedo Feedback")
dev.off()
```

[2] This transitional behavior of the albedo can also be modeled by a smooth function

$$\alpha(T) = (0.5 - 0.2 \times \tanh((T - 265)/10) \tag{5.19}$$

where 265 K is regarded as the ice formation temperature, and tanh is the hyperbolic tangent function, a smooth step function which is 1.0 at infinity and -1.0 at negative infinity. See Kaper and Engler (2013).

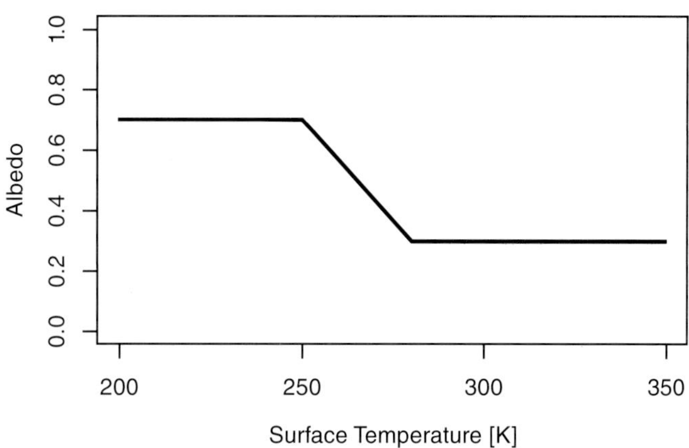

Fig. 5.5 Nonlinear dependence of albedo on the Earth's surface temperature: The variable albedo effect is a positive feedback on temperature, in the sense that the transitional albedo changes amplify the temperature changes, because a warming Earth becomes a darker Earth that reflects less solar energy and hence gets warmer; and a cooling Earth becomes a brighter Earth that reflects more solar energy and hence gets cooler.

```
#One can plot the albedo function directly without using a function
curve(ifelse(x < 260, 0.7, ifelse(x < 285,-0.016*x+ 4.86, 0.3)),
      from=200, to=350, lwd=3, ylim=c(0,1))
```

Then, the new EBM is

$$\epsilon\sigma T^4 = (1 - \alpha(T))(S/4). \tag{5.20}$$

This EBM has three solutions representing three different climate regimes:

- the snowball Earth for temperatures below 250 K,
- a temperate Earth for temperatures greater than 280 K, and
- a transitional regime for temperatures between these two.

The EBM in the snowball and temperate regimes can be solved analytically, while the EBM in the transitional regime has to be solved numerically. Computers can easily be programmed to solve it or one can solve it graphically as shown in Fig. 5.6.

Figure 5.6 can be produced using the following parameter values and R commands.

```
#Formulate and solve an EBM
S <- 1368
ep <- 0.6
sg <- 5.670373*10^(-8)
T <- seq(200,350, by=0.1)
Ein<-(1-ab(T))*(S/4)
Eout<- ep*sg*T^4
png("fig05-07.png",width=8,height=6, units = 'in', res = 300)
plot(T, Ein, xlim=c(200, 350), ylim=c(0,300),
     xaxp=c(200, 350, 15), yaxp=c(0, 300, 10),
     type="l",col="red", lwd=3,
     panel.first = grid(30, lty = "dotted", lwd = 1),
```

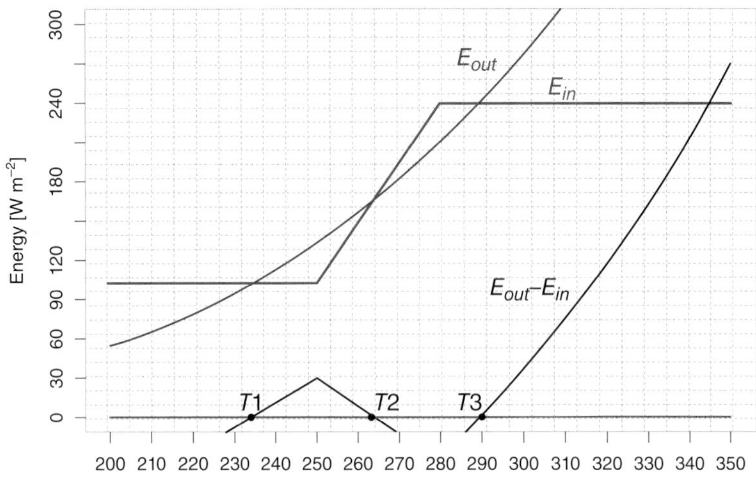

Fig. 5.6 Incoming solar energy E_{in} as a nonlinear smooth step function of temperature T due to albedo feedback, with outgoing radiation energy E_{out} based on the Stefan–Boltzmann law, and the graphical solution of an EBM: $E_{out} = E_{in}$ or $E_{out} - E_{in} = 0$.

```
     main="Simple Nonlinear EBM with Albedo Feedback: Eout = Ein",
     ylab="Energy [W/sq.m]",
     xlab="Surface temperature T [K]")
lines(T, Eout,col="blue",lwd=2.0)
lines(T, Eout-Ein,col="black",lwd=2.0)
y0<-0.0*T
lines(T, y0,col="purple")
text(310, 248, "Ein", col="red", cex=1.5)
text(290, 275, "Eout", col="blue", cex=1.5)
text(300, 100, "Eout - Ein", col="black", cex=1.5)
text(234,10, "T1", cex=1.5)
text(267,10, "T2", cex=1.5)
text(287,10, "T3", cex=1.5)
points(234, 0, pch=16)
points(263, 0, pch=16)
points(290, 0, pch=16)
dev.off()
# The three intersections of the red and blue lines
# are three solutions: T1=234, T2=263, T3=290 K.
```

For the snowball regime, the EBM is

$$\epsilon\sigma T^4 = (1-0.7)(S/4). \tag{5.21}$$

Hence,

$$T = \left(\frac{(1-0.7)(S/4)}{\epsilon\sigma}\right)^{1/4} = 234 \text{ K}. \tag{5.22}$$

Similarly, the temperate Earth has its temperature determined by

$$T = \left(\frac{(1 - 0.3)(S/4)}{\epsilon\sigma} \right)^{1/4} = 289 \text{ K}. \tag{5.23}$$

The numerical solution for the transitional regime can be found by the following R commands.

```
S<-1368
ep<-0.6
sg<-5.670373*10^(-8)
f <- function(T){return(ep*sg*T^4 - (1-ab(T))*(S/4))}
uniroot(f,c(260,275))
#$root
#[1] 263.4303
```

Thus, the solution is $T = 263$ K.

One can also use similar R commands to solve for the other two temperatures by

```
uniroot(f,c(275,295))
uniroot(f,c(220,240))
```

Consequently, the three solutions of the nonlinear EBM incorporating an albedo feedback are $T_1 = 234$, $T_2 = 263$, and $T_3 = 290$ K. The solution T_3 17°C is close to the current observed value for the Earth's average temperature. The solution $T_1 = -39$°C is a deeply frozen planet entirely covered with snow and ice, corresponding to a global ice age. The transitional solution $T_2 = -10$°C is also a snowball Earth, but will be shown to be unstable in the next section. An "unstable solution" means a state that cannot physically exist or be observed in nature, because it will immediately evolve into another state. This instability is explained below by considering time dependence of an EBM.

5.4 Time-Dependent Zero-Dimensional EBM for the Earth's Global Average Surface Temperature

At this point, we must stress a very important fact. As we have already pointed out, the B in EBM stands for balance, and any EBM is based on the crucial assumption of equilibrium. This assumption, that there is an exact balance between energy absorbed by the climate system and energy emitted by the climate system, is equivalent to saying that the climate system being modeled is energetically in a constant equilibrium state which does not change with time. That simplification can be a useful idealization in helping us to understand the system. However, we know that the climate of the Earth is definitely not in such an equilibrium now. The Earth's climate is now warming, and extensive research has shown that the dominant cause of the warming in recent decades is human activities, especially activities such as burning fossil fuels (coal, oil, and natural gas), which add heat-trapping substances such as carbon dioxide to the atmosphere, thus increasing the magnitude of the greenhouse effect. This phenomenon, popularly known as global warming, or man-made climate change, has many consequences for mankind and for ecosystems and is one of

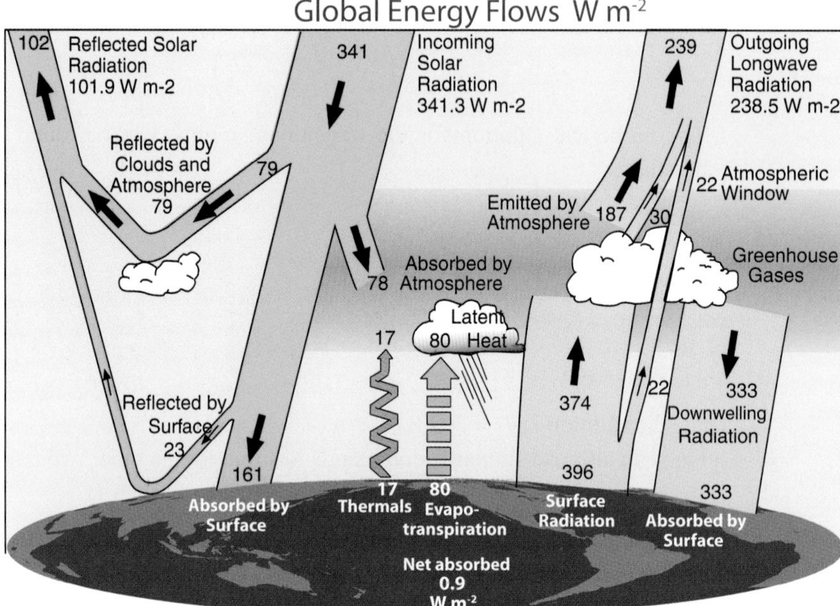

Fig. 5.7 Energy budget of the Earth. Incoming solar radiation and outgoing energy via radiation and other processes (Trenberth and Fasullo (2011)).

the most important challenges facing humanity today (IPCC 2013). An estimate of the actual energy budget of the Earth is shown in Fig. 5.7.

5.4.1 An EBM Including Time Dependence

In general, the Earth's temperature can vary with time. The rate of temperature change times the heat capacity is equal to the power needed to drive the change. This relation yields the time-dependent EBM

$$C\frac{dT}{dt} = (1-\alpha)(S/4) - \epsilon\sigma T^4 + P_i(t), \tag{5.24}$$

where $P_i(t)$ is the time-dependent internal variation of power, and C is the heat capacity of the climate system, including all air, ice, soil, and water, particularly the mixed layer of the ocean, assumed here to extend from the ocean surface to a depth of 70 meters. An estimated value of C for the current Earth is $C = 2.0983 \times 10^8$ [J K^{-1}m^{-2}].

This value $C = 2.0983 \times 10^8$ [J K^{-1}m^{-2}] is estimated by considering only the mixed layer ocean water with the following parameters: (i) depth of the mixed layer: $h =$ 70 [m]; (ii) Seawater density: $\rho = 1{,}023$ [kg m^{-3}], (iii) the specific heat of the seawater: $C_{Water} = 4{,}186$ [J kg^{-1} K^{-1}], and (iv) percentage of the Earth's surface being covered by the ocean: 70%. The heat capacity of the mixed layer ocean is $C_{Ocean} = \rho h C_{Water} =$ 299,759,460 [J K^{-1}m^{-2}]. The heat capacity of the land is denoted by C_{Land}. Then, the global heat capacity is

$$C = 0.7C_{Ocean} + 0.3C_{Land}.$$

We neglect the heat capacities of air and land in this approximation because they are very small, compared to the ocean heat capacity. Then,

$$C \approx 0.7C_{Ocean} = 2.0983 \times 10^8 \text{ [J K}^{-1}\text{m}^{-2}\text{]}.$$

The surface area of Earth is $S_{Earth} = 5.10072 \times 10^{14}$ [m^2]. The heat capacity for the entire Earth is then

$$C_{Earth} = C \times S_{Earth} = 1.07^{23} \text{ [J K}^{-1}\text{]}.$$

For more details, see Chapter 3 (Energy Balance Models) of the book by McGuffie and Henderson-Sellers (2014).

The above EBM (5.24) allows temperature T to change with respect to time and involves a derivative. An equation involving one or more derivatives is called a differential equation (often abbreviated as DE). Thus, this EBM is a differential equation model. A derivative is a slope of a curve, and it can be interpreted as the rate of change of a dependent variable with respect to an independent variable. In the above time-dependent EBM, the derivative $\frac{dT}{dt}$ is interpreted as the rate of temperature change with time. The rate of change is proportional to the magnitude of the power that causes the change, in this case the incoming absorbed shortwave solar radiation minus the outgoing emitted longwave terrestrial radiation. The rate of change is inversely proportional to the heat capacity C that quantifies the amount of heat needed to produce a given temperature change in the climate system. Heat capacity is an extensive property of matter, meaning that it is proportional to the size or extent of the system being considered. Under comparable conditions, air can be heated more readily than water, because air has a much smaller heat capacity than water.

The solution of a differential equation is a function. For example, the solution of $\frac{dy}{dt} = t - 1$ is $y = t^2/2 - t + D$ where D is an arbitrary constant. An initial condition $t = 0, y = 3$ can determine the constant D which is 3 in this example. Thus, the solution to this initial value problem of the differential equation is $y = t^2/2 - t + 3$.

Solving a differential equation analytically is often very difficult or impossible. For example, solving the above EBM analytically is impossible when the albedo feedback is included. Many computer program packages, such as R's `deSolve` package, can solve DEs for us. The study of DEs is itself a large and important area of mathematics. We will briefly discuss DEs later in this book, but we will limit ourselves to certain DE problems that are important from the perspective of calculus applications to climate sciences.

5.4.2 Stability Analysis of the Multiple Solutions of the EBM with a Nonlinear Albedo Feedback

Below we explore the stability of the three EBM solutions found earlier. Stability theory in mathematics has been extensively developed, and we treat here only a simple kind of stability. For our purposes, a solution of a DE is defined as stable if an initially small perturbation of this solution will evolve as time goes on in such a way as to remain small or even to decrease to zero, in other words, if the perturbed solution will always remain close to the unperturbed solution. If not, the solution is defined as unstable. This definition is consistent with certain intuitive stability properties encountered in daily life. An ordinary stepladder is said to be stable, for example, because a little accidental push to the ladder will not make the ladder fall down. However, a very slowly moving or stationary bicycle may be unstable, because a little push in a sideways direction applied to the rider can cause the bicycle and rider to fall.

When the right-hand side of the above EBM (5.24) is equal to zero, the derivative $\frac{dT}{dt}$ is zero. The model is then at an equilibrium state, in which the incoming power is equal to the outgoing power, if the internal random forcing is assumed to be zero. If such an equilibrium solution can remain at this state or in its nearby vicinity after it is subjected to a small power imbalance, called a small perturbation in mathematical terminology, then the equilibrium is said to be stable. The climate can then remain at this equilibrium state or in its nearby vicinity as time goes on. Otherwise, the equilibrium is said to be unstable. The equilibrium in the unstable case is still a mathematical solution of the time independent EBM, but the solution to the time-dependent EBM will not remain near the equilibrium state. Thus, an unstable solution of the time-independent EBM can be only a temporary or transient state of climate. The climate can therefore never remain in an unstable equilibrium state, and thus an unstable state is not observable.

We can argue that the largest solution (290 K) among the three solutions in the albedo feedback EBM described earlier is stable. For a small positive perturbation, $T > 290$ K, $C\frac{dT}{dt} < 0$ since the red line (incoming power) is below the blue (outgoing power) (see Fig. 5.6). The negative derivative means that T will decrease as time goes on, thus T will return to the equilibrium value of 290 K. For a small negative perturbation, $T < 290$ K, $C\frac{dT}{dt} > 0$ means that T will increase as time goes on, so in this case, T will also return to the equilibrium value of 290 K. Therefore, a small perturbation, either positive or negative, added to the equilibrium solution $T = 290$ K will result in the climate system returning to $T = 290$ K, i.e., the perturbation to the equilibrium solution will decrease to zero as time goes on. Therefore, the system is stable at $T = 290$ K.

One can use similar reasoning to show that the smallest of the three solutions, $T = 234$ K, is also stable, but the middle solution $T = 263$ K is unstable.

5.4.3 Energy Flow Budget and Greenhouse Effect for the Earth's Climate

The energy imbalance of the Earth is a significant driver of the climate change that is observed to be occurring now and that is projected to continue in coming decades. It is a subject of intensive research (von Schuckmann et al. 2016). Monitoring the energy flow imbalance is a complex scientific and technical challenge, requiring advances in both satellite remote sensing and *in situ* observational tools, as well as climate modeling. It is known that more than 90% of the heat gained by the Earth's climate system in recent decades because of this energy imbalance is stored in the ocean. Our ability to observe the global ocean and its heat content has recently been revolutionized by a technological advance: the development of autonomous Argo floats and the global deployment in the world ocean of thousands of these floats (Riser et al. 2016).

The present energy imbalance has multiple causes involving both natural and human-caused factors. The natural factors include the El Niño–La Niña phenomenon, the 11-year solar (or sunspot) cycle, and volcanic events. The human factors include the rapid increase of heat-trapping gases, especially the increase in atmospheric carbon dioxide concentration from about 315 parts per million by volume (ppmv) in 1958 to about 410 ppmv as this book is written in 2018.

The amount of water vapor in the Earth's atmosphere is highly variable, but the other atmospheric constituents are far less so. For that reason, it is useful when discussing the composition of the atmosphere to consider the dry atmosphere (with no water vapor) separately from water vapor. The dry atmosphere, excluding water vapor and small particles such as dust, is composed mainly of nitrogen (N_2, 78%), oxygen (O_2, 21%), and argon (Ar, 0.9%). The remaining 0.1% is made up of trace gases, so-called because they make up only a trace of the atmosphere. A more precise dataset from NASA
http://nssdc.gsfc.nasa.gov/planetary/factsheet/earthfact.html
lists the following values for dry air (percentage by volume): Major (99.03%) gases of nitrogen and oxygen, and minor (0.97%) gases of argon and trace gases.

The concentration of major gases (99.03%) is as follows:

- Nitrogen (N_2) 78.08%
- Oxygen (O_2) 20.95%

The concentration of minor gases (0.97%) in ppmv relative to the entire atmosphere is as follows:

- Argon (Ar) 9350
- Carbon dioxide (CO_2) 408
- Neon (Ne) 18.18
- Helium (He_2) 5.24
- Methane (CH_4) 1.7
- Krypton (Kr) 1.14
- Hydrogen (H_2) 0.55
- Ozone (O_3) less than 0.1

These numbers do not add up to exactly 100% due to round-off errors and observational uncertainties.

The atomic mass[3] of nitrogen is 14 daltons, that of oxygen is 16, carbon 12, hydrogen 1, and the noble gas argon 40; C, N, and O are next to each other in the periodic table and make up most of our atmosphere. The total mass of the atmosphere is about 5.1×10^{18} [kg] . The actual atmosphere includes water vapor, which varies in both space and time. Water vapor is highly variable, typically making up about 1% of the atmosphere, approximately about five Lake Superior-equivalents in terms of the amount of water in the air. Lake Superior is North America's largest fresh-water lake in terms of surface area (82,170 km^2). Its water volume is approximately 12,000 [km^3], about 13% of the Earth's total fresh lake water.

[3] Atomic mass is the mass of one atom. Its unit is the unified atomic unit, denoted by [u] or [Da] meaning dalton. One Da is defined as 1/12 of the mass of a single carbon-12 atom.

The atmospheric water vapor, mostly coming from oceanic evaporation, is the source of precipitation and, because of its radiative properties, water vapor also plays a major role as a regulator of the Earth's temperature. Aspects of global water vapor distribution in space and time are the subject of extensive research.

The trace gases include carbon dioxide (CO_2), methane (CH_4), nitrous oxide (NO_2), ozone (O_3), neon (Ne), and helium (He_2). Carbon dioxide (CO_2), methane (CH_4), nitrous oxide (NO_2), and ozone (O_3) have three or more atoms, unlike the simple two-atom molecules of nitrogen and oxygen, and they have more complex molecular structures. These complex gas molecules are bound together loosely and can be easily excited by radiation, leading to more molecular vibrations and hence higher temperature. Those molecules can absorb radiative energy and become warmer. A detailed explanation of why molecules with three or more atoms absorb infrared energy under the pressure and temperature conditions of the Earth's atmosphere is complicated and involves quantum mechanics, which is beyond the scope of this book. These gases are called greenhouse gases (GHG), because they absorb infrared energy. Water vapor (H_2O), a molecule with three atoms, is also a GHG. Thus, these GHG molecules are able to absorb some of the Earth's longwave radiation, hence becoming warmer, and they then come into contact with and consequently warm nearby molecules of nitrogen and oxygen surrounding them.

A physically insightful way to think about this ability of certain atmospheric gases to heat the Earth's surface is to consider that, in the absence of such gases, all of the radiation emitted to space by the Earth would originate at the Earth's surface, at the bottom of the atmosphere. Such an atmosphere with no GHGs would allow a portion of incoming sunlight to be absorbed, depending on the albedo, but would have no effect on outgoing infrared radiation. The GHGs, however, both absorb and re-emit infrared radiation, so that, seen from space, some of the planet's emitted energy if the atmosphere included GHGs would originate at an altitude above the Earth's surface. The effective radiating altitude of the planet thus will become higher, as the amount of such gases in the atmosphere increases. In the troposphere, atmospheric temperature decreases with height and is determined by various atmospheric processes. Consequently, the higher effective radiating altitude implies lower radiation temperature in the Stefan–Boltzmann law of blackbody radiation ($\epsilon\sigma T^4$), which then implies that less energy will be emitted. For the emitted energy to be able to balance the absorbed incoming energy, the temperature at the effective radiating altitude must increase. When this happens, the entire atmospheric column will warm, including the atmosphere at the surface. It is in this sense, because of these processes involving the dependence of the Earth's effective radiating altitude related to atmospheric GHG amounts, that GHGs may be said to "trap heat" like a blanket.

Although this atmospheric warming due to the GHG molecules is commonly called the *greenhouse effect*, the underlying physical mechanism differs from that operating in a typical gardening greenhouse. The transparent plastic or glass walls and roof in a typical greenhouse do allow much of the short-wave radiation to pass through unimpeded, and they do partially trap the outgoing long-wave radiation, or heat. However, the greenhouse walls and roof also greatly reduce the movement of air, or wind, so that in a greenhouse, the heated air cannot escape, and it is that wind reduction in a greenhouse that plays a dominant role in the resulting warming.

The atmospheric composition described above is for dry air. However, the actual atmosphere contains water vapor, and, in fact, much more water vapor than CO_2. Water vapor H_2O also has tri-atomic (three atoms) molecules, and is hence also a GHG. It has long been known that the amount of water vapor in the atmosphere is strongly determined by temperature, and a warm atmosphere contains significantly more water than a cool one. The relevant physics is described by the Clausius–Clapeyron relationship of classical thermodynamics, which expresses the dependence of saturation vapor pressure on temperature (Pierrehumbert 2010). On the other hand, CO_2 is not affected by phase change processes such as evaporation and condensation, which are involved in clouds and precipitation, and thus all the CO_2 in the atmosphere is always in the gaseous phase. Thus, adding CO_2 to the atmosphere warms the air, which has the effect of increasing the amount of water vapor in the air, a phenomenon that further increases the warming. The amount of CO_2 in the atmosphere may therefore be thought of as an effective "control knob" governing the Earth's surface temperature (Lacis et al. 2010). This control mechanism due to CO_2 can be incorporated in a rather crude way into the emissivity parameter of an EBM for the Earth's climate.

5.5 Increasing the Complexity of Climate Models

The EBMs discussed above for the Earth's climate are all very simple, and yet the time-varying albedo feedback EBM is already complex enough to prevent us from finding an analytical solution. Within the EBM model "family," the complexity of the models can easily be further increased in many ways, to consider latitude dependence, seasonality, land–ocean differentiation, ocean coupling, and many more kinds of physical processes. For example, a next level of complexity might be to introduce a latitude-band dependence of temperature, with higher temperatures in tropical regions and lower temperatures in polar regions. The addition of a latitudinal dependence of temperature means that the temperature will now be a function of both time and latitude $T(t,\phi)$, where ϕ is latitude. The EBM for this $T(t,\phi)$ involves derivatives with respect to both t and ϕ, which are partial derivatives. The resulting EBM is a time-dependent one-dimensional (1D) model, mathematically represented by a partial differential equation (PDE). In contrast to a PDE, an ordinary differential equation (ODE) involves only derivatives with respect to one variable, say t. An ODE is sometimes simply referred to as a DE.

When an EBM is formulated to also include the land and ocean differences, the temperature depends on at least three independent variables: time, latitude, and longitude, i.e., $T(t,\phi,\theta)$ where θ denotes longitude. The corresponding EBM is a nonlinear PDE of three variables. It is impossible to find an analytic solution to such a PDE, and it can even be a major effort to solve the PDE model numerically.

A comprehensive review including many members of the EBM model family is found in the book by North and Kim (2017). Although some of their EBMs involve solving multiple nonlinear partial differential equations, the models still do not include the circulation of either atmosphere or ocean. Although these fluid motions are not explicitly described in

EBMs, they are known to be critically important to weather and climate. For that reason, extensive effort since the 1950s has gone into the development of three-dimensional and time-dependent models known generically as GCMs. Originally the term GCM was an abbreviation of general circulation model. In meteorology, the term "general circulation" refers to the large-scale, time-averaged patterns of atmospheric motions, including, for example, the trade winds blowing from east to west in the tropics, and the westerly winds blowing from west to east in middle latitudes. In more modern usage, GCM is an abbreviation for global climate model (Donner et al. 2011; Goose 2015), which may refer to models of the ocean or the atmosphere or the coupled climate system including both ocean and atmosphere, as well as many other elements and processes, such as land surfaces, glaciers and ice sheets, and biogeochemical phenomena. The GCM equations are extremely complex PDEs often requiring high space and time resolution for their numerical solution. Depending on computer speeds, modern GCM resolutions can reach 1 km of horizontal space resolution and a few seconds of time resolution for global models and even finer resolution for regional climate models (RCMs). It is a major scientific team effort and a gigantic computational engineering task to develop a GCM or an RCM, which may involve hundreds of thousands of lines of code. Yet, many such models have been developed at research centers throughout the world, and several of these models, known generically as community models, are now freely available for use by scientists who have not been part of the teams which developed them. Thus, many climate scientists are now users rather than developers of the GCMs or RCMs, who may experiment with the models for specific purposes. For these users, it is usually not necessary to make major changes in the original PDEs in the climate models, or to rewrite the corresponding source codes. Instead, with today's easy and inexpensive access to supercomputing power, the well-documented and extensively tested community models have become popular and powerful research tools now readily available to a large number of researchers worldwide.

Despite the popular use of complex GCMs of high resolution, EBMs and other models of moderate or "intermediate" complexity can provide very useful estimates and insights that may guide research into issues such as the assessment of the overall greenhouse effect, the instability of a small ice cap on a polar region, and the sampling error estimation of climate observations (North and Kim 2017). Much experience has shown the great value of investigating a hierarchy of models, rather than restricting research only to GCMs or only to highly idealized models such as EBMs. Often the impressive realism of GCMs appears to come at a high cost in lack of understanding, because the complexity of the GCM can make it difficult to comprehend why the model behaves as it does. Then the more idealized models in the hierarchy may play a role in helping to understand both the GCM and the natural climate system.

Although the modeling portion of this book is limited to EBMs and does not include materials for developing GCMs, Chapters 10 and 11 of the book do supply R codes for analyzing and visualizing both GCM output and observational datasets, and these tools are valuable in meeting the needs of climate model users. Thus, this book has sufficient climate modeling materials for many climate science researchers and climate data users.

5.6 Chapter Summary

The Sun drives the climate system. The main energy in the climate system is the heat energy, and the leading measure of this quantity is temperature. The balance of the incoming energy and the outgoing energy forms an EBM equation with temperature as the unknown. Solving the EBM yields the temperature under specific conditions. This is the fundamental concept underlying the account of EBMs presented in this chapter. We have shown that the EBM is not only a useful concept in climate science, but it also has practical value in realistic applications, such as the accurate prediction of the Moon's surface temperature. The specific EBM topics included in this chapter are summarized as follows.

(i) The Moon's surface consists of an insulating regolith, and the Moon has no water and very little atmosphere, thus the lunar climate has little lateral heat exchange. This makes the lunar climate an ideal candidate to be modeled by an EBM. Except in the polar regions, our simple EBM prediction of the Moon's surface temperature at a given location and given time is impressively accurate, with an error of only a few kelvin, over the large lunar surface temperature range of 95 K–390 K. We discussed the evaluation of the EBM results by comparison with recent NASA Diviner satellite data.

(ii) The EBM we developed for a specific location on the lunar surface is based on the assumption that the incoming solar energy is balanced by blackbody radiation to space plus vertical conduction of heat in the regolith layer:

$$E_{in} = E_{bb} + E_{cond}. \tag{5.25}$$

Here, the heat conduction E_{cond} is downward in the regolith during the lunar day and upward at the lunar night. Although we customarily call this an equation of energy balance, strictly speaking it is actually a balance of power flux with a unit such as $[\mathrm{W\ m^{-2}}]$.

(iii) Because on Earth the atmosphere and ocean flows are responsible for lateral transfers of a massive amount of energy, an EBM for the Earth for a single location would be very complicated. As an introduction to the topic of EBMs, this chapter has been restricted to considering only the global average temperature averaged over the entire surface of the Earth. We have thus avoided the issues of lateral heat flow. This is the zero-dimensional EBM.

(iv) When we introduce an albedo feedback that makes a cold Earth more white and hence reflect more solar radiation, and makes a warm Earth less white and hence reflect less solar radiation, the zero-dimensional EBM becomes a nonlinear model and has three solutions, one of which corresponds approximately to the current climate of the Earth.

(v) If the assumption of balance is invalid, so there is more incoming energy and/or less outgoing energy, the energy surplus can heat the Earth. This may cause global warming and may alter other aspects of climate, including the circulation of the ocean and atmosphere. A time-dependent EBM might then be used to attempt to model the changing temperature of the Earth's surface.

(vi) However, the EBMs discussed in this chapter are very simple and have many limitations. Much more complicated global climate models (GCMs) have been developed at research centers around the world, and some of them have been made available to the climate research community so that researchers can use these community models as part of their research on climate change, such as estimating the response of the climate system to increasing the atmospheric concentration of carbon dioxide.
(vii) R codes have been written for the above steady-state EBMs. Exercise problems will give you a chance to use these codes.

The success of using a simple EBM to accurately predict the Moon's surface temperature may motivate you to apply it to predict the climate for other planets that do not have either atmosphere or water. Mercury is such a planet. An exercise problem about Mercury will give you a chance to carry out EBM modeling procedures similar to the ones we showed in this chapter for the Moon.

References and Further Readings

[1] Donner, L., W. Schubert, and R. C. J. Somerville (eds.), 2011: *The Development of Atmospheric General Circulation Models: Complexity, Synthesis, and Computation*, Cambridge University Press, New York.

> One reviewer called this book "A wonderful history of climate modeling science written by climate scientists" and went on to say, "This book contains some beautiful gems of writing about a field of research that has simultaneously emerged as the poster child of politicization, a subject of scorn by the observationalists in atmospheric and Earth system science, and an example of what science is capable of telling us about the future of the planet... whether we are ready to hear about it or not!" The book is a multi-authored treatment, and many of the authors are leading pioneers in the field of climate modeling.

[2] Goose, H., 2015: *Climate System Dynamics and Modeling*, Cambridge University Press, New York.

> This is a textbook, a reference book, and an excellent and very readable introduction to climate dynamics and climate modeling. It is aimed at graduate students and includes a glossary and exercises.

[3] Hayne, P. O., and co-authors, 2017: Global regolith thermophysical properties of the Moon from the Diviner Lunar Radiometer Experiment. *Journal of Geophysical Research: Planets*, 122, 2371–2400.

One of a series of important papers on the Diviner satellite experiment which provided observations of the Moon that we have used in conjunction with the research we describe in this book on developing energy budget models (EBMs) for the lunar climate.

[4] IPCC, 2013: *Climate Change 2013: The Physical Science Basis*. Contribution of Working Group I to the Fifth Assessment Report of the Intergovernmental Panel on Climate Change [Stocker, T. F., D. Qin, G.-K. Plattner, M. Tignor, S. K. Allen, J. Boschung, A. Nauels, Y. Xia, V. Bex and P. M. Midgley (eds.)], Cambridge University Press, New York.

This is an IPCC report, famous around the world. The IPCC assessment reports appear about every six years, and they are the definitive summary of our scientific understanding of climate change. This particular volume, the Working Group One report, deals with the physical science of the climate system. At some 1500 pages, with lots of charts, graphs and technical language, it is not easy reading. It and the many other IPCC reports are available for free at the IPCC website, www.ipcc.ch.

[5] Kaper, H. and H. Engler, 2013: *Mathematics and Climate*, SIAM Books, Philadelphia.

This book, written at a mathematically advanced level, is intended for research mathematicians or mathematics major students and aims to influence mathematicians to do research on climate science problems.

[6] Lacis, A. A., G. A. Schmidt, D. Rind, and R. A. Ruedy, 2010: Atmospheric CO_2: Principal control knob governing Earth's temperature. *Science*, 330, 356–359.

This paper states that "Ample physical evidence shows that carbon dioxide (CO_2) is the single most important climate-relevant greenhouse gas in Earths atmosphere. This is because CO_2, like ozone, N_2O, CH_4, and chlorofluorocarbons, does not condense and precipitate from the atmosphere at current climate temperatures, whereas water vapor can and does. Noncondensing greenhouse gases, which account for 25% of the total terrestrial greenhouse effect, thus serve to provide the stable temperature structure that sustains the current levels of atmospheric water vapor and clouds via feedback processes that account for the remaining 75% of the greenhouse effect. Without the radiative forcing supplied by CO2 and the other noncondensing greenhouse gases, the terrestrial greenhouse would collapse, plunging the global climate into an icebound Earth state."

[7] McGuffie, K. and A. Henderson-Sellers, 2014: *The Climate Modelling Primer*, 4th Edition, Wiley-Blackwell, New York.

This introductory text has the distinctly modern feature of providing students with interactive experiences in using Internet resources. The book has numerous colorful and insightful figures, covers both EBMs and GCMs, and presents the materials with a minimum of mathematical requirements.

[8] North, G. R. and K.-Y. Kim, 2017: *Energy Balance Climate Models*, Wiley, New York.

This book is a definitive overview of EBMs by two experts who both have many years of experience in EBM research. As they say in their preface, "The popularity of EBMs has been going up and down over the decades since they were introduced by Budyko and Sellers in the late 1960s. In the 1970s, they were in favor, but as general circulation models (GCMs) began to improve significantly, EBMs were often dismissed as too crude. But over time EBMs were recognized as important tools in the hierarchy of models. As GCMs include more and more processes and components (biology, carbon cycle, cryosphere, etc.) the output of their simulations becomes even more difficult to comprehend. EBMs and other simplified schemes can help to sort out some key processes in the system."

[9] Pierrehumbert, R. T., 2010: *Principles of Planetary Climate*. Cambridge University Press, New York.

A highly original and stimulating treatment beginning with the fundamental aspects of climate science and extending to the current research frontier. Highly recommended to all, including those interested in the climates of other planets as well as that of our Earth.

[10] Riser S. C., et al., 2016: Fifteen years of ocean observations with the global Argo array. *Nature Climate Change*, 6, 145–153.

A very readable overview of the remarkable advances in ocean observations attributable to the development of Argo, a global array of 3,800 free-drifting profiling floats that provide observational data for temperature, salinity, and velocity in the upper 2000 meters of the ocean. It used to be true that the atmosphere, thanks to the observational requirements of weather forecasting, was monitored, while the ocean was merely sampled. Not any longer. Argo is a giant step toward monitoring the ocean. At this writing (2018), Deep Argo, which will extend Argo's capabilities to observing the ocean at depths exceeding 2000 meters, is under development.

[11] Trenberth, K. E. and J. T. Fasullo, 2011: Tracking Earth's energy: From El Niño to global warming. *Survey in Geophysics*, 33, 413–426. DOI 10.1007/s10712-011-9150-2.

An excellent entry point to the scientific research literature on the topic of monitoring the global mean energy budget of the Earth. The authors are experts in this area and are also highly skilled science communicators.

[12] Vasavada, A. R., J. L. Bandfield, B. T. Greenhagen, P. O. Hayne, M. A. Siegler, J. P. Williams, and D. A. Paige, 2012: Lunar equatorial surface temperatures and regolith properties from the Diviner Lunar Radiometer Experiment. *Journal of Geophysical Research: Planets*, 117, E00H18. doi:10.1029/2011JE003987.

This paper explores the lunar temperature variation and the thermophysical properties of the lunar regolith from the surface to a depth of 0.4 meters. This work and its update (Hayne et al. (2017)) provided the values of some key parameters for our lunar EBM.

[13] von Schuckmann, K. et al., 2016: An imperative to monitor Earth's energy imbalance. *Nature Climate Change*, 6, 138–144.

An important review of a critical research topic in modern climate science. From the abstract, "The current Earth's energy imbalance (EEI) is mostly caused by human activity and is driving global warming. The absolute value of EEI represents the most fundamental metric defining the status of global climate change and will be more useful than using global surface temperature. EEI can best be estimated from changes in ocean heat content, complemented by radiation measurements from space. Sustained observations from the Argo array of autonomous profiling floats and further development of the ocean observing system to sample the deep ocean, marginal seas and sea ice regions are crucial to refining future estimates of EEI. Combining multiple measurements in an optimal way holds considerable promise for estimating EEI and thus assessing the status of global climate change, improving climate syntheses and models, and testing the effectiveness of mitigation actions. Progress can be achieved with a concerted international effort."

[14] Williams, J. P., D. A. Paige, B. T. Greenhagen, and E. Sefton-Nash, 2017: The global surface temperatures of the Moon as measured by the Diviner Lunar Radiometer Experiment. *Icarus*, 283, 300–325.

This paper displays many maps of the lunar surface temperature based on the Diviner satellite data. We used the same dataset to plot our lunar surface temperature and to verify our lunar energy budget models (EBMs).

Exercises

Exercises 5.1–5.6, 5.8, and 5.9 require the use of a computer.

5.1 Use the EBM and R to estimate the lunar surface temperature at lunar latitude 30° North and at 3:00 p.m., lunar local time.

5.2 Use the EBM and R to estimate the lunar surface temperature at 24 points uniformly distributed on the equator. List the results in a table of three columns. The first column is longitude, the second is temperature in kelvin, and third is temperature in degrees Celsius.

5.3 Use R to plot the results in the above table, then compare the EBM results with the Diviner observational data shown in this chapter, and discuss how the modeling parameters may affect the model output.

5.4 Similar to our Moon, Mercury is a planet without atmosphere and water. Use the Internet to find relevant EBM parameters for Mercury, and estimate Mercury's noon surface temperature at its equator.

5.5 Tune the "snowball" uniform Earth EBM parameters to find three types of climate conditions for the Earth. Discuss the numerical results generated by R.

5.6 Repeat the above problem for a given emissivity ϵ and for the case of the nonlinear albedo-feedback EBM, with the albedo modeled by a tanh function of temperature

$$\alpha = \alpha_1 - \alpha_2 \times \tanh((T - T_c)/T_s) \tag{5.26}$$

where α_1, α_2, and T_c are constants, and the temperature unit is K. Choose your own values for $\epsilon, \alpha_1, \alpha_2, T_c$, and T_s so that the EBM has three solutions, one of which is close to the global average temperature of the present Earth, i.e., around 15°C.

5.7 A simplified albedo-feedback nonlinear EBM can be solved "by hand." This model assumes the following:

(a) $\alpha = 0.7$ when $T < -8$°C, and $\alpha = 0.3$ when $T > -8$°C. Regard −273.15°C as the absolute zero temperature. Thus T [K] = 273.15 + T [°C]. The incoming solar radiation power is

$$E_{in} = (1 - \alpha)\frac{S}{4} \tag{5.27}$$

with $S = 1{,}368$ [W m^{-2}].

(b) The outgoing Earth radiation is approximated by a Budyko formula:

$$E_{out} = 189.4 + 2.77\ T\ [\mathrm{W\ m^{-2}}] \tag{5.28}$$

where the temperature T unit is [°C].

Find three equilibrium solutions for T of the EBM $E_{out} = E_{in}$ based on these two assumptions. Hand-plot relevant schematic figures about this EBM. Find analytic solutions by hand and using only calculators.

5.8 Use R to develop a table to document the relative differences between the Earth radiation computed by the Stefan–Boltzmann law and that by Budyko's linear approximation formula shown in (5.28), when T is in the range from $-60°$C to $+60°$C. Use the following parameters: $\epsilon = 0.6$, $S = 1,368$. Discuss and comment on your numerical results. Think about the real Earth whose radiation is large over the equatorial region and small over the polar regions. What is the meaning of our assumption of T being in the range $(-60, 60)°$C? Plot some figures to help you describe your numerical results and your ideas.

5.9 Explore how the Earth's surface temperature T depends on the solar radiation $Q = S/4$ using the EBM model.

(a) Derive that

$$Q = \frac{\epsilon\sigma T^4}{1-\alpha(T)}. \tag{5.29}$$

(b) Plot Q as a function of T using R.

(c) Use the figure to describe the variation of T as Q changes in the three climate regimes described in this chapter.

(d) Plot the Q–T dependence but use Q as the horizontal axis and T as the vertical axis. This is a so-called bifurcation diagram. *One may start with the following R code*

```
#T and solor constant Q relation
png("QT-relation.png",width=6,height=8, units = 'in', res = 200)
q = function(T){return(ep*sg*T^4/ (1-ab(T)))}
plot(q(T),T,type="l", lwd=2, xlim=c(200,700),ylim=c(200,350),
     main="Solar constant and temperature in an EBM",
     ylab="Temperature [K]",
     xlab="Solar radiaiton Q=S/4 [W/sq.m]")
Tm=seq(250,280)
lines(q(Tm),Tm,col="red", lwd=3)
dev.off()
```

6 Calculus Applications to Climate Science I: Derivatives

Let us take stock. In the preceding chapters, we have introduced the programming language R and discussed the mathematical topics of dimensional analysis, basic statistical methods, and matrices and linear algebra. We then surveyed energy balance models in Chapter 5. In the present chapter and the following one, we will discuss several topics involving climate science applications of calculus, focusing on derivatives in the present chapter and on integrals in the following chapter. Before starting these two chapters, we suggest reviewing Appendix D on calculus concepts and methods for climate science.

You may find our treatment of calculus somewhat different from the way you were first taught this subject. Re-learning calculus may thus resemble returning to a place you have visited before and seeing it from a new perspective. In Appendix D, we describe a simple and direct approach to calculus due to René Descartes (1596–1650), which is not at all the same as the conventional method found in most textbooks. Appendix D outlines our rationale for taking this pedagogically novel route. We hope that you will come to agree with our choice of approaches to calculus. In choosing this unusual approach, we have been mindful of a wise comment made by Ferdinand Porsche (1875–1951), the brilliant engineer who founded the Porsche automobile company. Dr. Porsche said, "Change is easy. Improvement is far more difficult."

In this chapter, we cover linear approximation, Newton's method, linearization of the Stefan–Boltzmann blackbody radiation law, Taylor expansion, and partial derivatives.

6.1 Stefan–Boltzmann Law and Budyko's Approximation

A blackbody is an idealized physical body having the property that it absorbs all the electromagnetic radiation that it receives, regardless of the angle of incidence or the frequency of that radiation. A blackbody also emits radiation which may be quantified according to the Stefan–Boltzmann law

$$E_{bb} = \sigma T^4 \tag{6.1}$$

where the temperature T is in kelvin, and σ is a known constant called the Stefan–Boltzmann constant. Note that a blackbody is different from a black hole which cannot emit radiation. Neither a blackbody nor a black hole reflects light.

Earth does not behave exactly as a blackbody would. It is more nearly what is known as a grey body, which is one for which its radiation is reduced from the value given by the Stefan–Boltzmann law by a numerical factor less than 1, called emissivity ϵ. Thus, the radiation of a grey body is less than the radiation of a blackbody, but it still retains the fourth-power dependence on temperature:

$$E_{gb} = \epsilon\sigma T^4. \tag{6.2}$$

In the previous discussion on energy balance models (EBMs), two values of emissivity ϵ were used: $\epsilon = 1.0$ (the blackbody value) and $\epsilon = 0.6$ (a grey body value). Our idealized EBM, often called a "toy model" in the climate research community due its extreme simplicity, yields a reasonable average surface temperature 16 °C for $\epsilon = 0.6$, and a snowball or ice ball Earth for $\epsilon = 1.0$. If we temporarily ignore the fact that a "toy model" omits many climatically important processes, and if we imagine that the "toy model" is adequate as a model of the actual Earth, then we might conclude that perhaps, during its long history, the Earth has experienced both cases: $\epsilon = 0.6$ and $\epsilon = 1.0$ (or ϵ might take on a value close to 1.0). In that case, we may ask what is the correct value for ϵ? Can we use observational data to determine the best ϵ value for the current conditions on the Earth surface?

In the 1960s, Mikhail I. Budyko (1920–2001) and his collaborators used a linear regression procedure to model the Earth's outgoing radiation (Budyko 1969). He deviated from the Stefan–Boltzmann law and instead assumed a linear relationship

$$E_{Bud} = A + BT \tag{6.3}$$

where the units of temperature T in Eq. (6.3) are Celsius degrees °C, not kelvins K as in the Stefan–Boltzmann law, and the units for the coefficients A and B are A [W m^{-2}] and B [W m^{-2}(°C)$^{-1}$]. These coefficients may be obtained by a linear regression from the observed data. This linear regression can be shown to be equivalent to a linear approximation of the nonlinear Stefan–Boltzmann law, which will be described in the next section.

McGuffie and Henderson-Sellers (2014) provided two estimates of parameters for the coefficients A and B: one attributed to Cess (1976)

$$A = 212\ [\text{W m}^{-2}], \qquad B = 1.6\ [\text{W m}^{-2}(°\text{C})^{-1}] \qquad \text{(Cess's estimate)}, \tag{6.4}$$

and another to Budyko for his original estimate

$$A = 202\ [\text{W m}^{-2}], \qquad B = 1.45\ [\text{W m}^{-2}(°\text{C})^{-1}] \qquad \text{(Budyko's estimate)} \tag{6.5}$$

The book by North and Kim (2017) also provides some estimated values of A and B and their justifications.

Figure 6.1 shows the curves of outgoing long-wave radiation energy when using $\epsilon = 0.6$ in the Stefan–Boltzmann radiation formula and using $A = 212$ [W m^{-2}], $B = 1.6$ [W m^{-2}(°C)$^{-1}$] for Budyko's radiation formula, over a surface temperature range from -50°C to $+50$°C. The two radiation formulas are

$$E_{SB} = 0.6 \times (5.670373 \times 10^{-8}) \times (273.15 + T)^4, \tag{6.6}$$

$$E_{Bud} = 212 + 1.6T. \tag{6.7}$$

Here, 273.15 comes from the conversion of the units from [K] to [°C], and -273.15 [°C] is equivalent to the zero of the absolute or Kelvin temperature scale [K].

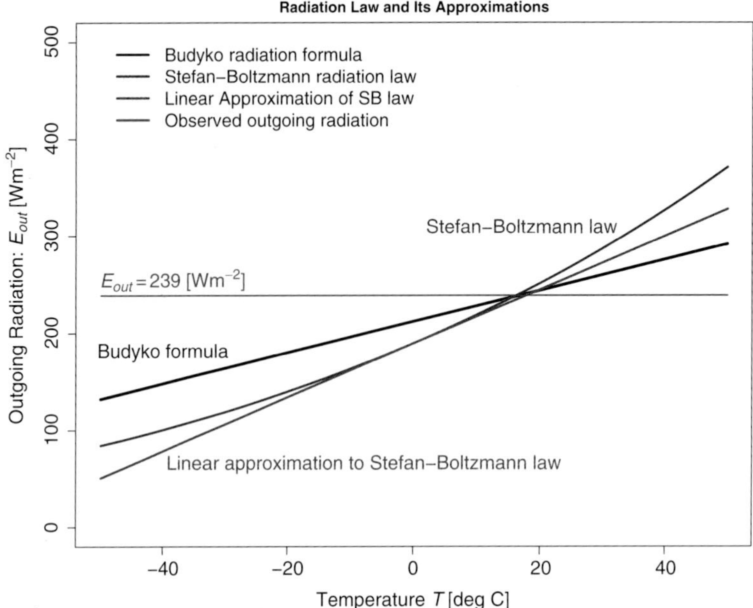

Fig. 6.1 Comparison of the Earth's outgoing radiation according to Stefan–Boltzmann (SB) law, Budyko's formula, and a linear approximation to the SB law.

Figure 6.1 is produced by the following R program

```
setEPS()
postscript("fig0601.eps", height=8, width=10)
#Cess' Budyko parameters
par(mar=c(4.5,5,2,0.5))
A = 212
B = 1.6
ep = 0.6
sg = 5.670373*10^(-8)
T = seq(-50,50, by=0.1)
S = 1365
alf = 0.30
plot(T,A+B*T, xlim=c(-50,50), ylim=c(0,500),
     type="l",lwd=4, cex.lab=1.5, cex.axis=1.5,
main="Radiation Law and Its Approximations",
xlab="Temperature T [deg C]",
ylab=expression(paste("Outgoing Radiation: ", E[out], " [W/sq. m]"))
)
lines(T, ep*sg*(273.15+T)^4,col="blue", lwd=3)
lines(T, 189.4 + 2.77*T,type="l", col="purple",lwd=3)
lines(T, (1-alf)*S/4*rep(1, length(T)), col="red",lwd=2)
legend(-50, 510, legend=c("Budyko radiation formula",
                          "Stefan-Boltzmann radiation law",
                          "Linear Approximation of SB law",
                          "Observed outgoing radiation"),
       col=c("black", "blue","purple","red"),
       lty=1, bty="n",lwd=3, cex=1.4)
text(-40,180, "Budyko formula",cex=1.5)
```

```
text(18,310, "Stefan-Boltzmann law", col="blue",cex=1.5)
text(-8,65, "Linear approximation to Stefan-Boltzmann law",
     col="purple",cex=1.5)
text(-37,253, expression(paste( E[out], "= 239 [W/sq. m]")),
     col="red", cex=1.5)
dev.off()
```

The figure shows a linear approximation for the Stefan–Boltzman law, and it also shows that there is a systematic difference or offset between the Budyko and the Stefan–Boltzmann formulas. The near-straight line property of the Stefan–Boltzmann law in this temperature range $[-50, 50]^\circ$C allows us to fit the regression better using improved data. For example, $A = 189.4$ [W m^{-2}], $B = 2.77$ [W m^{-2}($^\circ$C)$^{-1}$] yields a very close agreement between this new version of Budyko's radiation formula and the Stefan–Boltzmann radiation law (see Fig. 6.1).

With this R program, one can tune the two "knobs": A and B. We can find that $A = 189.4$ [W m^{-2}], $B = 2.77$ [W m^{-2}($^\circ$C)$^{-1}$] looks like the best fit. Why? It is the best linear approximation, which we may call simply "the linear approximation." In general, to approximate a curve by a tangent straight line is called a linear approximation. Thus, we now have a third outgoing radiation formula

$$E_{Lin} = 189.4 + 2.77T. \tag{6.8}$$

Each of these three different outgoing radiation formulas corresponds to an equilibrium: $E_{out} = E_{in} = (1-\alpha)S/4$. With $S = 1,365$ [W m^{-2}] and $\alpha = 0.30$, the solutions for the three cases of $E_{out} = E_{in}$ are

$$T_{SB} = 16.4^\circ\text{C}, \quad T_{Bud} = 16.8^\circ\text{C}, \quad T_{Lin} = 17.9^\circ\text{C}. \tag{6.9}$$

Each of these three solutions is indicated by one of the intersection points on the horizontal red line of Fig. 6.1.

If Budyko's original estimates for $A = 202$ [W m^{-2}] and $B = 1.45$ [W m^{-2}($^\circ$C)$^{-1}$] are used, then $T_{Bud} = 25^\circ$C. This is much higher than the result from Cess's estimate.

6.2 Linear Approximation

Clearly, the best linear approximation of the curve $y = f(x)$ at a point $x = a$ is the tangent line at $(a, f(a))$. The point-slope equation of the tangent line is

$$y - f(a) = f'(a)(x-a), \tag{6.10}$$

where $f'(a)$ is the derivative of the function $f(x)$ at $x = a$. The linear approximation is

$$y = L(x) = f(a) + f'(a)(x-a). \tag{6.11}$$

Example 6.1 Find the linear approximation of

$$f(x) = (1+x)^4 \tag{6.12}$$

near $x = 0$.

Solution: One can use WolframAlpha to find the derivative $f'(x) = 4(1+x)^3$. When $x = 0$, $f'(0) = 4(1+0)^3 = 4$. $f(0) = (1+0)^4 = 1$. Thus the linear approximation is

$$y - f(0) = f'(0)(x-0) \rightarrow y - 1 = 4 \times (x-0) \rightarrow y = 4x + 1. \tag{6.13}$$

This approximation is good when the absolute value $|x|$ is small. Table 6.1 shows that the error is relatively small, less than 25%, when x is in $[-0.15, 0.30]$.

Table 6.1 Linear approximation of $f(x) = (1+x)^4$ by $L(x) = 1 + 4x$

x	$f(x)$	$L(x)$	Error[%]
−0.3	0.2401	−0.2	183
−0.2	0.4096	0.2	51
−0.1	0.6561	0.6	9
0.0	1.0000	1.0	0
0.1	1.4641	1.4	4
0.2	2.0736	1.8	13
0.3	2.8561	2.2	23

Table 6.1 can be produced by the following R program

```
x <- seq(-0.3,0.3, by=0.1)
fx <- c(1:7)
lx <- c(1:7)
ex <- c(1:7)
for(i in 1:7){fx[i]=(1+x[i])^4
                  lx[i]=1+4*x[i]
      ex[i]=((1+x[i])^4-(1+4*x[i]))/((1+x[i])^4)*100
             }
round(cbind(x, fx, lx, ex), digits=4)
```

Example 6.1 can be applied to the Stefan–Boltzmann law in the following way:

$$\begin{aligned} E_{SB} &= 0.6 \times (5.670373 \times 10^{-8}) \times (273.15 + T)^4 \\ &= 0.6 \times (5.670373 \times 10^{-8}) \times 273.15^4 \left(1 + \frac{T}{273.15}\right)^4. \end{aligned} \tag{6.14}$$

The example $(1+x)^4$ applies to this formula with $x = T/273.15$ which is between -0.2 and 0.3 and can be considered small. This x interval $[-0.2, 0.3]$ corresponds to the T interval $[-41, 82]$°C. This range includes almost all the temperature regimes found on the Earth, except some parts of the Antarctic. Therefore, Budyko's approximation is applicable to almost any place on Earth.

When the linear approximation $(1+x)^4 = 1+4x$ is applied to the variable $x = T/273.15$ in formula (6.14), the linear approximation E_{Lin} given by formula (6.8) can be verified as follows:

$$E_{Lin} \approx 0.6 \times (5.670373 \times 10^{-8}) \times 273.15^4 \left(1+4\times\frac{T}{273.15}\right)$$
$$= 189.4 + 2.77T. \qquad (6.15)$$

6.3 Bisection Method for Solving Nonlinear Equations

Some simple equations can be easily solved. For example, the solution of $2x = 1$ is $x = 1/2$, and $x^2 + 2x + 1 = 0$ has the repeated root $x_1 = x_2 = -1$ which can be found by either the quadratic root formula or factorization. The root is also called a zero of the function $x^2 + 2x + 1$, or a solution of the equation $x^2 + 2x + 1 = 0$.

However, most equations arising from physical science are not so simple and cannot be solved analytically. For example, if we use the tanh function[1] to model the albedo in the time-independent EBM in Chapter 5, then the resulting EBM is

$$\epsilon\sigma T^4 = [1-(0.5-0.2\times\tanh((T-265)/10)]\,(S/4), \qquad (6.17)$$

which can only be solved numerically.

An R program solved the above equation (6.17) and yielded three numerical solutions: 234, 264, 289 K, which were found by an R command, such as,
`uniroot(f,c(220,240))` for 234.

Most equations in climate science require computers to solve them, either because the equations are nonlinear, or because there are too many of them, or both. Global climate models typically involve solving systems of equations of thousands of variables, and they may require finding numerical solutions of a large-dimensional matrix equation. Many of the numerical methods for solving such nonlinear equations are based on a linear approximation, called Newton's method.

Before describing Newton's method, we shall first examine a very simple method, the method of bisection. This "brute-force" method is used to find a root between a negative point and a positive point. Suppose that $f(x) = 0$ is the equation to be solved for x. We assume that $y = f(x)$ is a continuous function, meaning that the graph of the function is a continuous curve on the $xy-$plane. If $x = c$ is a root, then $f(c) = 0$. A frequent case is that $f(x)$ is negative on one side of c and positive on the other side. That is, if $f(a) < 0$

[1] The function

$$\tanh(x) = \frac{e^x - e^{-x}}{e^x + e^{-x}} \qquad (6.16)$$

is called the hyperbolic tangent function. It is a smooth step function that smoothly varies from the -1 level as $x \to -\infty$ up to the $+1$ level as $x \to \infty$.

and $f(b) > 0$, then there is a root c between $x = a$ and $x = b$ under the assumption that the function $f(x)$ is continuous, i.e., the curve $y = f(x)$ does not have jumps, holes, or extremely rapid oscillations.[2] The bisection method is a procedure to search for c. The algorithm is as follows.

Make an initial guess that c is in $[a,b]$ and compute $f(a)$ and $f(b)$. If $f(a)$ and $f(b)$ have different signs, then take this pair a and b as the first successful guess. Otherwise, make another initial guess. The first successful guess is called a_1 and b_1. The bisection of this pair is $(a_1 + b_1)/2$. Compute $f((a_1 + b_1)/2)$. From the signs of $f(a_1), f(b_1)$, and $f((a_1 + b_1)/2)$, one can determine which half of the bisected interval $[a_1, (a_1 + b_1)/2]$ or $[(a_1 + b_1)/2, b_1]$ has a root c. For the chosen half, we have the root c in $[a_2, b_2]$. Similarly, one can find $[a_3, b_3]$. Since the interval $[a,b]$ gets bisected each time, the length of the remaining half is $(b-a)/2^{n-1}$ which approaches zero as n goes to infinity. Therefore, the two end points $\{a_n\}_{n=1}^{\infty}$ and $\{b_n\}_{n=1}^{\infty}$ approach the same point, which is the root c.

This method is very simple and easy to understand, but it typically requires many iterations to find a solution, which means that the convergence rate is very slow.

Another drawback is that it cannot find a double root, on both sides of which $f(x)$ has the same sign.

The R package `NLRoot` has a command for the bisection method: `BFfzero (func, a, b)`, where `func` is the function whose zeros are to be found, and `a` and `b` are the end points. For example, for the function $f(x) = x^3 - x - 1$, we have $f(0) = -1 < 0$ and $f(2) = 5 > 0$; thus, the function $f(x)$ must have a zero in $[0,2]$. To find the zero by the bisection method, one can use the following R code:

```
install.packages("NLRoot")
library(NLRoot)
func<-function(x){x^3-x-1}
BFfzero(func, 0, 2)
#[1] 1.324716 is the solution
```

Example 6.2 Use the bisection routine to find a zero of the function $f(x) = (1+x)^4 - (2+x)$ between $x = 0$ and $x = 1$.

The problem can be solved by the following R code:

```
f1<-function(x){(1+x)^4-(2+x)}
BFfzero(f1,0,2)
#[1] 0.2207428 is the solution
```

Example 6.3 Find a solution of the following EBM equation

$$\epsilon\sigma T^4 = \left\{1 - \left[0.5 - 0.2 \times \tanh\left(\frac{T-265}{10}\right)\right]\right\}\frac{S}{4} \tag{6.18}$$

in the interval $[270, 300]$ by the bisection method.

[2] This statement can be treated as the definition of continuity of a function. A more rigorous statement is that the function $y = f(x)$ is continuous at a given point if sufficiently small changes of x about this point result in arbitrarily small changes of y. Most functions in climate science are continuous. Examples of discontinuous functions include the air pressure function across a shock and the heat capacity function across a continent–ocean boundary.

The problem can be solved by the following R code:

```
S <- 1368
ep <- 0.6
sg <- 5.670373*10^(-8)
f2 <- function(T){ep*sg*T^4 - (1-(0.5 - 0.2*tanh((T-265)/10)))*(S/4)}
BFfzero(f2,270,300)
#[1] 289.3097 is the solution
```

6.4 Newton's Method

Newton's method is much faster than the bisection method and uses the tangent line linear approximation. To find a root of $f(x)=0$, we make an initial guess x_0, then compute $f(x_0)$. From the point $(x_0, f(x_0))$, we draw a tangent line

$$y - f(x_0) = f'(x_0)(x - x_0) \tag{6.19}$$

and compute the intersection point of this line and the x-axis, i.e., we find the root for this linear equation, which is very easy. The solution of this linear equation $y=0$ is x_1

$$x_1 = x_0 - f(x_0)/f'(x_0). \tag{6.20}$$

This x_1 is regarded as the second guess, and it is a calculated guess. Next, one can calculate the next guess, x_2, and so on. Therefore, we have a convenient iteration algorithm for root-finding:

$$x_{n+1} = x_n - f(x_n)/f'(x_n), \qquad n = 0,1,2,3,\ldots \tag{6.21}$$

The above procedure for finding a root by Newton's method is illustrated by Fig. 6.2.

An R routine for Newton's method is as follows:

```
newton <- function(f, tol=1E-12,x0=1,N=20) {
        h <- 0.001
        i <- 1; x1 <- x0
        p <- numeric(N)
        while (i<=N) {
                df.dx <- (f(x0+h)-f(x0))/h
                x1 <- (x0 - (f(x0)/df.dx))
                p[i] <- x1
                i <- i + 1
                if (abs(x1-x0) < tol) break
                x0 <- x1
        }
        return(p[1:(i-1)])
}
```

To use this routine, called `newton`, we need to specify the function, the error tolerance, the initial guess, and the maximum number of steps allowed. For example, to find the roots for $x^3+4x^2-10=0$ near 1.0, we can write the following code

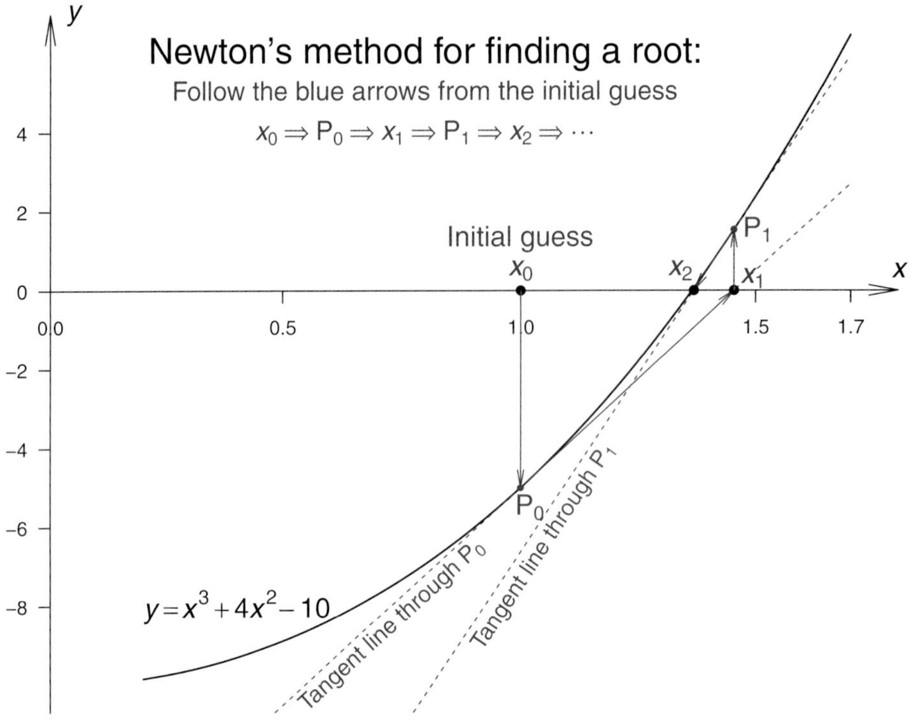

Fig. 6.2 Illustration of Newton's method for finding a root from the initial guess $x_0 = 1.0$.

```
f <- function(x) { x^3 + 4*x^2 -10 }
root <- newton(f, tol=1E-12, x0=1, N=10)
root
#[1] 1.454256 1.368917 1.365238 1.365230 1.365230 1.365230 1.365230
```

One can easily verify this solution: `f(1.365230)`, which is $-2.215123e-07$, very close to zero. Only seven iterations are used to find this good approximate solution $x = 1.365230$. Figure 6.2 illustrates the method.

The R command `uniroot(f,c(1,2), tol=10^(-6))` to find a solution of the nonlinear equation $f(x) = 0$ in Sections 5.1 and 5.3 also uses Newton's method and hence yields the same result `1.365230`. Here, `tol=10^(-6)` is the desired accuracy, i.e., the residual is less than or equal to an error of the order `tol=10^(-6)`.

Figure 6.2 can be produced by the following R code.

```
#Illustration of Newton's method
setEPS()
postscript("fig0602.eps", height=6, width=8)
par(mar=c(0,1,0,0))
x = seq(0.2,1.7, len=30)
f = function(x) { x^3 + 4*x^2 -10 }
g = function(x) { 3*x^2 + 8*x }
plot(x,f(x), type='l', lwd=1.5, bty="n",
     xaxt="n",yaxt ="n", xlab="", ylab="",
     xlim=c(0,1.8),ylim=c(-10,7))
```

```
axis(1, at=c(0, 0.5, 1.0, 1.5, 1.7), pos=0, cex.lab=1.3)
axis(2, at=c(-8, -6, -4, -2, 0, 2, 4), pos=0, cex.lab=1.3, las=1)
arrows(0,0, 1.8,0, angle=10, length=0.25)
arrows(0,-11, 0,7, angle=10, length=0.25)
text(1.8, 0.5, expression(x), cex=1.4)
text(0.05, 7, expression(y), cex=1.4)
text(0.4, -8, expression(y == x^3 + 4*x^2 -10), cex=1.4)
#Plot the initial points and then follow tangent lines
x0 = 1.0
x1 = 1.4543
x2 = 1.3689
lines(x,f(x0) + g(x0)*(x - x0),type="l", lty=2, col="red")
text(0.73,-8.5, expression(paste("Tangent line through ", P[0])),
     col="red", cex=1.2, srt=42)
points(1,0, pch=19)
text(1.0, 1.3, "Initial guess", cex=1.4, col="blue")
text(1.0, 0.5, expression(x[0]), cex=1.4, col="blue")
arrows(1,0,1,-5, angle=10, length=0.15, col="blue")
points(x0, f(x0), col="red", pch=20)
text(x0+ 0.02, f(x0) - 0.5, expression(P[0]), cex=1.4, col="red")
arrows(x0,f(x0), x1,0, angle=10, length=0.15, col="blue")
points(x1,0, pch=19)
text(x1+0.04, 0.3, expression(x[1]), cex=1.4, col="blue")
arrows(x1,0,x1,f(x1), angle=10, length=0.15, col="blue")
points(x1, f(x1), col="red", pch=20)
text(x1+ 0.05, f(x1), expression(P[1]), cex=1.4, col="red")
lines(x,f(x1) + g(x1)*(x - x1),type="l", lty=2, col="red")
text(1.05,-6.5, expression(paste("Tangent line through ", P[1])),
     col="red", cex=1.2, srt=56)
arrows(x1,f(x1),x2,0, angle=10, length=0.15, col="blue")
points(x2,0, pch=19)
text(1.34,0.5, expression(x[2]), cex=1.4, col="blue")
text(0.8, 6, "Newton's method for finding a root:", cex=1.8)
text(0.8, 5,"Follow the blue arrows from the initial guess",
    cex=1.3, col="blue")
text(0.8, 4, expression(x[0] %=>% P[0] %=>% x[1] %=>%
                P[1] %=>% x[2] %=>% ...), cex=1.3, col="blue")
dev.off()
```

One can next try to use Newton's method to find a root of

$$f(x) = (1+x)^4 - (2+x) = 0$$

between $x = 0$ and $x = 1$ by starting at $x_0 = 1$. The result is $x = 0.2207441$.

```
f = function(x) { (x+1)^4 -(2+x) }
root <- newton(f, tol=1E-12, x0=1, N=10)
root
#[1] 0.5809697 0.3335992 0.2359959 0.2210877 0.2207447
```

We can also use Newton's method to find the equilibrium solutions of the albedo-feedback EBM solutions for temperature T. Define the parameters and the function f whose roots are to be found. Then call the R routine `newton`. These three steps are illustrated by the following example.

```
#Define parameters
S <- 1365
```

```
ep <- 0.6
sg <- 5.670373*10^(-8)
#Define the function for energy
f3 <- function(T){return(ep*sg*T^4 -
(1-(0.5 - 0.2 * tanh ((T-265)/10)))*(S/4))}
```

To find the three equilibrium solutions using the R routine `newton`, we begin with different initial guesses `x0`, each of which will converge to a solution for a given error tolerance `tol` and the maximum number of iterations `N`. Our numerical experiments are below:

```
#Compute the solutions
root1 <- newton(f3, tol=1E-12, x0=220,N=20)
root1
#  [1] 235.6965 234.3860 234.3817 234.3817 234.3817
root2 <- newton(f3, tol=1E-12, x0=270,N=20)
root2
#[1] 262.0567 264.5071 264.3378 264.3377 264.3377 264.3377
root3 <- newton(f3, tol=1E-12, x0=300,N=20)
root3
#[1] 289.9086 289.1469 289.1401 289.1401 289.1401 289.1401
```

The three equilibrium solutions are: 234, 264, and 289 K. The solutions converge quickly, in only five or six steps with a high precision of 1×10^{-12}. The first solution 234 K is the same as the `uniroot(f3,c(220,240))` solution in Section 5.3. The other two solutions 264 and 289 K are close to but not the same as those in Section 5.3, because the albedo transition models are different, and slightly different parameters are used. As pointed out earlier, the R command `uniroot` also uses Newton's method. Our R command `newton(f3, tol=1E-12, x0=220,N=20)` here explicitly uses the algorithm for Newton's method.

Newton's method is very efficient when one can estimate the reasonable range of possible solutions in advance by, for example, using physical reasoning. For the EBM, we certainly should limit our first guess to be in the range from 200 to 340 kelvin, because the Earth's surface air temperature is most probably in the range of $(-70,70)^{\circ}$C. If our initial guess is far outside this range, Newton's method may still yield a solution, which, however, may not be what we expected. For example, when we take the initial guess to be $T_0 = 100$,

```
root5 <- newton(f3, tol=1E-12, x0=100, N=20)
root5
 #[1] 827.2544 623.5417 474.8968 372.5619 313.3648 292.0552 289.2231
 #[8] 289.1402 289.1401 289.1401 289.1401
```

Newton's method still gives a solution, but it takes more steps to do so. In this case, instead of yielding the smallest solution 234 K, it also produces the largest solution 289 K.

6.5 Examples of Higher-Order Derivatives

Higher-order derivatives, or simply higher derivatives, are briefly described in Appendix D, and are described in this section with additional application examples and as preparation for Section 6.8 on Taylor series.

A slope, i.e., a derivative, can be an indicator of a mountain path's changing rate of height with respect to horizontal distance, commonly known as "rise over run" in daily life. Such a slope can vary and has its derivative, the second derivative of height with respect to horizontal distance, that indicates how rapidly the slope changes with distance. Geometrically, the second derivative determines the concavity of a curve and is relevant to the curvature of a curve.

The second derivative of a parabolic path is a constant. If $y = ax^2$ where a is a constant, then the second derivative is $y'' = 2a$. When a is positive, then $y'' > 0$ and the parabola opens upward, i.e., is concave up. When $y'' < 0$, the curve is concave down.

Speed is the rate of change of position with respect to time. When a car accelerates from a stop sign, starting with speed equal to zero and ending at a constant cruising speed of, say, 65 miles per hour, the speed changes with time, and the car experiences an acceleration in order to reach a constant cruising speed. The acceleration measures the changing rate of speed, and is a derivative of the first derivative (speed), i.e., acceleration is the second derivative of position with respect to time.

Consider the gravitational force from the free-fall distance formula

$$h = -9.8t^2/2 + 20t + 1 \text{ [m]}, \tag{6.22}$$

where h is the height of a ball being thrown upward from an initial height 1 [m] with an initial speed 20 [m s^{-1}]. The speed is $s = dh/dt = -9.8t + 20$ [m s^{-1}]. The acceleration is $a = ds/dt = -9.8$ [m s^{-2}], which is the gravitational constant, also called the Earth's gravitational acceleration. Due to this negative acceleration, gravity acts downward on the ball to slow the ball, which has an initial speed of 20 [m s^{-1}]. When the ball reaches its highest point, at which its speed becomes zero: $-9.8t + 20 = 0$ yields $t = 20/9.8$, approximately 2 seconds. The highest point is approximately $H = -9.8 \times 2^2/2 + 20 \times 2 + 1 = 21$ [m], which, if we compare it with the height of a typical building, is about 5 stories high. After the ball reaches its highest point, it starts to drop, i.e., its speed becomes negative. Thus, the Earth's gravity exerts a force on the ball and causes the negative acceleration, according to Newton's second law $F = ma$, of the ball whose initial positive speed first decreases to zero, and then becomes negative. In this simple theoretical analysis, air resistance and other forces are neglected.

The unit of force is [newton] or [N], which is the force needed to make a 1 [kg] ball have 1 [m s^{-2}] acceleration. The gravitational force acting on a 0.1 [kg] ball is thus 0.1 [kg] $\times$ 9.8 [m s^{-2}] = 0.98 [kg m s^{-2}] or 0.98 [N].

Mathematically, one can extend the concept of a second derivative, also called second-order derivative, to a third derivative, a fourth derivative, and so on. The notations are below.

(i) The first derivative, or simply derivative: $f'(x), f', D[f,x], D[f], df/dx, \frac{df}{dx}, y', \dot{y}, \frac{dy}{dx}$

(ii) The second derivative, or second-order derivative: $f''(x), f'', D_2[f,x], D_2[f], d^2f/dx^2, \frac{d^2f}{dx^2}, y'', \ddot{y}, \frac{d^2y}{dx^2}$

(iii) The third derivative, or third order-derivative: $f'''(x), f''', D_3[f,x], D_3[f], d^3f/dx^3, \frac{d^3f}{dx^3}, y''', \frac{d^3y}{dx^3}$

(iv) The fourth derivative, or fourth-order derivative: $f^{(4)}(x), f^{(4)}, D_4[f,x], D_4[f]$, $d^4f/dx^4, \frac{d^4f}{dx^4}, y^{(4)}, \frac{d^4y}{dx^4}$

(v) The fifth derivative, or fifth-order derivative: $f^{(5)}(x), f^{(5)}, D_5[f,x], D_5[f]$, $d^5f/dx^5, \frac{d^5f}{dx^5}, y^{(5)}, \frac{d^5y}{dx^5}$

(vi) The nth derivative, or nth-order derivative: $f^{(n)}(x), f^{(n)}, D_n[f,x], D_n[f]$, $d^nf/dx^n, \frac{d^nf}{dx^n}, y^{(n)}, \frac{d^ny}{dx^n}$

One can use the derivative–antiderivative table and differential rules in Appendix D to find the derivative and higher derivatives of a function. For example, we use $x = 0.5\sin(2\pi t)$ [m] to model the displacement of a pendulum's center of mass from its equilibrium rest position, where t [s] is time. Then, its derivative and higher derivatives are below.

$$\frac{dx}{dt} = \pi\cos(2\pi t) \quad [\text{m s}^{-1}] \text{ (speed)}, \tag{6.23}$$

$$\frac{d^2x}{dt^2} = -2\pi^2\sin(2\pi t) \quad [\text{m s}^{-2}] \text{ (acceleration)}, \tag{6.24}$$

$$\frac{d^3x}{dt^3} = -4\pi^3\cos(2\pi t), \tag{6.25}$$

$$\frac{d^4x}{dt^4} = 8\pi^4\sin(2\pi t) \tag{6.26}$$

We can also use an R program to find the derivatives. For example, to find $D[x^2]$, the R command is

```
D(expression(x^2),"x")
# 2 * x
```

More examples are below

```
D(expression(exp(-x^2)),"x")
# -(exp(-x^2) * (2 * x))
D(expression(sin(-3*t)-2*cos(4*t-0.3*pi)),"t")
# -(cos(-3 * t) * 3 - 2 * (sin(4 * t - 0.3 * pi) * 4))
```

To find the nth-order derivatives, we simply apply the D command n times. For example, the following R commands can find the second-order derivative for the ball-throwing problem:

```
D(expression(-g*t^2/2 + v0*t + h0),"t")
# v0 - g * (2 * t)/2
# Find derivative of this result function to find the second derivative
D(expression(v0 - g * (2 * t)/2),"t")
# -(g * 2/2)
#or simply
D(D(expression(-g*t^2/2 + v0*t + h0),"t"),"t")
# -(g * 2/2)
#The third-order derivative
D(D(D(expression(-g*t^2/2 + v0*t + h0),"t"),"t"),"t")
#[1] 0
```

R derivative function notations are more rigid than those of WolframAlpha, which allow many different variations of the differentiation symbols.

6.6 Pressure Gradient Force and Coriolis Force

An air parcel in the atmosphere is subject to many types of forces: a pressure gradient force (PGF), an apparent force due to the Earth's rotation, called the Coriolis force in oceanography and meteorology, the gravitational force, frictional forces, and others. The PGF is a force caused by the differences in pressure across a surface. The pressure difference results in a force directed toward lower pressure, which can result in an acceleration according to Newton's second law, $F = ma$. Let $p(x, y)$ be the pressure field at a given time. The PGF is

$$\mathbf{F}_p = -\nabla p/\rho \tag{6.27}$$

where ρ is the density of the air parcel. This formula shows the dimension of

$$\begin{aligned}[\nabla p/\rho] &= [p]L^{-1}/[\rho] = ([Force]/L^2)L^{-1}/[ML^-3] \\ &= [Force]/M = MLT^{-2}/M = LT^{-2}.\end{aligned} \tag{6.28}$$

So, the PGF is actually an acceleration, and has units of force per unit mass, such as [newton/kg].

The condition in which the PGF and the Coriolis force due to Earth's rotation are the only important forces and are in balance with one another is known as geostrophic balance. Many large-scale flows in both the atmosphere and the ocean are approximately in geostrophic balance. See Figs. 6.3 and 6.4 for two atmospheric examples. Figure 6.3 illustrates in an idealized way how an air parcel starting from rest might initially move from higher pressure to lower pressure in response to the PGF, then would begin curving to the right (in the Northern Hemisphere) as the Coriolis force becomes stronger with increasing wind speed, and finally would move parallel to the isobars (lines of constant pressure) as geostrophic balance becomes established. Figure 6.4 illustrates that the wind directions in a local region may be approximately parallel to the isobars, a sign of geostrophic balance.

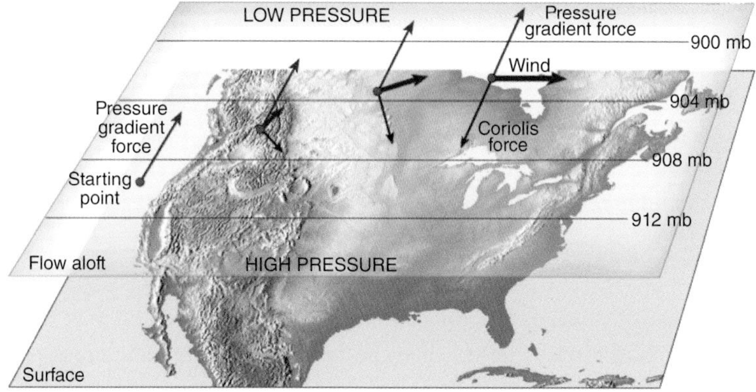

Fig. 6.3 Wind direction and the action of the pressure gradient force and Coriolis force on an air parcel: Time evolution of the motion of an idealized air parcel initially at rest, then moving with increasing speed and ultimately establishing geostrophic balance. *Credit: National Weather Service of the United States. Source:* www.weather.gov

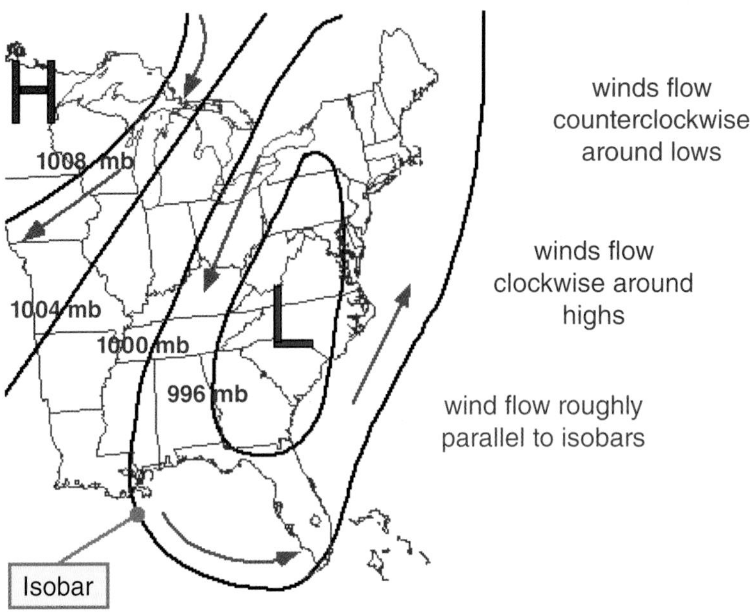

Fig. 6.4 Wind along the isobars in the counterclockwise direction around a lower pressure center over the Eastern United States. In general, the wind directions over a local region in the Northern Hemisphere, in an approximately geostrophic flow, are nearly parallel to the isobars and have the lower pressure on the left and the higher pressure on the right. Thus, the air circulates counterclockwise around an area of low pressure and clockwise around an area of high pressure in the Northern Hemisphere. *Credit: University of Illinois. Source:* https://mrcc.illinois.edu/resources/glossary/images/isobar.png

Newton's laws of motion are valid in an inertial frame of reference, that is, a frame of reference which does not accelerate or rotate. When Newton's laws are expressed in a form suitable for a rotating frame of reference, the Coriolis force appears. The rotating reference frame that we have in mind here is the Earth itself, which rotates. However, the Earth requires an entire day to complete one rotation, so effects due to the Coriolis force generally are most significant when the motions occur over relatively large distances and long time periods. Examples of such motions include the large-scale winds in the atmosphere and currents in the ocean.

The mathematical expression of the Coriolis force on a mass m in a frame rotating with angular velocity $\boldsymbol{\Omega}$ is

$$\mathbf{F}_\mathrm{C} = -2m\boldsymbol{\Omega} \times \mathbf{v}, \tag{6.29}$$

where $\times$ means the cross product of two vectors and $\mathbf{v}$ is the velocity of mass m. The Coriolis force is large when the velocity is large. The cross product of two vectors is a vector and follows the right-hand rule. See Appendix B.

The amplitude of the Coriolis force is

$$|\mathbf{F}_\mathrm{C}| = 2m|\boldsymbol{\Omega}|\ |\mathbf{v}|\ \sin\gamma, \tag{6.30}$$

where γ is the angle between $\mathbf{\Omega}$ and $\mathbf{v}$. The notation $|\mathbf{v}|$ means the length or magnitude of the vector $\mathbf{v}$. See Appendix B for more information on the cross product.

6.7 Spatiotemporal Variations of the Atmospheric and Oceanic Temperature Fields

The lowest portion of the atmosphere is called the troposphere. It extends from the surface up to an altitude that can range from about 6 km to 20 km. This altitude is greatest in the tropics and is least in the polar regions in winter. Within the troposphere, the temperature of the atmosphere generally decreases with increasing height. The higher you are, the colder it is. When a passenger airplane cruises at about 10 km altitude, the atmospheric temperature outside the airplane can easily be 70 degrees Celsius colder than at ground level.

Figure 6.5 shows the average observed atmospheric temperature, averaged over the year and over all longitudes, as a function of pressure, the vertical coordinate, and latitude. This figure implies that the atmospheric temperature varies with respect to both time and space. The rate of vertical change is generally much greater than the rate of horizontal change.

The atmospheric temperature $T(x,y,z,t)$ is a function in a four-dimensional (4D) space: three dimensions in space and one dimension in time. The horizontal change of $T(x,y,z,t)$ is typically small, often with a near-zero spatial derivative, while the vertical change is usually large, that is, $|\frac{\partial T}{\partial x}| << |\frac{\partial T}{\partial z}|$ in most cases.

The average vertical variation of temperature is shown in Fig. 6.6. Note that the dependent value temperature is shown on the horizontal axis, and the independent

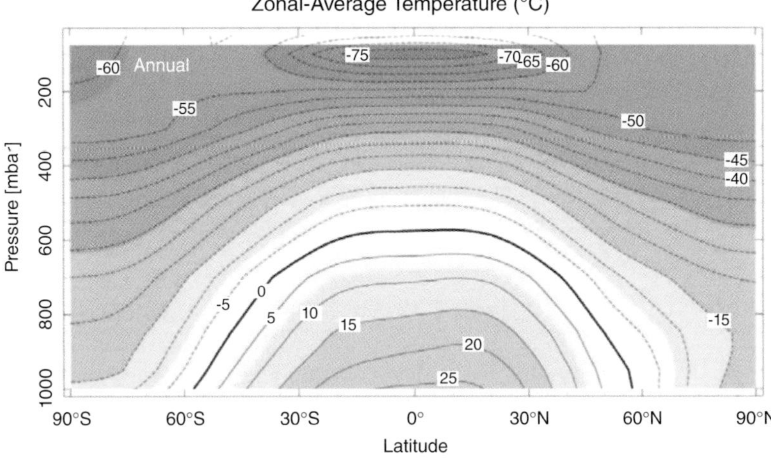

Fig. 6.5 The observed annual mean zonal mean temperature in the troposphere. This figure is obtained by averaging over all longitudes and over the year. This average temperature is plotted as a function of latitude from the South Pole at the left to the North Pole at the right, and as a function of altitude, expressed as atmospheric pressure, from the surface at a pressure of about 1,000 mb to an altitude at a pressure of somewhat less than 200 mb. *Source: Atmosphere, Ocean and Climate Dynamics, An Introductory Text*, by John Marshall and R. Alan Plumb, Elsevier, 2008.

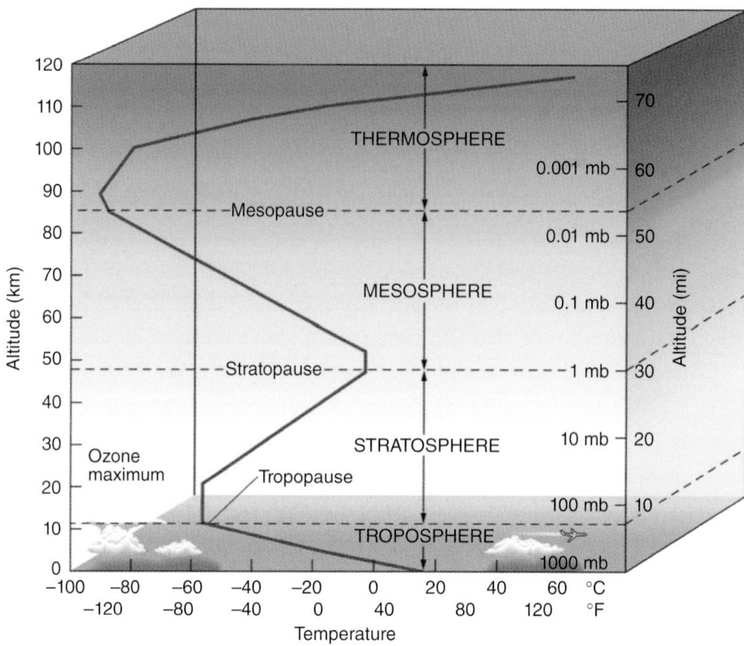

Fig. 6.6 Vertical profile of atmospheric temperature. *Credit: NOAA. Source:* www.scienzafisica.it/pressione-atmosferica-2/pressione-atmosferica-quote

variable height is on the vertical axis. The derivative $\partial T/\partial z$ is usually negative in the troposphere. It becomes positive in the stratosphere, the region above the troposphere. The boundary between troposphere and stratosphere is called the tropopause, which is located at about 10 km above sea level, lower over the polar regions and higher over the equatorial regions. Almost all weather is confined to the troposphere, and climate science is most concerned with the troposphere and is rarely concerned with regions above the stratosphere, which extends from about 10 km above the sea level to about 50 km.

The rate of change of atmospheric temperature with respect to height is called the lapse rate. Note the sign convention. The lapse rate is defined as the negative of the derivative of temperature with respect to height:

$$\Gamma = -\frac{\partial T}{\partial z}. \tag{6.31}$$

If $\Gamma < 0$ in an atmosphere layer, so that temperature increases with height, then this layer is said to have a temperature inversion. Local temperature inversions may occur for many reasons (Wallace and Hobbs 2006, and Goose 2015). A temperature inversion in portions of the troposphere may occur in the polar winter, for example, when the surface temperature may be quite cold, while the air temperature above the surface may be somewhat warmer.

Ocean water properties also vary with depth. Figure 6.7 shows typical average profiles of temperature, salinity, and density.

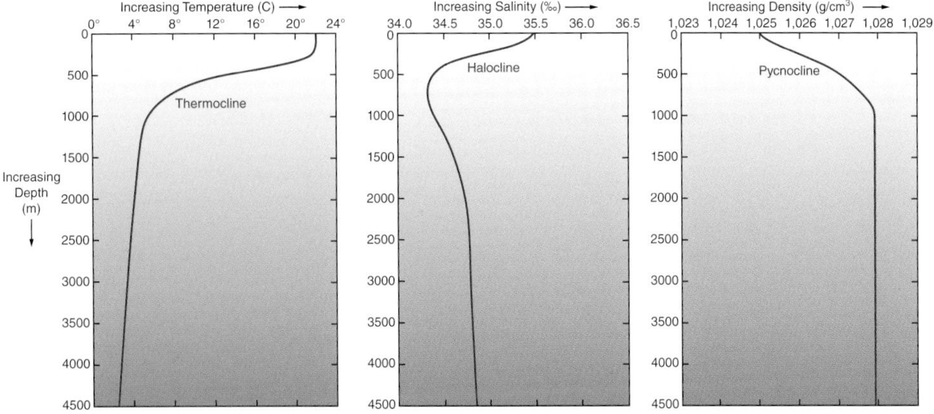

Fig. 6.7 Vertical profiles of ocean water temperature, salinity, and density. *Credit: NOAA. Source:* www.hurricanescience.org/science/basic/water

If we use a z-axis pointing upward to measure the depth from the sea level, then $\partial T/\partial z \approx 0$ in the top isothermal layer, and $\partial T/\partial z$ has a large positive value in the thermocline, and is still positive but with a very small value in the deep ocean.

6.8 Taylor Polynomial as a High-Order Approximation

Section 6.2 describes the linear approximation of a function, which uses a tangent line

$$L(x) = f(a) + f'(a)(x-a) \tag{6.32}$$

to approximate a curve $y = f(x)$ when x is close to a. Can we extend this approximation from a straight line to a quadratic curve, or a cubic curve? Can we generalize the same idea to general polynomial functions of order n:

$$Q_n(x) = a_0 + a_1x + a_2x^2 + \cdots + a_nx^n. \tag{6.33}$$

This is an nth-order polynomial with coefficients $a_i (i = 0, 1, 2, \ldots, n)$. In Section 6.2, we have already demonstrated that we may form a good approximation using a straight line. We may expect that a quadratic or higher-order approximation might be even better. Then, what are the errors of these approximations? The above questions may be answered by Taylor's theorem, named after the English mathematician Brook Taylor (1685–1731).

On the other hand, the functional relationship $y = f(x)$ is often unknown in climate science. We then may use climate data to fit a polynomial to approximate the unknown function (see Fig. 11.10 for an example of fitting the global average annual mean surface air temperature by 9th- and 20th-order polynomials). When, the polynomial is of order one, the data fitting is the linear regression. The slope of the linear function is the linear trend, frequently used in climate change studies.

6.8.1 Taylor's Theorem

Theorem 6.1 *Taylor's theorem: If a function $f(x)$ has an $(n+1)$th-order derivative which is continuous in an open interval containing $x=0$, then*

$$f(x) = P_n(x) + R_n(x), \tag{6.34}$$

where

$$P_n(x) = f(0) + f'(0)x + \frac{f''(0)}{2!}x^2 + \frac{f'''(0)}{3!}x^3 + \cdots + \frac{f^{(n)}(0)}{n!}x^n \tag{6.35}$$

is called the Taylor polynomial, and

$$R_n(x) = f^{(n+1)}(\xi)\frac{x^{(n+1)}}{(n+1)!} \tag{6.36}$$

is called the remainder and has ξ in the interval $(0,x)$. The value ξ is to be determined and depends on both the function f and the interval $(0,x)$.

Although Taylor's theorem holds for any value x, the Taylor polynomial approximation is particularly good when $|x| < 1$, because then x^n approaches zero as $|x|$ goes to zero, and $1/n!$ approaches zero very fast as n goes to infinity. For example, when $n=4$ and $x=0.3$, the remainder

$$R_n(x) = f^{(n+1)}(\xi)\frac{0.3^{(n+1)}}{(n+1)!} = 0.00002025 f^{(n+1)}(\xi), \tag{6.37}$$

is very close to zero if $f^{(n+1)}(\xi)$ is bounded by a given number C.

Therefore, the Taylor polynomial can be used to approximate a function, and the approximation error is the remainder which is small when $|x| < 1$ and n is large. When $|x|$ is close to zero, then the approximation is still very accurate even if n is small, such as $n=1$ for the linear approximation, and $n=2$ for the quadratic approximation. Figure 6.10 visually shows the approximations by the first-order, second-order, and third-order Taylor polynomials to the function e^x when $|x| < 1$.

To make $|x| < 1$ in applications, one can normalize the original variable, with division by a large standard value. For example, in Section 6.3, for a linear approximation we normalized the surface air temperature T [°C] by the Celsius equivalent value -273.15 [°C] of the absolute (kelvin) zero temperature:

$$x = \frac{T}{273.15} \tag{6.38}$$

which makes x in the interval $(-0.3, 0.3)$ for T in $(-82, 82)$ [°C].

When $|x| \geq 1$, one can still use Taylor polynomial $P_n(x)$ for an approximation although the convergence is not as fast, i.e., one may need a large n to get a good approximation, because $|x|^{(n+1)}$ can be a very big number. However, $x^{n+1}/(n+1)!$ still goes to zero as n goes to infinity for any x; hence $P_n(x)$ can still be a good approximation for $f(x)$ even for $|x| \geq 1$ as long as n is sufficiently large.

An extreme case is that n is infinity, then the Taylor polynomial becomes the Taylor series

$$P_s(x) = f(0) + f'(0)x + \frac{f''(0)}{2!}x^2 + \frac{f'''(0)}{3!}x^3 + \cdots, \tag{6.39}$$

or

$$P_s(x) = \sum_{n=0}^{\infty} \frac{f^{(n)}(0)}{n!}x^n. \tag{6.40}$$

The Taylor polynomial approximation, also called Taylor series approximation or Taylor series expansion, is used widely in natural science and engineering applications, such as finding numerical solutions of climate models.

Taylor's theorem may be regarded as an extension of the differential mean value theorem (see Appendix D)

$$\frac{f(x) - f(0)}{x - 0} = f'(\xi), \tag{6.41}$$

or

$$f(x) = f(0) + f'(\xi)x \tag{6.42}$$

which is the zeroth-order Taylor's theorem. One can prove Taylor's theorem using the mean value theorem.

In this book, we choose to prove Taylor's theorem by the repeated use of the integral mean value theorem and the DA pair definition (i.e., the height increment is the integral of slope) (see Appendix D)

$$f(x) - f(0) = \int_0^x f'(t)dt. \tag{6.43}$$

This method of proof is straightforward and has a clear geometric meaning.

Proof Replacing the dummy variable t by t_1, the above DA pair definition can be written as follows:

$$f(x) = f(0) + \int_0^x dt_1\, f'(t_1) \tag{6.44}$$

$$= f(0) + \int_0^x dt_1 \left[f'(0) + \int_0^{t_1} dt_2\, f''(t_2) \right] \tag{6.45}$$

$$= f(0) + \int_0^x dt_1 f'(0) + \int_0^x dt_1 \int_0^{t_1} dt_2\, f''(t_2) \tag{6.46}$$

$$= f(0) + f'(0)x + \int_0^x dt_1 \int_0^{t_1} dt_2\, f''(t_2). \tag{6.47}$$

This is the linear approximation:

$$f(x) = P_1(x) + R_1(x), \tag{6.48}$$

where

$$P_1(x) = a + bx \tag{6.49}$$

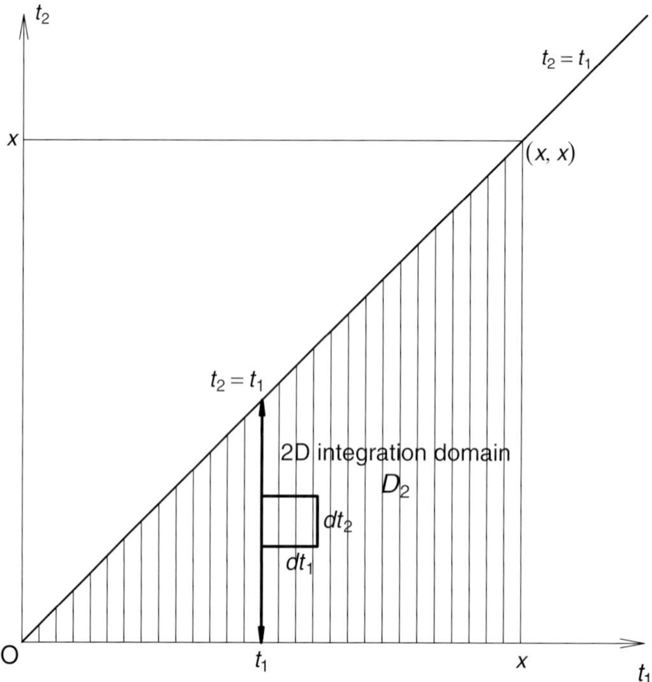

Fig. 6.8 The 2D triangle integration domain for the remainder term $R_1(x)$ in the first-order Taylor expansion $P_1(x)$.

is the tangent line at $x = 0$ with intercept $a = f(0)$ and slope $b = f'(0)$, and the error term $R_1(x)$ is related to the second-order derivative:

$$R_1(x) = \int_0^x dt_1 \int_0^{t_1} dt_2\, f''(t_2). \tag{6.50}$$

This can be written as a double integral

$$R_1(x) = \iint_{D_2} f''(t_2)\, dt_1 dt_2, \tag{6.51}$$

where the domain D_2 of the double integral is a right triangle resulting from a diagonal cut of a square of side equal to x, as shown in Fig. 6.8.

A single integral is equal to the mean function value times the length of the integral domain, and a double integral is equal to the mean function value times the area of the integral domain. This is the mean value theorem of calculus, as discussed in Appendix D. Thus,

$$R_1(x) = f''(\xi_2)\frac{1}{2}x^2, \tag{6.52}$$

where $f''(\xi_2)$ is the mean value and $\frac{1}{2}x^2$ is the area of the triangular integration domain shown in Fig. 6.8.

We carry out the above procedure to the next order:

$$f(x) = f(0) + f'(0)x + \int_0^x dt_1 \int_0^{t_1} dt_2 \left[f''(0) + \int_0^{t_2} dt_3\, f'''(t_3) \right] \tag{6.53}$$

$$\begin{aligned} &= f(0) + f'(0)x + \int_0^x dt_1 \int_0^{t_1} dt_2\, f''(0) \\ &\quad + \int_0^x dt_1 \int_0^{t_1} dt_2 \int_0^{t_2} dt_3\, f'''(t_3). \end{aligned} \tag{6.54}$$

The first set of integrals above can be calculated as follows:

$$\int_0^x dt_1 \int_0^{t_1} dt_2\, f''(0) \tag{6.55}$$

$$= \int_0^x dt_1 \left[\int_0^{t_1} dt_2\, f''(0) \right] \tag{6.56}$$

$$= \int_0^x dt_1 f''(0) t_1 \tag{6.57}$$

$$= f''(0) \int_0^x dt_1 t_1 \tag{6.58}$$

$$= f''(0) \frac{1}{2} x^2. \tag{6.59}$$

Substitution of this back into the previous set of integrals yields

$$\begin{aligned} f(x) &= f(0) + f'(0)x + f''(0)\frac{1}{2}x^2 + \int_0^x dt_1 \int_0^{t_1} dt_2 \int_0^{t_2} dt_3\, f'''(t_3). \\ &= P_2(x) + R_2(x). \end{aligned} \tag{6.60}$$

This is the quadratic approximation $b_0 + b_1 x + b_2 x^2$ to $f(x)$. The remainder term can be written as a triple integral

$$R_2(x) = \iiint_{D_3} f'''(t_3)\, dt_1 dt_2 dt_3, \tag{6.61}$$

where the integration domain D_3 is the right tetrahedron $OABC$ in the 3D space (t_1, t_2, t_3) shown in Fig. 6.9. The four vertices of the tetrahedron are $O(0,0,0)$, $A(x,0,0)$, $B(x,x,0)$, and $C(x,x,x)$. The base of the tetrahedron is the triangular region OAB in the previous approximation. The height of the tetrahedron is x, which is the length of edge BC. The area of the base is $\frac{1}{2}x^2$, which is the triangle's area in the previous approximation. The volume of a right tetrahedron in the 3D space is one third of the base area times height, and is thus

$$V = \frac{1}{2}x^2 \times \frac{x}{3} = \frac{x^3}{3!}. \tag{6.62}$$

By the integral form of the mean value theorem, the triple integral for the remainder $R(2)$ in Eq. (6.61) is

$$R_2(x) = \iiint_{D_3} f'''(t_3)\, dt_1 dt_2 dt_3 = f'''(\xi_2) \times V = f'''(\xi_2) \frac{x^3}{3!} \tag{6.63}$$

where $f'''(\xi_2)$ is the mean value of the function $f(t_3)$ in the right tetrahedron $OABC$.

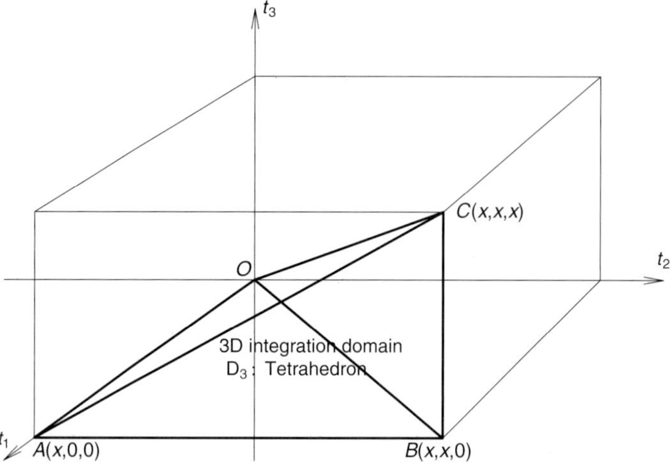

Fig. 6.9 The 3D right tetrahedron integration domain for the remainder term $R_2(x)$ in the second-order Taylor expansion $P_2(x)$.

If the third-order derivative is small, then the error R_2 is small. The extreme case is that the third-order derivative is zero. Then $R_2 = 0$, because the quadratic approximation to a quadratic function is 100% accurate.

The above procedures have shown that the nth-order Taylor approximation itself needs only the condition that all the needed derivatives exist and are continuous all the way until $f^{(n+1)}(x)$. The estimation of the error term has used the mean value theorem that requires $f^{(n+1)}(x)$ to be a continuous function.

Carrying out the same procedure for

$$f'''(t_3) = f'''(0) + \int_0^{t_3} dt_4\, f^{(4)}(t_4) \tag{6.64}$$

and calculating the following integral, we obtain

$$\int_0^x dt_1 \int_0^{t_1} dt_2 \int_0^{t_2} dt_3\, f'''(0) \tag{6.65}$$

$$= \int_0^x dt_1 \int_0^{t_1} dt_2\, \left[f'''(0)t_2\right] \tag{6.66}$$

$$= \left[f'''(0)\frac{1}{2}\int_0^x dt_1\, t_1^2\right] \tag{6.67}$$

$$= f'''(0)\frac{1}{2}\frac{1}{3}x^3 \tag{6.68}$$

$$= \frac{f'''(0)}{3!}x^3. \tag{6.69}$$

Substitution of this integral back into the previous set of integrals yields

$$f(x) = f(0) + \frac{f'(0)}{1!}x + \frac{f''(0)}{2!}x^2 + \frac{f'''(0)}{3!}x^3 + \int_0^x dt_1 \int_0^{t_1} dt_2 \int_0^{t_2} dt_3 \int_0^{t_3} dt_4\, f^{(4)}(t_4)$$
$$= P_3(x) + R_3(x). \tag{6.70}$$

This yields the third-order approximation $P_3(x)$:

$$P_3(x) = f(0) + \frac{f'(0)}{1!}x + \frac{f''(0)}{2!}x^2 + \frac{f'''(0)}{3!}x^3. \tag{6.71}$$

Its error term is

$$R_3(x) = f^{(4)}(\xi_3)\frac{x^4}{4!}, \tag{6.72}$$

where $f^{(4)}(\xi_3)$ is the mean value of the fourth-order derivative in a 4D right tetrahedron.

Carrying out the same procedure, one can reach the nth-order polynomial approximation of Taylor's theorem.

6.8.2 Taylor Series Example: Exponential Function

All the derivatives of e^x are e^x which is equal to 1.0 at $x = 0$. Namely, $D[e^x] = e^x$, $D^{(n)}[e^x] = e^x$ and $e^0 = 1$ at $x = 0$. Taylor's theorem implies that

$$P_n(x) = 1 + x + \frac{x^2}{2!} + \frac{x^3}{3!} + \cdots + \frac{x^n}{n!} \tag{6.73}$$

is the nth-order polynomial approximation to the exponential function e^x around $x = 0$. The above is true because, as we have stated, all the derivatives of e^x are e^x which is equal to 1.0 at $x = 0$.

Figure 6.10 shows the Taylor series approximation to e^x around $x = 0$ in the interval $(-1, 1)$. When $n = 1$, it is a linear approximation. When $n = 2$, it is a

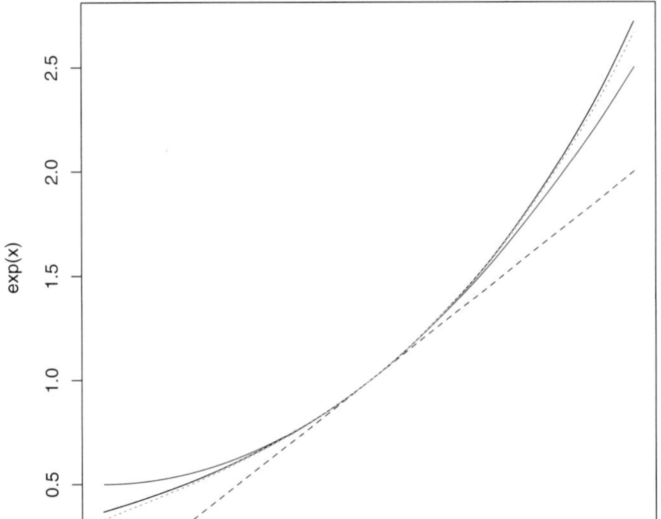

Fig. 6.10 Taylor series approximations to e^x (black line) around $x = 0$: the first order (blue), second order (red), and third order (green).

quadratic approximation, more accurate than the linear approximation. When $n = 3$, it is the cubic approximation (the green line in the figure), which can barely be distinguished from the original function e^x when $x \in (-0.5, 0.5)$.[3] The approximation is extremely good. This example clearly shows the efficiency of the Taylor polynomial approximation.

6.8.3 Numerical Integration Using Taylor Expansion

For the zeroth-order approximation,

$$f(x) = f(0) + R_0(x). \tag{6.74}$$

An approximation of an integral is

$$\int_a^b f(x)dx \approx f(0)(b-a). \tag{6.75}$$

This is the rectangular rule of integration often used in calculus to illustrate the definition of an integral.

A more accurate approximation is the linear approximation: noting that the line from point $(a, f(a))$ to point $(b, f(b))$ is a line with nonzero slope, and it thus forms a trapezoid with bases $f(b)$ and $f(a)$ and height $b-a$. Thus, the trapezoid's area

$$I_T = \frac{f(a)+f(b))}{2} \times (b-a) \tag{6.76}$$

may be used to approximate the integral:

$$\int_a^b f(x)dx \approx \left[\frac{1}{2}f(a) + \frac{1}{2}f(b)\right](b-a). \tag{6.77}$$

This is the trapezoidal rule of integration. $\left[\frac{1}{2}f(a) + \frac{1}{2}f(b)\right]$ is a weighted average of $f(x)$ with two weights 1/2 and 1/2.

If the domain $[a, b]$ is cut into two equal intervals, the trapezoid rule is applied to two trapezoids over two equal intervals: $[a, (a+b)/2]$ and $[\frac{a+b}{2}, b]$. Then

$$\begin{aligned}
\int_a^b f(x)dx &= \int_a^{\frac{a+b}{2}} f(x)dx + \int_{\frac{a+b}{2}}^b f(x)dx \\
&\approx \left[\frac{1}{2}f(a) + \frac{1}{2}f\left(\frac{a+b}{2}\right)\right]\left(\frac{a+b}{2} - a\right) \\
&\quad + \left[\frac{1}{2}f\left(\frac{a+b}{2}\right) + \frac{1}{2}f(b)\right]\left(b - \frac{a+b}{2}\right) \\
&= \left[\frac{1}{4}f(a) + \frac{1}{2}f\left(\frac{a+b}{2}\right) + \frac{1}{4}f(b)\right](b-a).
\end{aligned} \tag{6.78}$$

The square bracket may be considered a weighted average of $f(x)$ since the sum of the weights $1/4, 1/2, 1/4$ is one.

[3] The symbol $\in$ stands for "belongs to the set." Thus, $x \in (-0.5, 0.5)$ means that x belongs to the set $(-0.5, 0.5)$.

One may extend this weighted average idea to develop higher-order numerical integration methods by finding appropriate weights for appropriate points in $[a,b]$.

The following weighted average

$$\int_a^b f(x)dx \approx \left[\frac{1}{6}f(a)+\frac{4}{6}f\left(\frac{a+b}{2}\right)+\frac{1}{6}f(b)\right](b-a) \tag{6.79}$$

is called Simpson's rule of integration. Again, the sum of the three weights $1/6, 4/6, 1/6$ is one. This is equivalent to approximating $f(x)$ by a quadratic function $P_2(x)$, and is hence more accurate than the trapezoidal rule, which comes from a linear approximation.

Simpson's rule improves the integration accuracy without increasing the computational work for an interval $[a,b]$. However, when cutting the interval into n subintervals and adding together the Simpson's rule results for every interval, it is clear that Simpson's rule requires more computational effort, as one would naturally expect.

An alternative to Simpson's rule might be to divide the interval into three parts using four points $a, (2a+b)/3, (a+2b)/3, b$, and to assign weights as $1/8, 3/8, 3/8, 1/8$, whose sum is still one. Thus,

$$\int_a^b f(x)dx \approx \left[\frac{1}{8}f(a)+\frac{3}{8}f\left(\frac{2a+b}{3}\right)+\frac{3}{8}f\left(\frac{a+2b}{3}\right)+\frac{1}{8}f(b)\right](b-a). \tag{6.80}$$

Similarly, one can easily invent new numerical integration rules by following the principle of using a weighted average. Often a symmetric rule can be reasonable and useful. In that case, one can calculate the order of approximation according to the Taylor expansion.

6.9 Chapter Summary

This chapter describes several commonly used mathematical methods in climate sciences involving derivative applications. The methods and their main ideas are summarized below.

(i) Linear approximation $L(x)$ of a function $f(x)$ at $x=a$ is the tangent line of this curve at the point $(a, f(a))$. Thus, the linear approximation function is given by the linear equation (6.11), i.e.,

$$L(x) = f(a) + f'(a)(x-a). \tag{6.81}$$

The linear approximation can be applied to the Stefan–Boltzmann law of blackbody radiation

$$\sigma(273.15+T)^4 = \sigma 273.15^4\left(1+\frac{T}{273.15}\right)^4 \tag{6.82}$$

for the variable $T/273.15$ since $T/273.15$ is small for the Earth's surface temperature between $-50\,^\circ$C and $50\,^\circ$C. The linearized radiation law is

$$E_{out} = A + BT. \tag{6.83}$$

(ii) Budyko's approximate law of radiation is such a linear function of temperature T, in which the coefficients A and B can be obtained by a regression approach based on the observed data.

(iii) Newton's method is based on the linear approximation

$$L(x_1) = f(x_0) + f'(x_0)(x_1 - x_0) = 0 \tag{6.84}$$

that determines the first approximation to a root of a nonlinear equation from an initial guess of the root x_0. The next step is to treat x_1 as the guess and find the second approximation x_2. This procedure continues iteratively until the guess and the next approximation differ by only a small amount. If so, the procedure is said to be convergent.

(iv) Atmospheric pressure varies spatially, and its gradient vector ∇p divided by the air density is a force per unit mass [N kg^{-1}], which, in large-scale atmospheric and oceanic flows, is often approximately balanced by the Coriolis force. This state is called geostrophic balance. The gradient is computed by partial derivatives

$$\nabla p = \left(\frac{\partial p}{\partial x}, \frac{\partial p}{\partial y}, \frac{\partial p}{\partial z} \right). \tag{6.85}$$

(v) A derivative of a derivative yields the second derivative, which means acceleration if the dependent variable is displacement and the independent variable is time. Climate models typically use only the first and second derivatives in their fundamental governing differential equations. However, when exploring the approximate solutions to climate models or analyzing climate data, the third or higher-order derivatives may be used, such as the Taylor polynomial approximation. When assuming the existence of the $(n+1)$th-order derivative, a function can be approximated by an nth-order polynomial whose coefficients can be represented by the derivatives as shown in Eq. (6.35):

$$f(x) = f(0) + f'(0)x + \frac{f''(0)}{2!}x^2 + \frac{f'''(0)}{3!}x^3 + \cdots + \frac{f^{(n)}(0)}{n!}x^n + O(x^{n+1}). \tag{6.86}$$

This is Taylor's theorem and can be proved by repeatedly applying multiple integrals.

(vi) The Taylor polynomial approximation is an extension of the linear approximation from the first order to nth order. An integration based on the linear approximation yields the trapezoidal rule of numerical integration. An integration based on a higher-order polynomial can yield a more accurate numerical integration, such as Simpson's rule of numerical integration based on the approximation of the second-order polynomial.

The applications of derivatives included in this chapter follow the philosophy of direct calculus presented in Appendix D: the derivative is the slope of a curve, and the integral is the height of the curve from point P_1 to point P_2 when treating the height as an accumulation of the elevations of many small steps determined by the local slope times the small step's horizontal distance. If the derivative is considered to be the rate of change, then the integral is the total change from time τ_1 to time τ_2 when treating the total change as an accumulation of the small changes determined by the instantaneous rate times the size of the time step. These ideas and methods of direct calculus invented in the early seventeenth century turned out to be quite practical and made many seemingly impossible calculations easy, such as the calculation of the area underneath the curve $y = x^2$ from $x = 0$ to $x = 1$, the area of a polar cap, and the distance traveled by a free-falling body in a given time interval. Our motivation for presenting calculus and its applications is the same as that of René Descartes, Pierre de Fermat, and Isaac Newton to develop a method to solve practical problems using the concept of variable slopes, and small steps and their accumulations. Almost four centuries after Descartes and others developed the beginnings of calculus, this powerful method has now become a core course in many high schools and a mandatory course for the STEM (science, technology, engineering, and mathematics) majors in many universities. Its use in climate science has been extensive, and we are confident that it will play an even more important role in climate modeling and climate data analysis in the new era of big data and machine learning.

References and Further Readings

[1] Budyko, M. I., 1969: The effect of solar radiation variations on the climate of the Earth. *Tellus*, 21, 611619.

> This is one of the earliest papers on quantitative studies using an energy balance model. The now well-known Budyko model of long-wave radiation emitted by the Earth was developed in this paper.

[2] Cess, R. D., 1976: Climate change: An appraisal of atmospheric feedback mechanisms employing zonal climatology. *Journal of the Atmospheric Sciences*, 33, 1831–1843.

> This paper published a numerical estimate of the key parameters used in Budyko's EBM. The zonal dependence of the parameters was considered as well.

[3] Goose, H., 2015: *Climate System Dynamics and Modeling*. Cambridge University Press, New York.

> This is a textbook, a reference book, and an excellent and very readable introduction to climate dynamics and climate modeling. It is aimed at graduate students and includes a glossary and exercises.

[4] McGuffie, K. and A. Henderson-Sellers, 2014: *The Climate Modelling Primer*, 4th Edition, Wiley-Blackwell, New York.

> This text has the distinctly modern feature of providing students with interactive experiences using Internet resources.

[5] North, G. R. and K.-Y. Kim, 2017: *Energy Balance Climate Models*, Wiley, New York.

This book is a definitive treatment of EBMs by two experts both of whom have many years of experience in EBM research.

[6] Wallace, J. M. and P. V. Hobbs, 2006: *Atmospheric Science: An Introductory Survey*. 2nd Edition, Academic Press, New York.

A classic standard textbook, much loved by generations of atmospheric science students.

Exercises

Exercises 6.4–6.7 and 6.11–6.14 require the use of a computer.

6.1 A ball has been shot straight up at 10 $[\mathrm{m\ s^{-1}}]$ from an initial height 2 $[\mathrm{m}]$. How long it will take the ball to reach its maximum height? Hint: Choose the gravitational acceleration to be 9.8 $[\mathrm{m\ s^{-2}}]$, and find an approximate answer. The general formula is $h = -gt^2/2 + v_0 t + h_0$.

6.2 Find the second-order polynomial approximation of the Stefan–Boltzmann law of radiation by using Taylor's theorem for the variable $x = T/273.15$:

$$\sigma(273.15 + T)^4 = \sigma 273.15^4 \left(1 + \frac{T}{273.15}\right)^4. \tag{6.87}$$

6.3 Find the linear approximation of $f(x) = x^2 - 1$ at $x = 1.5$. If this 1.5 is used as the first guess, find the next approximate root of $f(x) = 0$ by Newton's method.

6.4 (a) Use Newton's method and the R code shown in Section 6.4 to find all the solutions of the following equation

$$50\sin(x) = 4x^2 - 0.1x^4 - 1. \tag{6.88}$$

Show your R codes and describe your procedures as in the examples shown in Section 6.4.

(b) Plot the tangent line, similar to that shown in Fig. 6.2, at the initial guess x_0 for the first root. Use this tangent line to find x_1 by hand either by a formula or from the plot.

(c) Use the R command `uniroot` to find all the solutions, and compare the result with the solutions in (a).

6.5 (a) Make a linear approximation table similar to Table 6.1 for the first root in the above problem. Use seven points as in the table, but you may choose your own step size, i.e., the distance between each pair of points. Your final results should be in the form of a table. You should also give the details of your R code and supply a text describing your procedure.

(b) Discuss the above table of results.

6.6 (a) Use Fig. 6.5 and the derivative concept to approximately calculate $\frac{\partial T}{\partial p}$ for three points on the figure, where T is temperature and p is pressure. To calculate an approximate value of a derivative, you can use either Descartes' method of drawing tangent lines, or Fermat's method of small increments.

(b) Identify the regions with very large absolute values of the derivative $\frac{\partial T}{\partial p}$.

(c) Identify the regions with very small absolute values of the derivative $\frac{\partial T}{\partial p}$.

(d) If p has an exponential dependence on altitude z, what can you say about the temperature gradient with respect to altitude? Use Fig. 6.6 as a reference.

6.7 (a) Find the fourth-order Taylor polynomial expansion $P_4(x)$ of the function $f(x) = e^x + e^{-x}$.

(b) Plot $P_4(x)$ and $f(x)$ on the same figure for $x \in (-1, 1)$.

6.8 Find the Taylor series of the function $f(x) = \sin(x-1)$.

6.9 Use the differential mean value theorem to prove Taylor's theorem.

6.10 (a) Calculate the approximate derivatives with respect to depth z in Fig. 6.7 for temperature, salinity, and density. Select two points for each variable: temperature, salinity, and density.

(b) Discuss where the vertical gradient is the largest for each variable.

6.11 Using a fifth-order Taylor series and regression, develop a mathematical model for the temperature profile in ocean water as a function of depth, i.e., find a fifth-order polynomial such that $T = P_5(s)$ where $s = z/h_0$ is the standardized depth variable, where z is depth, and h_0 is the characteristic depth and is a constant. Use the online data that you can find.

6.12 Similar to the Taylor series of polynomials, a continuous function $f(x)$ in the interval $[0, T]$ can also be decomposed into an infinite series of sine or cosine functions, such as

$$f(x) = A_0 + \sum_{n=1}^{\infty} A_n \cos(2\pi n\omega x + \phi_n) \tag{6.89}$$

where A_n are called amplitude spectra, ϕ_n are called phases, and $\omega = 1/T$ is called fundamental frequency with T being the period of the function $f(x)$. This kind of trigonometric series is called Fourier series and is named after Joseph Fourier (1768–1830), a French mathematician and physicist.

(a) Given that

$$A_0 = 0.25, \tag{6.90}$$

$$A_n = \frac{2}{n^2}, \quad n = 1, 2, \ldots, \tag{6.91}$$

$$\phi_n = \frac{(2n-1)\pi}{4} \quad n = 1, 2, \ldots, \tag{6.92}$$

use R to plot the following functions on the same figure

$$S_N(x) = A_0 + \sum_{n=1}^{N} A_n \cos(\pi n x/4 + \phi_n) \tag{6.93}$$

for $N = 2, 6, 10, 14, 18$, and 22.

(b) Use R to plot the points (n, A_n) for $n = 0, 1, 2, \ldots, 22$. This figure is called the amplitude spectrogram.

(c) What is the fundamental frequency of this Fourier series?

6.13 Carry out the same procedures as the previous problem but for the following data

$$A_0 = 0, \tag{6.94}$$

$$A_n = (-1)^n \frac{2}{n\pi}, \quad n = 1, 2, \ldots, \tag{6.95}$$

$$\phi_n = 0 \quad n = 1, 2, \ldots, \tag{6.96}$$

6.14 In climate data analysis, one often uses the method of Fast Fourier Transform (FFT) to calculate a spectrogram. For a given data vector $x = (x_1, x_2, \ldots, x_n)$, application of the FFT algorithm to this data vector yields a vector of spectral data $X = (X_1, X_2, \ldots, X_n)$, which are complex numbers. From the spectral data, an inverse transform can recover the original data. The FFT and the inverse FFT (IFT) can be carried out using the following R commands

```
fft(x)
fft(X, inverse=TRUE)/length(X)
```

(a) Randomly generate 20 data entries by R using `rnorm(20)` and apply the above R commands of FFT and IFT to the data.

(b) Verify that the real part of the IFT result can approximately recover the original data.

7 Calculus Applications to Climate Science II: Integrals

We now move on to illustrate several ways in which integrals are used in climate science. We focus first on hydrostatic balance, the equilibrium between the forces of gravity and the vertical pressure gradient that is found in both the ocean and the atmosphere for phenomena characterized by large horizontal length scales. We then introduce the important concept of geopotential, the gravitational potential energy per unit mass at a given atmospheric altitude z with respect to sea level. We provide a general derivation of the hypsometric equation showing that atmospheric pressure decreases exponentially with increasing elevation, and we discuss approximations to this equation. We also show how the hypsometric relationship was used in the 1800s to determine the height of mountains with surprising accuracy, long before modern technology was developed.

We next explore the role of integrals in several fundamental topics in thermodynamics, including work done when a system consisting of air is expanded or compressed into a different volume. We discuss internal energy, enthalpy, entropy, the ideal gas law, and the first law of thermodynamics. We show how, as an application of integration, the Stefan–Boltzmann law can be derived from Planck's law of radiation. Because our focus is on selected mathematical aspects of these climate subjects rather than on details of the physics, readers interested in learning more about these topics may wish to also consult standard textbook references listed at the end of this chapter, such as the books by Curry and Webster (1990), Feynman et al. (2013), Pierrehumbert (2010), and Wallace and Hobbs (2006).

7.1 Geopotential and Atmospheric Pressure

This section uses the method of integration to analyze the concepts of geopotential and atmospheric pressure. The integration method is to divide the object to be analyzed into many small pieces, use differentials to make relevant approximations, and finally integrate the approximations to derive an integrated result. Figure 7.1 shows a small piece of air, called a small parcel of the atmosphere in climate science. We will build differentials and integrals based on the small parcel.

7.1.1 Vertical Forces on a Small Parcel of Atmosphere

Figure 7.1 shows the balance of forces on a small parcel of atmospheric air or ocean water, which is assumed to be not moving. Consider the forces acting in the vertical direction (parallel to gravity) on such a parcel. The parcel has mass, so we know that the force

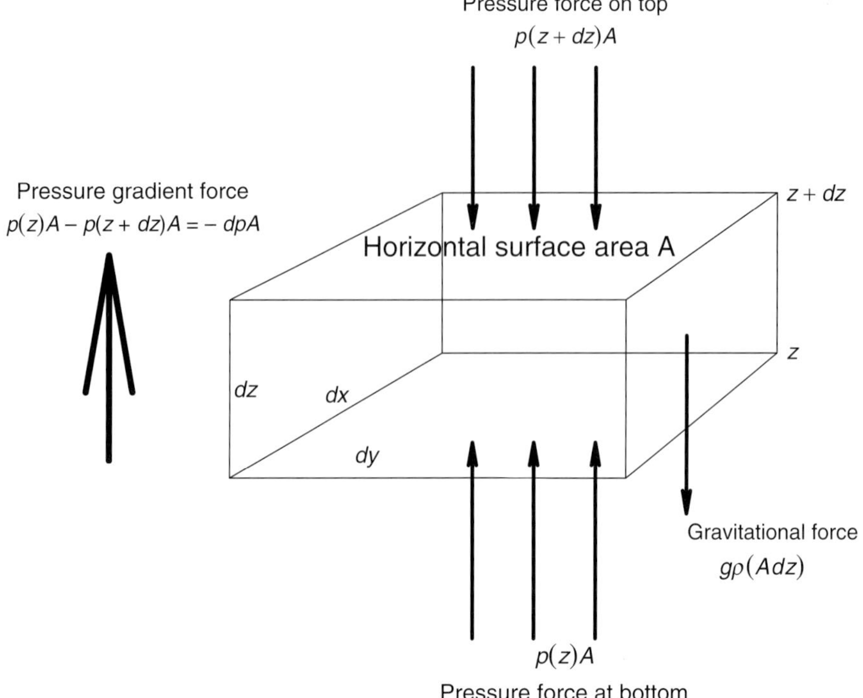

Fig. 7.1 A small parcel of atmosphere with a horizontal base area A, height dz, and mass $\rho(Adz)$, and the vertical forces acting on it: A small pressure decrease due to a small altitude increment (dz) causes a pressure gradient force ($-dpA$) to balance the gravitational force ($g\rho(Adz)$).

of gravity acts on it, tending to accelerate it downwards toward the center of the Earth. Because the parcel is not moving, there must be an equal and opposite force to balance that of gravity. This is the pressure gradient force (PGF).

Strictly speaking, the balance between the downward force of gravity and the upward pressure gradient force holds only when the fluid (air or water) is not moving. This balance is called hydrostatic balance, and the term "static" occurring in the word "hydrostatic" refers to a motionless state. In both the atmosphere and the ocean, however, vertical variations of pressure are almost always observed to be quantitatively much larger than either horizontal variations or time variations. Thus, the atmosphere and the ocean are almost always nearly in hydrostatic balance in the vertical direction. The main exceptions are the relatively rare cases where vertical accelerations are large, usually found in small regions, such as air motions in tornadoes or violent thunderstorms, and in regions of intense convection in the ocean. In the atmosphere, hydrostatic balance is generally found wherever the characteristic horizontal length scale of the motions is much larger than their characteristic vertical length scale. Thus, the large-scale circulation of the atmosphere may safely be assumed to be hydrostatic, but relatively small-scale phenomena such as fronts and convective clouds are likely to be non-hydrostatic.

We may also note here a major difference between the atmosphere and the ocean. Both are fluids, and both are held to the Earth by gravity, but seawater, being a liquid, is nearly incompressible, so its density variation is small in both space and time, and thus the ocean water has an upper surface. The atmosphere, however, is almost entirely a gas (strictly speaking, a mixture of gases) and is thus compressible, and atmospheric density varies with pressure, so the atmosphere lacks a top. Rather than having a finite depth, the atmosphere instead gradually becomes less and less dense as altitude increases, and the very thin atmosphere at great altitudes finally blends into interplanetary space.

To develop a mathematical expression for hydrostatic balance, consider a small volume of the fluid (which may be either atmospheric air or ocean water), as shown in Fig. 7.1. Let the small volume have a horizontal base area A and a vertical extent dz, so its volume is Adz, and its mass is ρAdz, where ρ is density. Newton's second law $F = ma$ then tells us that the gravitational force on this parcel is $g\rho Adz$, where g is the gravitational acceleration. Opposing this force is a vertical pressure gradient force (PGF) illustrated by the thick upward arrow in Fig. 7.1, and this force is equal to the pressure force at the bottom surface $p(z)A$ minus the pressure force on the top surface $p(z+dz)A$, i.e., $p(z)A - p(z+dz)A \approx -Adp$, where dp is the differential of p with respect to z. Because the pressure is a decreasing function of the vertical coordinate z, dp is negative, hence the PGF $-Adp$ is positive. The hydrostatic balance means that the gravitational force $g\rho Adz$ is equal to the PGF $-Adp$, i.e.,

$$g\rho Adz = -Adp, \tag{7.1}$$

which can be simplified to

$$-dp/dz = g\rho. \tag{7.2}$$

This is called the hydrostatic equation, expressing the balance between the downward gravitational force and the upward pressure gradient force.

According to Figure 7.1, this hydrostatic balance can also be explained in another way using the balance of three forces in the vertical direction: (i) the upward pressure force from the bottom of the parcel $p(z)A$, (ii) the downward pressure force from the top $-p(z+dz)A$, and (iii) the downward gravitational force $-g\rho Adz$. The sum of these three forces is zero for the condition of vertical static balance:

$$p(z)A - p(z+dz)A - g\rho Adz = 0, \tag{7.3}$$

which can be written as

$$-g\rho Adz = (p(z+dz) - p(z))A. \tag{7.4}$$

Denote

$$dp \approx p(z+dz) - p(z) \tag{7.5}$$

as the differential of p with respect to z, i.e., a small pressure change due to the small increment of the vertical position. Then,

$$dp = -g\rho dz, \tag{7.6}$$

$$\frac{dp}{dz} = -g\rho. \tag{7.7}$$

when the non-zero area A is canceled.

We may write the derivative with respect to the vertical coordinate z as a partial derivative, in recognition of the fact that pressure may also vary in the horizontal dimensions and time:

$$\frac{\partial p}{\partial z} = -g\rho. \tag{7.8}$$

This is another form of Eq. (7.2).

We may integrate the hydrostatic equation to yield a relationship between pressure and either depth in the ocean or height in the atmosphere. For example, the pressure at altitude z may be expressed mathematically as an integral of $dp = -g\rho dz$ from the infinite altitude where the pressure is zero to z:

$$\int_0^p dp = -\int_\infty^z g\rho dz. \tag{7.9}$$

This yields

$$p(z) = -\int_\infty^z g\rho dz = \int_z^\infty g\rho dz. \tag{7.10}$$

This is the mathematical expression of the physics definition of hydrostatic pressure that is the sum of the gravitational force per unit area acting on all the mass above the given altitude z. The pressure measured by a barometer is called the barometric pressure or atmospheric pressure, which includes the effect of acceleration of the atmospheric motion. In most climate science analyses, the hydrostatic pressure is a very good approximation to the atmospheric pressure.

To carry out this integration, it is often assumed that g is constant, but this is not strictly true. For example, g varies with distance from the center of the Earth, and therefore g also varies with geographical location, because the Earth is not exactly a sphere. In fact, the Earth is more nearly an oblate spheroid, or oblate ellipsoid, and the polar radius of the Earth is about 21 km smaller than the equatorial radius.

A further complication is introduced when we recognize that the rotation of the Earth gives rise to two apparent forces, the Coriolis force and the centrifugal force. As Vallis (2017) points out, these forces are not true forces but arise because, when we choose to represent motions in a coordinate system that rotates with the Earth, then bodies in such a non-inertial coordinate system behave as if other forces are present that affect their motions. The centrifugal force and the gravitational force are both potential forces, and so it is convenient to modify our definition of g, making it an effective gravity equal to the sum of true gravity plus the centrifugal force. Then true gravity will be directed approximately toward the center of the Earth, with small deviations due primarily to the Earth's oblate shape mentioned above, and the effective gravity will deviate slightly from this direction.

A convenient way to account for these variations in g is to employ the concept of geopotential, introduced in the next subsection. We note also that in middle latitudes, the centrifugal force is about an order of magnitude larger than the Coriolis force. With the introduction of the geopotential height as a coordinate in the climate model equations, the centrifugal force will disappear from the equations. Then, the climate model equations can explicitly show a very important dynamical balance between the Coriolis force and the pressure gradient force, called geostrophic balance. This balance can help model and explain many large-scale horizontal flows in both atmosphere and ocean.

7.1.2 Geopotential

The geopotential Φ is the gravitational potential energy per unit mass at a given altitude z with respect to sea level. Geopotential at sea level is conventionally taken to be zero. The mathematical expression for geopotential is then an integral

$$\Phi(\phi, \theta, z, t) = \int_0^z g dz, \tag{7.11}$$

where the gravitational acceleration g is a function of latitude, longitude, geometric elevation from sea level, and time. The unit of geopotential Φ is $[\mathrm{m}^2\ \mathrm{s}^{-2}]$ in the SI system, which is the work done to raise 1 [kg] mass from sea level to level z.

It turns out that in constructing a convenient coordinate system, spherical coordinates are natural, and there are several reasons to define the vertical direction as the direction of the geopotential force, and then to make the approximation of treating surfaces of constant geopotential Φ as though they were exactly spherical. As Vallis (2017) states, "The horizontal component of effective gravity is then identically zero, and we have traded a potentially large dynamical error for a very small geometric error." We are fortunate that the oblate spheroid shape of the Earth is very nearly spherical, because over geological time, the Earth has developed an equatorial bulge of just the right magnitude to make the Earth's surface a spheroid on which geopotential Φ is nearly constant. For a further discussion, see Vallis (2017) and references therein.

The geopotential height is

$$Z = \frac{1}{g_0} \int_0^z g(\phi, \theta, z) dz, \tag{7.12}$$

where g_0 is the standard gravity at the mean sea level, and is 9.80665 $[\mathrm{m\ s}^{-2}]$, or 32.174 ft s^{-2}. The unit of Z is [gpm], the geopotential meter, whose dimension is the same as length. One [gpm] is very close to one meter in the lower atmosphere. However, 1.0 gpm $>$ 1 m in the higher atmosphere, say, at an airliner's cruising altitude.

Thus, $Z = f(\phi, \theta, z)$ is a function of latitude, longitude, and altitude. There are accurate mathematical theories and detailed algorithms to convert the geometric height z to the geopotential height Z and vice versa. However, this topic is beyond the scope of this book, and interested readers may find relevant materials in the literature (e.g., Lewis (2007)).

In terms of the geopotential height Z, the hydrostatic balance equation can be expressed in a way not explicitly dependent on $g(\phi, \theta, z)$. This can be shown as follows.

The chain rule of differentiation leads to

$$\frac{\partial p}{\partial z} = \frac{\partial p}{\partial Z}\frac{\partial Z}{\partial z} = \frac{\partial p}{\partial Z}\frac{g}{g_0}. \tag{7.13}$$

The hydrostatic balance equation (7.8) now becomes

$$\frac{\partial p}{\partial Z}\frac{g}{g_0} = -g\rho, \tag{7.14}$$

or

$$\frac{\partial p}{\partial Z} = -g_0\rho. \tag{7.15}$$

This hydrostatic balance equation depends on the standard gravity g_0, which is a constant, and does not explicitly depend on the variable gravitational acceleration $g(\phi,\theta,z)$. This equation has many important applications, such as a convenient hydrostatic approximation in climate models, the theoretical basis for designing the radiosonde instruments to determine the height of the radiosonde balloons, and a modern approach to make accurate and alternative observation of atmospheric pressure with the help of GPS data. We will discuss these in the next section.

In the lower atmosphere below 2 km elevation, the geopotential height Z and geometric height z (i.e., elevation from sea level) are very close and have a difference less than one meter. However, this difference can be as much as several meters at an altitude of about 10 km.

A common meteorological use of geopotential height is the Z value corresponding to a certain atmospheric pressure, $Z = f(\phi,\theta,p,t)$. The pascal (Pa) is the unit of pressure in the SI system. A pascal is one newton per meter squared. In meteorology, the traditional unit of pressure is the millibar (mb) which is 100 pascals or 1 hectopascal. A typical value of sea level pressure is about 1,000 mb which is one bar, defined as 100,000 pascals. For a given pressure, we may ask, what is the geopotential height? Thus, $Z = f(\phi,\theta,500\text{ mb},t)$ may be used to describe the surface on which a given pressure, say, 500 mb, is constant. See Fig. 7.2 for an example.

In the example depicted in this figure, we can see that the 500 mb constant-pressure surface is located at an altitude of 588 [dam] (decametre =10 meters), i.e. 5,880 meters, over the Gulf of Mexico, but it is at an altitude of only 572 [dam] over the northeastern United States. Thus, a map of geopotential height at a given time, or averaged over a given time interval, is an intuitively attractive way to represent the spatial variations in the atmospheric pressure field.

In the idealized case of geostrophic flow with no wind acceleration, small vertical velocity, and no friction, the pressure difference between a point in the atmosphere (x,y,z) and the corresponding sea level point $(x,y,0)$ is the integration of the total amount of the gravitational force in the vertical interval $[0,Z]$:

$$p_0(x,y) - p(x,y,Z,t) = \int_0^Z \rho g dz \tag{7.16}$$

where ρ is the air density, and 0 in the integral denotes sea level.

For a given pressure, say $p(x,y,z,t) = 500$ mb, Z must vary with respect to location and time (x,y,t), which defines the geopotential height surface $z = Z(x,y,t)$ evolving with time. A map of mean 500 mb geopotential height field during a particular interval of time is shown in Fig. 7.2.

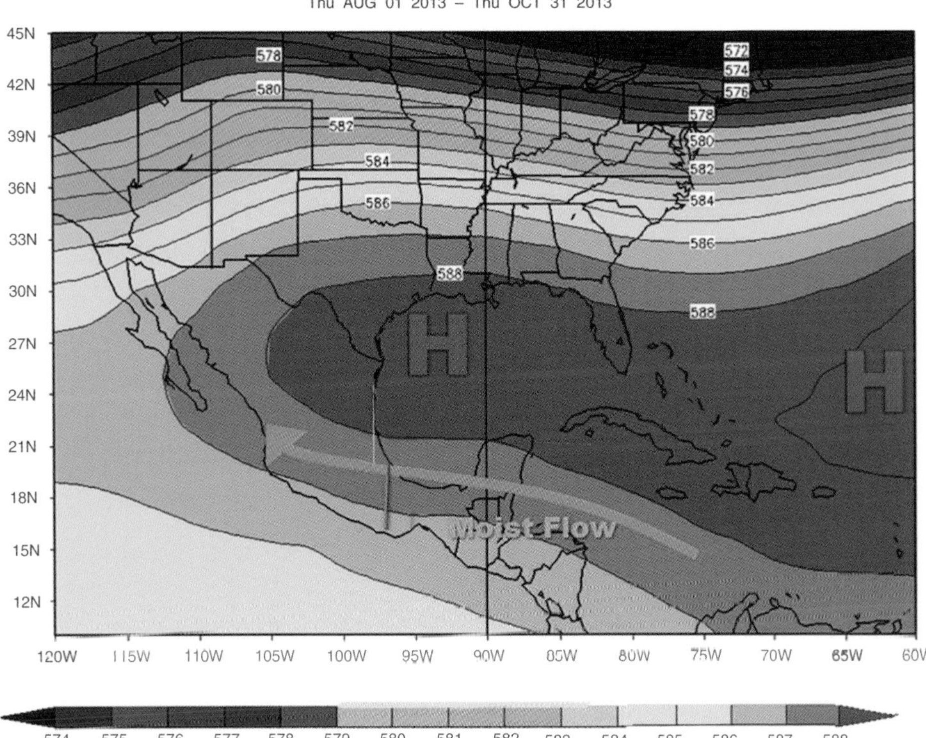

Fig. 7.2 Geopotential height of 500 mb for the Gulf of Mexico region and the adjacent Atlantic Ocean during the 2013 hurricane season. Hurricane Ingrid: September 12–17, 2013, peak wind 85 mph, minimum surface pressure 983 mb. Tropical Storm Fernand: August 25–26, 2013, peak wind 60 mph, minimum surface pressure 1001 mb. *Credit: The United States National Weather Service. Source:* www.weather.gov/bro/2013event_hurricaneseasonwrap

In general, a pressure of 500 mb corresponds to an altitude of about 5,000 meters, and to temperatures around $-20\,^\circ$C, while a pressure of 850 mb corresponds to an altitude of about 1,500 meters and to temperatures around $5\,^\circ$C, thus a little above the freezing point of water which is $0\,^\circ$C. Jet passenger aircraft typically operate at altitudes of approximately 10 km, corresponding to an atmospheric pressure of about 250 mb and air temperatures of about $-50\,^\circ$C.

7.2 Hypsometric Equation: Exponential Decrease of Pressure with Respect to Elevation

7.2.1 The General Hypsometric Equation

In the atmosphere, the ideal gas law, or equation of state for an ideal gas, is very accurate and is usually an excellent approximation to physical reality. Given the gas law and the hydrostatic assumption, atmospheric pressure can be shown to decrease exponentially with increasing elevation, i.e., the elevation is in a logarithmic relationship with the pressure. This relationship is referred to as the hypsometric equation. The word "hypsometric" is derived from the Greek word "hypsos" meaning height and another Greek word "metre" meaning measure, because the equation can be used to calculate elevation from atmospheric pressure data. We next derive the hypsometric equation.

We use the hydrostatic equation derived in the previous section:

$$dp = p(z+dz) - p(z) = -\rho g dz. \tag{7.17}$$

The ideal gas law is

$$pV = nR^*T, \tag{7.18}$$

where n is the number of moles of gas in the volume V, and R^* is the ideal gas constant, also called the universal gas constant, which is 8.314462 $[\mathrm{J(mol)^{-1}K^{-1})}]$.

The law can also be written as

$$p/T = (n/V)R^*, \tag{7.19}$$

in which

$$\rho^* = \frac{n}{V} \tag{7.20}$$

is the gas density in the unit of $[(\mathrm{mol})\mathrm{m}^{-3}]$.

The density in the SI system often uses units $[(\mathrm{kg})\ \mathrm{m}^{-3}]$ or $[\mathrm{g(cm)}^{-3}]$. One can then express the gas constant in terms of weight, rather than moles. The gas constant in weight is called the specific gas constant, denoted by R, since it is specific to a given gas. For example, the weight of a mole of dry air, known as the molar mass, is $M = 28.9647 \times 10^{-3}$ $[\mathrm{kg\ mol^{-1}}]$. The specific gas constant R for the dry air is then

$$R = \frac{R^*}{M} = \frac{8.314462[\mathrm{J\ (mol)^{-1}\ K^{-1})}]}{28.9647 \times 10^{-3}[\mathrm{kg\ mol^{-1}}]} = 287.055\ [\mathrm{J\ (kg)^{-1}\ K^{-1})}]. \tag{7.21}$$

In terms of the specific gas constant, the ideal gas law can be written as

$$pV = nMRT, \tag{7.22}$$

or

$$\frac{nM}{V} = \frac{p}{RT}. \tag{7.23}$$

Here,

$$\rho = \frac{nM}{V} \tag{7.24}$$

is the density for a specific gas with an SI unit, such as $[\mathrm{kg\ m^{-3}}]$.

Thus, the gas law in terms of specific gas constant R and SI density ρ can be written as

$$\rho = \frac{p}{TR}. \tag{7.25}$$

Substituting this relation into Eq. (7.17) yields

$$dp = -\frac{p}{TR} g dz, \tag{7.26}$$

which can be written as

$$\frac{1}{p} dp = -F dz, \tag{7.27}$$

with

$$F = \frac{g}{TR} \tag{7.28}$$

as a new quantity whose dimension is L^{-1} and varies with respect to elevation z as well as latitude, longitude, and time since the air temperature T [K] does. At altitudes less than about 9 km, g varies by less than about 0.3% and may be regarded as a constant. For a very high elevation, g's variation with respect to z should also be considered. The quantity

$$H = \frac{1}{F} = \frac{TR}{g} \tag{7.29}$$

is called the scale height and has a unit [m] if R is expressed using the unit [J (kg K)$^{-1}$]. When using the Earth's surface air temperature data to calculate the H field, one obtains a spatial field roughly corresponding to the geopotential height field of 350 mb, higher in the tropics and lower in the polar regions.

The scale height H is the height of a constant-density (also called homogeneous) atmosphere. If we integrate the hydrostatic equation (7.17), assuming a constant density, the resulting atmosphere has a finite depth with a top at $z = H$.

Also, the scale height H is the e-folding height for pressure in a constant-temperature (also called isothermal) atmosphere. As we shall see below, if we integrate the hydrostatic equation (7.27) with $T = const$, we obtain the solution

$$p = p_0 \exp(-z/H), \tag{7.30}$$

where p_0 is the pressure at the surface where $z = 0$. The e-folding value of p occurs when $z = H$.

Radiosonde data from the NOAA/ESRL Radiosonde Database https://ruc.noaa.gov/raobs/ show that F as a function of elevation z varies slowly and even monotonically within the troposphere, as shown in Fig. 7.3 for Corpus Christi, Texas, United States. The radiosonde stations in other locations around the world show similar properties: F slowly varies with respect to z. See Free et al. (2005) for the radiosonde data and the atmospheric temperature profile with respect to elevation and pressure.

For a given location on Earth and a given time, F is only a function of z. We integrate both sides of the equation (7.27) from elevation z_1, corresponding to a pressure p_1, to another elevation z_2, corresponding to a pressure p_2:

$$\int_{p_1}^{p_2} \frac{1}{p} dp = \int_{z_1}^{z_2} -F(z) dz. \tag{7.31}$$

The antiderivative of $1/p$ is $\ln p$. Thus, the left-hand side is

$$\int_{p_1}^{p_2} \frac{1}{p} dp = \ln p_2 - \ln p_1 = \ln(p_2/p_1). \quad (7.32)$$

The above two expressions imply that

$$\ln(p_2/p_1) = -\int_{z_1}^{z_2} F(z)dz, \quad (7.33)$$

or

$$p_2 = p_1 \exp\left(-\int_{z_1}^{z_2} F(z)dz\right). \quad (7.34)$$

This is the general equation for pressure under the hydrostatic balance assumption.

An analytic solution of the integral on the right-hand side of the above equation is likely to be impossible since we do not even know the functional expression for $F(z)$. However, it is known that $F(z)$ varies slowly with respect to z. Using the geometric meaning of an integral, the following integral

$$\int_{z_1}^{z_2} F(z)dz \quad (7.35)$$

is equal to the area underneath the red $F(z)$ curve in Fig. 7.3 and above the z-axis between z_1 and z_2. Because of the slow variation of $F(z)$, and its monotonic variation within the troposphere, the area can be approximated accurately by the area of the blue trapezoid

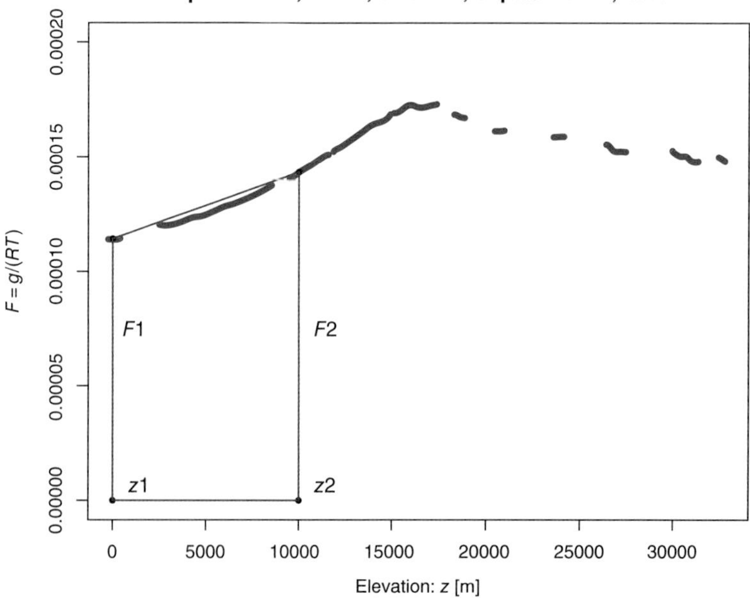

Fig. 7.3 The quantity F [m^{-1}] as a function of elevation z for Corpus Christi, Texas, United States.

which is

$$A_{trapezoid} = \frac{F_1 + F_2}{2}(z_2 - z_1), \tag{7.36}$$

where $F_1 = F(z_1)$ and $F_2 = F(z_2)$.

With this approximation, Eq. (7.31) becomes

$$\ln(p_2/p_1) = -\frac{F_1 + F_2}{2}(z_2 - z_1), \tag{7.37}$$

or

$$\ln(p_2/p_1) = -\bar{F}(z_2 - z_1), \tag{7.38}$$

where

$$\bar{F} = \frac{F_1 + F_2}{2} \tag{7.39}$$

is the average of F_1 and F_2.

Solving this equation for pressure at elevation z_2, we have

$$p_2 = p_1 \exp(-\bar{F}(z_2 - z_1)). \tag{7.40}$$

Equation (7.40) shows the atmospheric pressure decays exponentially with respect to elevation in the atmosphere.

One can also solve Eq. (7.38) for elevation z_2 with given pressure data:

$$z_2 = z_1 + \frac{1}{\bar{F}} \ln(p_1/p_2). \tag{7.41}$$

In this step, we have used the property of a logarithmic function $-\ln(p_2/p_1) = \ln(p_1/p_2)$.

When a very high elevation is considered and when the temperature lapse rate changes sign, the above trapezoidal rule of integration needs to be broken into different sections. Within each section, the temperature has only a monotonic change with respect to elevation. In this case, the gravitational constant g's variation with respect to elevation must be considered

$$g_2 = g_1 \left(\frac{6,400,000 + z_1}{6,400,000 + z_2} \right)^2, \tag{7.42}$$

where $6,400,000$ [m] is the approximate radius of Earth. This formula is derived from Newton's law of universal gravitation

$$g = G\frac{M}{r^2} \tag{7.43}$$

where r is the distance between the point of elevation z and the Earth's center, $M = 5.9722 \times 10^{24}$ [kg] is the mass of Earth, and $G = 6.67384 \times 10^{-11}$ [m^3 (kg s^2)$^{-1}$] is the universal gravitational constant. Do not confuse G with the Earth's gravitational acceleration g [m s^{-2}]. The above law implies that

$$gr^2 = GM \tag{7.44}$$

is constant; hence

$$g_1 r_1^2 = g_2 r_2^2 \tag{7.45}$$

for the two points of different elevations z_1 and z_2 with $r_1 \approx 6{,}400{,}000 + z_1$ and $r_2 \approx 6{,}400{,}000 + z_2$.

Substituting $F = g/(TR)$ into Eq. (7.41) and making some algebraic simplification, we obtain the following

$$z_2 = z_1 + \frac{2R_1R_2T_1T_2}{g_1R_2T_2 + g_2R_1T_1} \ln\left(\frac{p_1}{p_2}\right), \tag{7.46}$$

where g_i [m s^{-2}], R_i [J (kg K)$^{-1}$], and T_i [K] are the gravitational constant, gas constant, and temperature at elevation $z_i (i = 1, 2)$. This is the general hypsometric equation, which can be used to calculate elevation z_2 when data z_1, p_1, p_2, T_1, T_2, g_1, g_2, R_1, and R_2 are given. The hypsometric equation, also known as the thickness equation, relates an atmospheric pressure ratio to the equivalent thickness of an atmospheric layer.

However, z_2 is unknown. Thus, Eq. (7.46) for z_2 is nonlinear. One can use an iterative method to solve this equation. Continue the same cycle until reaching a point when successive z_2 values change only very little. Newton's method described in Chapter 6 is an iterative method and can be used to solve this equation.

Within the troposphere, the gravitational constant g changes very little, less than 0.4%. The atmosphere is well-mixed within the troposphere, and hence the gas constant R also changes very little in the troposphere. With the possible difficulty that the water vapor content of the air, and thus the value of R, may change, we thus can assume that R and g are constant in the troposphere. At very high altitudes in the troposphere, because the temperature there is very low compared to the surface temperature, the water vapor content is also very small, because it is limited by the saturation vapor pressure being a strong monotonic function of temperature, a relationship known as the Clausius–Clapeyron equation (Curry and Webster 1999). In brief, because the troposphere is cold at high altitudes, it is also dry at high altitudes, so R is nearly constant at high altitudes. Then, the hypsometric equation for the troposphere is reduced to

$$z_2 = z_1 + \frac{2RT_1T_2}{g(T_1 + T_2)} \ln\left(\frac{p_1}{p_2}\right). \tag{7.47}$$

This is why in the pre-GPS years a pilot or a mountaineer could use a barometer (i.e., an air pressure gauge) and a thermometer to approximately determine his elevation. This is significant since a barometer measurement for pressure and thermometer measurement for temperature are very easy to obtain and are very accurate. In contrast, a direct survey of the elevation of a mountain peak by measuring height directly could take days or months of hard work or might even be dangerous or effectively impossible in the case of some hostile mountain environments.

The hypsometric equation holds under two assumptions:

(a) the hydrostatic approximation is valid for the real atmosphere, and
(b) the ideal gas law with constant R is a good approximation to the real atmosphere.

These two conditions can be approximately valid when the atmosphere is relatively calm and dry. The time period of early morning calm may provide excellent conditions for the hypsometric equation. Several application examples, together with their atmospheric conditions, will be presented below.

7.2.2 An Application of the Hypsometric Equation: Calculate the Elevation of Mount Mitchell

Mount Mitchell in North Carolina has a peak elevation of 6,684 ft (2,037 m) and is the highest point in the eastern United States. Professor Elisha Mitchell of the University of North Carolina conducted several expeditions beginning in 1835 to measure the height of this mountain. One of his calculation methods was to use the hypsometric equation. He used barometer and thermometer readings and calculated the peak elevation to be 6,476 ft, not far from the correct modern value. The mountain was named after him. Mitchell fell to his death on the mountain in 1857, having returned to verify his earlier measurements.

We have repeated Elisha Mitchell's calculation using modern observational data of atmospheric pressure and temperature taken at two stations at 9:30 a.m., March 3, 2018. The observed data were from the North Carolina Climate Office http://climate.ncsu.edu/.

- Station MITC: Mount Mitchell State Park Station: Location (35.7585°N, 82.2712°W); Elevation: 6,200 feet above sea level; Temperature −6.1°C; Pressure 810.8 mb.
- Station BURN: Burnville Tower Station: Location (35.9189°N, 82.2604°W); Elevation: 2,702 feet above sea level; Temperature 0.7°C; Pressure 929.9 mb.

The nearby lower elevation Station BURN is only 18 km away and is used as the base location. Station MITC is the target location to be calculated. The hypsometric equation (7.47) yields the Station MITC's elevation as

$$
\begin{aligned}
z_2 &= z_1 + \frac{2RT_1T_2}{g(T_1+T_2)} \ln\left(\frac{p_1}{p_2}\right) \\
&= 2,702 \times 0.3048 \\
&\quad + \frac{2 \times 287.055 \times (273.15+0.7)(273.15+(-6.1))}{9.80665 \times ((273.15+0.7)+(273.15+(-6.1)))} \times \ln\left(\frac{929.9}{810.8}\right) \\
&= 1,908.395\ [m],
\end{aligned}
\tag{7.48}
$$

or $1,908.395/0.3048 = 6,261$ [ft], only 61 feet different from the correct value 6,200 feet, about only 1% error. This error might be caused by the strong wind of 26 miles per hour (mph), which may indicate an invalid assumption of hydrostatic equilibrium, due to inevitable vertical acceleration when a strong wind was blowing over a high mountain. The relative humidity at MITC station was 24% and that at BURN station was 36%, which are relatively dry in both locations and make the ideal gas law a very good approximation. The observational data may also have some instrumental errors.

The gas constant used here is that for dry air 287.055 [J kg^{-1} K^{-1}]. The gas constant's unit was converted to [J kg^{-1} K^{-1}] for unit consistency in the hypsometric equation.

We made the same calculation for a calm condition at 8:00 p.m., March 4, 2018. At this time, at the MITC station, the wind speed was only 7 mph and the relative humidity was 33%. At the BURN station, the wind speed was 3 mph and relative humidity was 49%.

The observed data were $p_1 = 926.9$ [mb], $p_2 = 812.0$ [mb], $T_1 = 3.4\,^\circ\mathrm{C}, T_2 = 1.1\,^\circ\mathrm{C}$. Because the hydrostatic and ideal gas assumptions were better satisfied, a more accurate result is expected. As expected, the general hypsometric equation yields a very accurate solution of $z_2 = 6,202$ feet, almost equal to the true value of 6,200 feet. This level of accuracy is truly remarkable! However, we cannot rule out that some sources of error may have canceled each other.

We tested another type of weather condition: a moderate wind speed. This was 1:00 p.m., March 5, 2018 when the wind speed at MITC was 14 mph and relative humidity was 31%. The observed data were $p_1 = 925.9$ [mb], $p_2 = 810.8$ [mb], $T_1 = -2.3\,^\circ\mathrm{C}, T_2 = 1.8\,^\circ\mathrm{C}$. The general hypsometric equation yields $z_2 = 6,180.773$ feet, again very close to the true value 6,200 feet with a difference of only 19 ft. This implies a relative error of only 0.3%!

We have tested several other sets of data. All the results indicate small errors and show that a slower wind speed, i.e., a nearly calm atmosphere, yields a more accurate elevation result, while the relative humidity does not have much influence on the accuracy.

7.2.3 Hypsometric Equation for an Isothermal Layer

Many textbooks derive the hypsometric equation based on the isothermal assumption, which means that the atmospheric layer under consideration is assumed to have the same temperature everywhere in the layer. The derivation we have given does not require this assumption and is thus more general. Our result without the isothermal assumption is not much more complicated than that with the assumption. However, in the unit of K, the isothermal assumption is a reasonable approximation, because the temperature below the stratopause (about 50 km above sea level) typically varies between 210 K and 310 K. The relative variation is $(310 - 210)/[(210 + 310)/2] = 38\%$ which may be considered "small." Further, in many applications at less than 6 km elevation, the temperature is often between 260 K and 285 K. The relative variation is even smaller and is 9%. Thus, the isothermal assumption can still yield reasonably good results, which, however, are certainly not as accurate as the general hypsometric equation (7.46) or (7.47). Section 7.2.4 will quantify the errors due to the isothermal assumption.

In an isothermal layer with the same temperature $T = T_1 = T_2$, the hypsometric equation (7.47) is reduced to

$$z_2 = z_1 + \frac{RT}{g} \ln\left(\frac{p_1}{p_2}\right). \tag{7.49}$$

Realizing that the isothermal condition is unrealistic, because temperature does have an apparent change with respect to elevation, some textbooks then replace the isothermal temperature T by the average of the temperatures at z_1 and z_2

$$T = \frac{T_1 + T_2}{2} \tag{7.50}$$

and obtain a good estimate of the thickness of the layer of atmosphere $z_2 - z_1$ within the troposphere.

We claim using the average temperature to replace the isothermal condition is equivalent to using a linear approximation of the hypsometric equation (7.47). The linear approximation of this type is a method we described in the last chapter. It uses the tangent line to approximate a curve. Our claim is mathematically justifiable, as shown below.

We regard the temperature at z_i as a "small" perturbation S_i from the base temperature $T_0 = 273.15$ K:

$$T_i = T_0 + S_i = T_0(1 + r_i) \tag{7.51}$$

where $r_i = S_i/T_0$ have small absolute values for many applications in the troposphere $(i = 1, 2)$. According to Fig. 6.6, the Earth's atmosphere temperature S_i is between -100 and 80°C, which yields

$$-0.37 < r_i < 0.29. \tag{7.52}$$

The temperature factor in the hypsometric equation (7.47) can be written in the following ways

$$\begin{aligned}
\frac{2T_1T_2}{T_1+T_2} &= T_0\frac{(1+r_1)(1+r_2)}{1+(r_1+r_2)/2} \\
&\approx T_0(1+r_1+r_2+r_1r_2)\times[1-(r_1+r_2)/2] \\
&\qquad \text{(Drop off the high-order terms in the second factor)} \\
&\approx T_0[1+(r_1+r_2)/2] \quad \text{(Drop off the high-order terms in the first factor)} \\
&= T_0(1/2+1/2+r_1/2+r_2/2) \\
&= \frac{T_0(1+r_1)+T_0(1+r_2)}{2} \\
&= \frac{T_1+T_2}{2}.
\end{aligned} \tag{7.53}$$

In the second step above, we used the following linear approximation

$$\frac{1}{1+x} \approx 1 - x \quad \text{when } |x| << 1. \tag{7.54}$$

The higher order terms $x^n (n \geq 2)$ are omitted.

For Mount Mitchell data at 9:30 a.m., March 3, 2018, the result computed from the hypsometric equation with the average temperature and that from the original general equation have a negligible difference of less than one foot. This means that omission of higher-order terms does not affect the results of the hypsometric equation at lower elevations, particularly below 5 km. Two examples are presented below.

Example 7.1 We can calculate the thickness of a given layer of atmosphere with an average temperature $T = 7$°C, bottom pressure $p_1 = 850$ [mb], and top pressure $p_2 = 700$ [mb]. Then the hypsometric equation yields the thickness $\Delta z = z_2 - z_1 = 1{,}591$ [m]. This might be regarded as a reasonable estimate of cloud thickness for a cloudy summer day somewhere over the southern United States.

Example 7.2 One can use the hypsometric equation to calculate the thickness of a layer of atmosphere. Suppose that (i) the average temperature is 10°C, (ii) the layer's bottom is at sea level with $z_1 = 0$ and $p_1 = 1{,}009$ [mb], and (iii) the layer's top pressure is $p_2 = 500$ [mb]. Calculate the layer's top elevation z_2:

$$z_2 = z_1 - \frac{TR}{g}\ln(p_2/p_1) = 0 - \frac{283 \times 286.9}{9.8}\ln(500/1{,}009)\ [\mathrm{m}] = 5{,}817\ [\mathrm{m}]. \tag{7.55}$$

This result might be a reasonable estimation for a real atmosphere over the southern United States in summer. The height is around 5,800 meters at 500 mb. See Fig. 7.2 for the observational data of geopotential height of the 500 mb pressure surface in a three-month period from August through October of 2013.

Two other heights which may be significant are 10,000 meters (about the cruising level of jet airliners) at 250 mb, and 1,500 meters (a typical cloud and rain level) at 850 mb. They are also supported by the hypsometric equation when the temperature is again taken to be the average temperature in the entire layer of atmosphere being considered.

7.2.4 Error Estimate of the Linear Approximation to the Hypsometric Equation

This subsection finds the second-order approximation to the hypsometric equation, and also estimates the errors of the linear approximation. We can thus explain why the average temperature yields an elevation error less than one foot for the Mount Mitchell application of the hypsometric equation. In fact, the error of the linear approximation in the previous subsection can be calculated exactly. Although being able to calculate the error of such an approximation is rare in science, it works out beautifully here.

The error of the linear approximation $(T_1+T_2)/2$ to $2T_1T_2/(T_1+T_2)$ in the previous subsection is defined as their difference

$$\begin{aligned} E(r_1,r_2) &= \frac{2T_1T_2}{T_1+T_2} - \frac{T_1+T_2}{2} \\ &= T_0\frac{(1+r_1)(1+r_2)}{1+(r_1+r_2)/2} - T_0\left[1+\frac{r_1+r_2}{2}\right]. \end{aligned} \tag{7.56}$$

Several steps of algebraic simplification lead to the following tidy expression of the linear approximation error

$$E(r_1,r_2) = -\frac{(r_1-r_2)^2}{2(2+r_1+r_2)}T_0. \tag{7.57}$$

Consequently, the elevation error due to the linear approximation is

$$\delta z = -\frac{RT_0}{g}\frac{(r_1-r_2)^2}{2(2+r_1+r_2)}\ln\left(\frac{p_1}{p_2}\right). \tag{7.58}$$

For the Mount Mitchell data at 9:30 a.m., March 3, 2018, this formula yields $\delta z = -0.563$ feet. This is exactly the error due to the approximation by an average temperature: the result from the general hypsometric equation (i.e., 6,261.137 feet) minus the average temperature approximation (the result from the above subsection: 6,261.700 feet). The difference is 0.563 feet. The average temperature approximation over-estimates the elevation of the Mount Mitchell weather station in the North Carolina ECONet.

Since $-0.37 < r_i < 0.29$ according to formula (7.52), the error $E(r_1, r_2) \leq 0$ following Eq. (7.57) because $2 + r_1 + r_2 > 0$. This means that the approximation by the average temperature always over-estimates elevation z_2.

The error formula (7.57) can be explained as the second-order approximation to $2T_1T_2/(T_1 + T_2)$, when we interpret formula (7.57) in the following way

$$E_2 = -\frac{(r_1 - r_2)^2}{4} T_0. \tag{7.59}$$

Here, we have omitted the third- or higher-order quantities of r_1 and r_2 in $E(r_1, r_2)$.

This formula can be expressed in terms of the temperature data in K:

$$E_2 = -\frac{(T_1 - T_2)^2}{4T_0}. \tag{7.60}$$

Thus, the error is proportional to the square difference of the temperatures at the two elevations. When T_2 is much higher than T_1, the average temperature approximation can lead to sizable errors.

7.2.5 Applications of Geopotential Height in Radiosonde Measurements

Besides the error due to the temperature approximation in the hypsometric equation, the variation of gravitational acceleration $g(\phi, \theta, z)$ yields errors if g is treated as a constant. This error can be eliminated if we use the conversion between the geometric height z and the geopotential height Z, as discussed in the last section. The hydrostatic balance equation (7.15) $\partial p/\partial Z = -g_0\rho$ and the ideal gas law (7.25) $\rho = p/(RT)$ imply that

$$\frac{\partial p}{\partial Z} = -g_0 \frac{p}{TR}. \tag{7.61}$$

Then, all the derivations for the general hypsometric equation are still valid when replacing the vertical coordinate z by Z and g by g_0. The general equation (7.34) for pressure is now as follows

$$p_2 = p_1 \exp\left(- \int_{Z_1}^{Z_2} F_0(Z) dZ \right), \tag{7.62}$$

where

$$F_0 = \frac{g_0}{RT}. \tag{7.63}$$

These equations in the geopotential height Z coordinate have eliminated the variability of the gravitational acceleration $g(\phi, \theta, z)$ and retained only the variation of temperature $T(\phi, \theta, Z, t)$. This mathematical result is important not only for climate model equations, but also for climate observations, such as radiosonde measurements.

Radiosonde balloons can reach altitudes as high as 20 to 40 km. In the upper atmosphere, the variation of the gravitational acceleration $g(\phi, \theta, z)$ is significant for radiosonde measurements of meteorological parameters, such as pressure, temperature, and humidity. Equation (7.62) provides a way to accurately measure the atmospheric pressure p_2 in the upper atmosphere in the modern era when the geometric height z can be accurately determined by GPS satellites. The radiosonde sensors will measure temperature

$T(\phi,\theta,z,t)$. GPS can measure the geometric height z from the surface to z_2, which can be converted into the geopotential height Z. Based on these data, a numerical method can be used to compute the integral in Eq. (7.62), which determines p_2. This computed p_2 can be compared with the pressure data measured by the radiosonde pressure sensors, which have observational errors mainly due to environmental changes, such as the variations of temperature, humidity, wind, and solar radiation at different elevations from the surface to a height of 20–40 km. The pressure error for a modern sensor is within the range of ± 0.4 hPa (i.e., 0.4 mb). The GPS-based radiosonde since 2000 has generally reduced the error range, particularly in the stratosphere. Further information is readily available on the GPS-based radiosonde instruments, such as the Vaisala Radiosonde RS41 introduced in 2013 (see website: www.vaisala.com).

If the integral in Eq. (7.62) is computed by the mean value theorem of calculus (see the theorem in Appendix D)

$$\int_{Z_1}^{Z_2} F_0(Z)dZ = \frac{g_0}{RT_c}(Z_2 - Z_1), \tag{7.64}$$

where $1/T_c$ is the mean value of the function $1/T(\phi,\theta,Z,t)$ in the interval (Z_1,Z_2), then the general hypsometric equations (7.40) and (7.41) become

$$p_2 = p_1 \exp\left(-\frac{g_0}{RT_c}(Z_2 - Z_1)\right), \tag{7.65}$$

$$Z_2 = Z_1 + \frac{RT_c}{g_0} \ln\left(\frac{p_1}{p_2}\right). \tag{7.66}$$

However, it was difficult, if not impossible, to estimate the mean value T_c in the pre-GPS era, because the functional profile $T(\phi,\theta,Z,t)$ vs. Z was unknown.

Another approximation is to estimate the integral in Eq. (7.62) by the trapezoidal rule using only two points Z_1 and Z_2. Then, the general hypsometric equations (7.40) and (7.41) become

$$p_2 = p_1 \exp(-\bar{F}_0(Z_2 - Z_1)), \tag{7.67}$$

$$Z_2 = Z_1 + \frac{1}{\bar{F}_0} \ln\left(\frac{p_1}{p_2}\right), \tag{7.68}$$

where

$$\bar{F}_0 = \frac{g_0(T_1 + T_2)}{2RT_1T_2}. \tag{7.69}$$

In this case, one can use Equations (7.68) and (7.69) to compute the geopotential height Z_2 when the pressure data p_1 and p_2, the temperature data T_1 and T_2, and the initial geopotential height Z_1 are given. The resulting Z_2 can be converted into the geometric height z_2. This theory could be used in the pre-GPS era to determine the height of a radiosonde balloon.

This approach and the method presented earlier in this section yield almost no difference in results in the troposphere, because of the small variation of $g(\phi,\theta,z)$. This difference becomes significant in the stratosphere.

7.3 Work Done by an Air Mass in Expansion

Suppose a parcel consisting of air with a volume V_1 is expanded to a volume V_2. How can one express the work done by the air parcel on its surroundings, or on its environment?

Suppose that the air parcel has a cylindrical shape with a cross-sectional area A and the expansion occurs through an extension of the cylinder's length. When the cylinder is expanded by a small increment dx in length, the force is pA and the distance of the force is dx. Thus, the small amount of work done by the air parcel is

$$dW = pAdx. \tag{7.70}$$

Here Adx can be regarded as a small volume increment dV. Thus, the work is

$$dW = p(V)dV, \tag{7.71}$$

where pressure is a function of volume. The total amount of work done by the air parcel on its surrounding environment when the volume is increased from V_1 to V_2 is

$$W = \int_{V_1}^{V_2} pdV \tag{7.72}$$

If compression occurs, i.e., $V_2 < V_1$, then the work is negative, meaning the environment has done work on the air parcel. The sign convention here is arbitrary, so it does not matter if we define work done on the air parcel by the environment positive or negative, and correspondingly work done by the parcel on the environment negative or positive, or if we reverse these signs, so long as we choose one sign convention or the other and always use the chosen sign convention consistently.

When a parcel or system goes through a cycle of expanding from V_1 to $V_2 > V_1$ via a pressure path $p_{12}(V)$ and compressing from V_2 to V_1 via another path $p_{21}(V)$ (see Fig. 7.4), the work done by the parcel or system on its environment may not be zero when the two paths are not the same:

$$\Delta W = \int_{V_1}^{V_2} p_{12}(V)dV + \int_{V_2}^{V_1} p_{21}(V)dV = \oint p(V)dV \neq 0. \tag{7.73}$$

Here, the contour interval symbol $\oint$ denotes an integral along a closed path (also called closed contour). A consequence of this dependence of work on the path of the expansion or compression is that the p–V relationship is not an *exact differential*, because the increment of work pdV cannot be found simply by the differential of a function of the state of a system. The work done by the system on its surroundings depends on not only the initial and final states of the process, but also the path in the p–V plane. The Carnot heat engine is an example of this kind of closed contour integral. Nicolas Léonard Sadi Carnot (1796–1832) was a French military engineer and physicist and is often called the "father of thermodynamics."

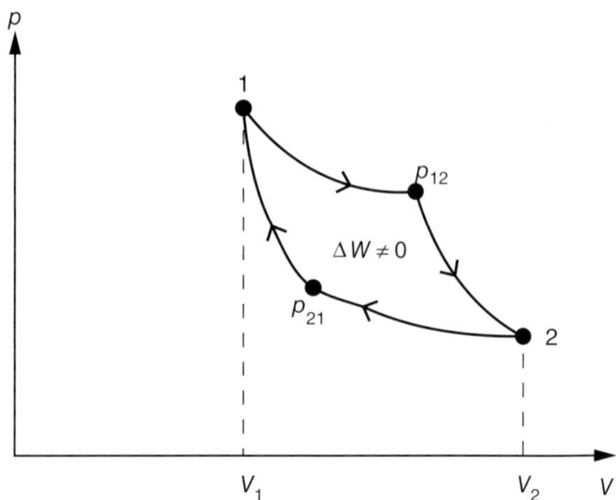

Fig. 7.4 An integral along a closed path in a p–V system may result in a non-zero work on the system's surroundings.

7.4 Internal Energy, Enthalpy, and Entropy

7.4.1 Internal Energy and Enthalpy

When it is heated from temperature T_1 to T_2, a system consisting of dry air gains energy. This gain in energy may take different forms. It may increase the temperature (molecular speeds) of the air, and/or it may increase the internal motions (vibration and rotation) of the air molecules. If the system consisted of water instead of dry air, some or all of the energy added might go to changing the state of the system, such as from liquid water to water vapor, by decreasing the forces attracting the water molecules to one another. All of these changes are examples of changes to the *internal energy* of the system. The change or increment in internal energy dU will increase if an increment of heat dQ is added to the system or if an increment of work dW is done on the system. This is a conservation law, expressed mathematically as

$$dU = dQ + dW. \tag{7.74}$$

This law is a form of the first law of thermodynamics. A way to state the first law, due to Clausius in 1850, is "In a thermodynamic process involving a closed system, the increment in the internal energy is equal to the difference between the heat accumulated by the system and the work done by it." Rudolf Clausius (1822–1888) was a German mathematician and physicist who was another important founder of the science of thermodynamics.

If the system's volume is unchanged, the gain is the change in the system's internal energy from u_1 to u_2. It can be shown that the change in internal energy Δu is an integration of specific heat capacity with respect to temperature:

$$\Delta u = u_2 - u_1 = \int_{T_1}^{T_2} c_v(T)dT, \tag{7.75}$$

where c_v is the specific heat capacity of dry air at the state of constant volume, and the unit of internal energy is [joule/unit mass].

When the system's pressure is unchanged, the gain is enthalpy from h_1 to h_2. Here, enthalpy for the system is defined by

$$H = U + pV, \tag{7.76}$$

where U is the internal energy, p pressure, and V is volume. H has the same unit as U. Enthalpy includes both internal energy gain and the work done on the system, because the fact that the volume changes at constant pressure means work is done on the system. It can be shown that this change in enthalpy per unit mass Δh is also an integration of specific heat capacity with respect to temperature:

$$\Delta h = h_2 - h_1 = \int_{T_1}^{T_2} c_p(T)dT \tag{7.77}$$

where c_p is the specific heat capacity of dry air at the state of constant pressure, and the unit of enthalpy is also [joule/unit mass].

Intuition suggests that for the same increment of heat, the constant pressure process (7.77) is more efficient to allow absorption of more heat energy into the system due to the volume expansion, while the constant volume process (7.75) is comparably less efficient. In other words, some of the heat added in a constant-pressure process will be used to do work on the environment, but in a constant-volume process, all of the heat added to the system will be used in increasing the temperature of the air comprising the system. Thus,

$$c_p(T) \geq c_v(T). \tag{7.78}$$

The difference between the specific heat at constant volume and the specific heat at constant pressure is an important property of the gas in the system. For an ideal gas, it can be shown that this difference is the specific gas constant, denoted by R, which is the same constant that occurs in the ideal gas law (7.18) and Eq. (7.21):

$$R = c_p - c_v. \tag{7.79}$$

This constant R does not change with temperature T and is an intrinsic property of the gas. Different ideal gases have different R values, which is why this R is called the specific gas constant. The gas constant with a unit $[\mathrm{J\,K^{-1}(kg)^{-1}}]$ for some gases in the atmosphere is listed below:

- Ordinary air: $R = 287$
- Nitrogen gas: $R = 297$
- Oxygen gas: $R = 260$
- Water vapor: $R = 462$
- Carbon dioxide gas: $R = 189$
- Methane gas: $R = 310$.

It can be shown that the specific gas constant is a universal constant divided by the molecular weight of the gas. The universal constant is the product of Avogadro's number and the Boltzmann constant. The appropriate molecular weight for a mixture of gases, such as dry air, is found by using Dalton's law, which states that the total pressure exerted by the mixture is the sum of the partial pressures that would be exerted by each of the gases in the mixture individually, if they each filled the entire volume at the temperature of the mixture of gases. For a more complete discussion of these aspects of thermodynamics, the reader may consult the references cited at the beginning of this chapter.

For an ideal gas, the equation of state or ideal gas law may also be written as

$$pV = nR^*T \tag{7.80}$$

where the universal gas constant $R^* = 8.3145$ [J mol^{-1}K^{-1}], and n [unit: moles] denotes the amount of the gas, in terms of the number of moles of the gas. Because the molecular weight of dry air is 0.02896 [kg mol^{-1}], we can compute the specific gas constant for ordinary dry air, just as we did in Eq. (7.21):

$$R = \frac{8.3145\ [\text{J mol}^{-1}\ \text{K}^{-1}]}{0.02896\ [\text{kg mol}^{-1}]} = 287\ [\text{J K}^{-1}\ \text{kg}^{-1}]. \tag{7.81}$$

This is the value listed on p. 189.

The heat capacity most commonly referred to is the specific heat capacity at constant volume. Water has a much larger heat capacity than air. Per unit mass, the heat capacity of water is 4.18 [J g^{-1} K^{-1}], about four times that of dry air 1.01 [J g^{-1} K^{-1}]. The total amount of water on the Earth's surface is about $1{,}350 \times 10^{18}$ [kg], which is about 260 times the mass of all the air in the Earth's atmosphere, about 5.1×10^{18} [kg]. Thus to warm all of the air in the atmosphere uniformly by 1°C would require $5{,}151 \times 10^{18}$ [J] of heat. To warm all the water in the world by the same 1°C would require about 1,000 times more heat, $10^3 \times 5{,}643 \times 10^{18}$ [J]. This is thus much more heat than would be needed to warm the global atmosphere by the same temperature increment.

Enthalpy is a very useful parameter in thermodynamics that measures both internal energy and the work done to a system. The differential of enthalpy dH denotes a small increment of H when all the relevant independent variables are also subject to a small change. This statement is denoted by the differential of Eq. (7.76):

$$dH = dU + pdV + Vdp. \tag{7.82}$$

Note that the differential of a product $d(uv) = udv + vdu$ is a sum of two differentials, each being the increment in one parameter while another is fixed. This relationship is called the product rule of differentiation, or the product rule of derivatives.

The first two terms on the right-hand side are the total energy increment of the system: the internal energy increment dU plus the mechanical energy pdV. These energies both come from the heat dQ put into the system. Thus, another form of the first law of thermodynamics is

$$dH = dQ + Vdp. \tag{7.83}$$

A system may expand its volume from V_1 to V_2 in many ways. An important way is for the temperature of the system to remain unchanged. Such a process is called an isothermal process. Another important way is for no heat to be added to or removed from the system. This kind of process is called an adiabatic process. A Carnot heat engine includes both isothermal and adiabatic processes, and forms a closed path in the p–V plane. These processes can lead to many kinds of relationships among p, V, and T which can be expressed in terms of integrals and differentials. More details on these thermodynamic processes can be found in the Further Readings listed at the end of this chapter.

7.4.2 Entropy

Entropy increase for a gas is defined by a reversible change from one state A to another state B

$$\Delta S_{AB} = \int_A^B \frac{dq}{T}, \tag{7.84}$$

where T is temperature and $dq = c_p dT - v dp$ from the first law of thermodynamics with c_p being the specific heat under a constant pressure, v the volume of unit mass, and p pressure.

The second law of thermodynamics states that in any cyclic process the entropy will either increase or remain the same. It will not decrease. The mathematical expression of this statement is

$$\oint \frac{dq}{T} \geq 0, \tag{7.85}$$

where $\oint$ denotes an integral of a closed path. The closed path is a description of the cyclic process, which means a system goes from state A to B and then comes back to A.

7.5 Use of Integrals to Derive Stefan–Boltzmann's Blackbody Radiation Formula from Planck's Law of Radiation

In Chapter 5, we repeatedly used the Stefan–Boltzmann law of blackbody radiation

$$E_{bb} = \sigma T^4 \ [\text{unit} : \text{W m}^{-2}] \tag{7.86}$$

to quantify the amount of radiation energy emitted to space from the Earth. Here T in the unit of K is the Earth's global average surface temperature, and σ is the Stefan–Boltzmann constant which is

$$\sigma = 5.670367 \times 10^{-8} \ [\text{W m}^{-2}\ \text{K}^{-4}]. \tag{7.87}$$

As an application of integration, the Stefan–Boltzmann law can be derived from Planck's law of radiation, which quantifies the radiation at different frequency emitted by a blackbody in thermal equilibrium at a given temperature. Max K. Planck (1858–1947) was a German theoretical physicist who won the 1918 Nobel Prize in Physics for his discovery of electromagnetic wave energy emission in quantized form. Planck proposed a law of wave energy radiation from a blackbody at given temperature and wavelength:

$$E(\lambda, T) = \frac{2hc^2}{\lambda^5} \times \frac{1}{\exp[hc/(k_b \lambda T)] - 1}, \tag{7.88}$$

where

- $h = 6.626070040(81) \times 10^{-34}$ [J sec] is the Planck constant,
- $k_b = 1.38064852(79) \times 10^{-23}$ [J K^{-1}] is the Boltzmann constant,
- $c = 300,000,000$ [m s^{-1}] is the approximate speed of light in vacuum,
- T [K] is temperature,
- λ [m] is wavelength, and
- $E(\lambda, T)$ is the spectral flux of the radiation power, and its SI unit is W m^{-2} m^{-1}.

The wavelength of the radiation from the Sun and the Earth is in the range of micrometers: λ [μm]. The spectral flux of the radiation power $E(\lambda, T)$ from the solar surface is in the range of kW m^{-2} nm^{-1}, and that from the Earth's surface is 10^{-6}kW m^{-2} nm^{-1}.

Figure 7.5 shows the plots of $E(\lambda, T)$ as a function of wavelength λ [μm] for given temperatures T. The figure shows that the peak radiation moves to the shorter wavelength zone (i.e., the higher frequency zone) as the temperature increases. This agrees with our experience and intuition. For example, a very high-temperature body, such as a burning arc welding rod (around 6,000°C), shows a bright purple color, which has a shorter wavelength (or higher frequency) than liquid iron that emits bright red light (around 1,200°C).

The figure shows that spectral power flux has a maximum for a given temperature. This maximum can be found by taking a derivative:

$$\frac{d}{d\lambda} E(\lambda, T) = 0. \tag{7.89}$$

This equation gives the wavelength of maximum emission denoted by λ_{max}:

$$\lambda_{max} \approx \frac{hc}{4.965114 \times k_b T}, \tag{7.90}$$

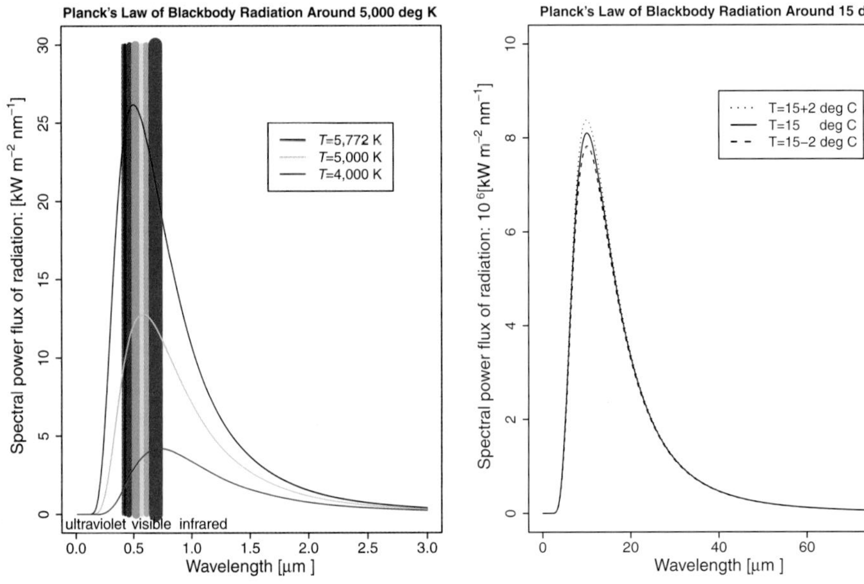

Fig. 7.5 Planck's law of blackbody radiation: Spectral power flux of radiation at (a) 5,000 K, and (b) 15°C.

or

$$\lambda_{max} \approx \frac{2,898}{T} \; [\mu\text{m}]. \tag{7.91}$$

The average temperature of the Sun's surface is approximately $T = 5{,}772$ K, so the wavelength of maximum emission is $\lambda_{max} = 0.50$ [μm] and is in the range of visible light. The average temperature of the Earth's surface is approximately $T = 15°$C, or 288 K; the wavelength of maximum emission is $\lambda_{max} = 10.06$ [μm] and is in the infrared range.

For a given T, the total amount of energy emitted by a blackbody throughout the entire range of wavelength is an integration of $E(\lambda, T)$ with respect to wavelength λ from zero wavelength to infinity:

$$E_T = \int_0^\infty E(\lambda, T)d\lambda = \int_0^\infty \frac{2hc^2}{\lambda^5} \times \frac{1}{\exp[hc/(k_b\lambda T)] - 1} d\lambda. \tag{7.92}$$

Wave frequency ν and wavelength λ are related by

$$\nu\lambda = c. \tag{7.93}$$

We can integrate by substitution with

$$\lambda = \frac{c}{\nu}, \quad d\lambda = -\frac{c}{\nu^2} d\nu \tag{7.94}$$

and with $\nu = \infty$ when $\lambda = 0$ and $\nu = 0$ when $\lambda = \infty$:

$$E_T = \int_\infty^0 \frac{2hc^2\nu^5}{c^5} \times \frac{1}{\exp[h\nu/(k_bT)] - 1} \left(-\frac{c}{\nu^2}\right) d\nu, \tag{7.95}$$

or

$$E_T = \int_0^\infty \frac{2h\nu^3}{c^2} \times \frac{1}{\exp[h\nu/(k_bT)] - 1} d\nu, \tag{7.96}$$

Further integration by substitution can be made to convert the above integral into a standard integral that can be found from an integration table.

Let

$$x = h\nu/(k_bT), \tag{7.97}$$

then

$$dx = hd\nu/(k_bT), \tag{7.98}$$

or

$$d\nu = (k_bT/h)dx. \tag{7.99}$$

Substitution of these into the above E_T formula yields

$$E_T = 2\frac{k_b^4T^4}{h^3c^2}I, \tag{7.100}$$

where I is a standard integral

$$I_3 = \int_0^\infty \frac{x^3}{e^x - 1} dx \tag{7.101}$$

whose value can be found from an integration table

$$I_n = \int_0^\infty \frac{x^{n-1}}{e^x - 1} dx = \Gamma(n)\zeta(n). \tag{7.102}$$

Here, $\Gamma(n)$ is called the gamma function which has a simple formula when n is a positive integer:

$$\Gamma(n) = (n-1)!; \tag{7.103}$$

and $\zeta(n)$ is called the Riemann zeta function, which also has simple formulas when n is a small positive integer, such as $n = 4$:[1]

$$\zeta(4) = \sum_{i=1}^{\infty} \frac{1}{i^4} = \frac{\pi^4}{90}. \tag{7.104}$$

Thus,

$$E_T = 2\frac{k_b^4 T^4}{h^3 c^2}\frac{\pi^4}{15} = \frac{2\pi^4 k_b^4}{15h^3c^2}T^4. \tag{7.105}$$

Its SI unit is [W m^{-2} sr^{-1}]. This power flux of radiation is per solid angle over a hemisphere covering the surface of the radiation source. Here, sr stands for steradian, which is a measure of a solid angle on a sphere. Similar to degree or radian for measuring an angle, steradian *sr* for measuring a solid angle is also dimensionless. A steradian can be defined as the solid angle subtended at the center of a unit sphere by a unit area on its surface. For a general sphere of radius r, any portion of its surface with area $A = r^2$ subtends one steradian. A solid angle thus measures the size of a cone with its vertex at the sphere's center and its top on the sphere, defined by the sphere's area inside the cone A and divided by the square of the sphere's radius r^2: $\Omega_s = A/r^2$ [sr]. Thus, the solid angle for the entire sphere is 4π [sr], that for a hemisphere is 2π [sr], and that for the Arctic Circle, defined by the parallel of latitude 67°N, is 0.5 [sr].

[1] One can derive this formula using integration by parts and sum of infinite series. Let us consider this integration by series

$$\begin{aligned} I &= \int_0^\infty \frac{x^3}{e^x - 1} dx \\ &= \int_0^\infty \frac{x^3 e^{-x}}{1 - e^{-x}} dx. \end{aligned} \tag{7.106}$$

Part of the integrand can be expanded into a convergent series

$$\frac{e^{-x}}{1 - e^{-x}} = \sum_{n=1}^{\infty} e^{-nx}. \tag{7.107}$$

Integration by parts for the following integral

$$\int_0^\infty x^3 e^{-nx} dx \tag{7.108}$$

leads to

$$\sum_{n=1}^{\infty} \int_0^\infty x^3 e^{-nx} dx = \sum_{n=1}^{\infty} \frac{6}{n^4} = \frac{\pi^4}{15}. \tag{7.109}$$

We used an online table for the commonly used infinite series in the last step to compute the sum.

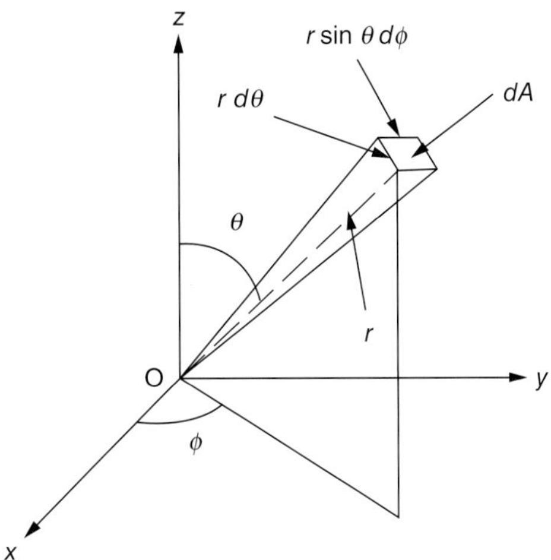

Fig. 7.6 Zenith angle θ, azimuth angle ϕ, and solid angle covered by the small area $dA = r^2 \sin\theta d\theta d\phi$ over a hemisphere. The power flux to space from the Earth's surface is in the z-direction. This figure is adapted from Figure 10.3 of Jacobson (1999).

For the Earth's radiation to space, we need the power flux through a spherical shell enclosing the Earth's surface. If a point on the hemisphere has a zenith angle θ and azimuth angle ϕ (see Fig. 7.6) then the Earth's radiation power flux per solid angle [sr] to space along the zenith axis is

$$E_T \cos\theta. \tag{7.110}$$

The total radiation power flux is an integration of $E_T \cos\theta$ with respect to the solid angle over a hemisphere:

$$E_{bb} = \int_0^{\pi/2} d\theta \int_0^{2\pi} d\phi \ \sin\theta E_T \cos\theta = \pi E_T, \tag{7.111}$$

or

$$E_{bb} = \sigma_1 T^4, \tag{7.112}$$

with

$$\sigma_1 = \frac{2\pi^5 k_b^4}{15h^3c^2}. \tag{7.113}$$

This formula can yield the theoretical value of the Stefan–Boltzmann constant

$$\sigma_1 = \frac{2\pi^5 k_b^4}{15h^3c^2} = 5.670373 \times 10^{-8} \ [\text{W m}^{-2} \text{ K}^{-4}]. \tag{7.114}$$

which can be computed from the parameter values given earlier.

7.6 Chapter Summary

Climate science uses calculus extensively. This chapter illustrates the use of integrals, while the previous chapter showed the use of derivatives. The main ideas and results of the integral applications we have discussed can be summarized as follows.

(i) The hydrostatic pressure of air is defined as the weight of the air per unit area above the point (x,y,z) at time t. Its mathematical expression is below (i.e., Eq. (7.10)):

$$p(x,y,z,t) = \int_z^\infty g(x,y,\xi)\rho(x,y,\xi,t)d\xi. \tag{7.115}$$

This hydrostatic pressure is often used to approximate the barometric pressure in climate science.

(ii) Geopotential height Z is an important concept in weather analysis, climate modeling, and instrumental design, such as the radiosonde. The geopotential height Z is defined as

$$Z = \frac{1}{g_0}\int_0^z g(\phi,\theta,z)dz. \tag{7.116}$$

The relationship between the geopotential height Z and the geometric height z allows the climate model equations to use Z as a coordinate with a constant gravitational acceleration g_0. For example, the pressure derivative with respect to Z is independent of g:

$$\frac{\partial p}{\partial Z} = -g_0\rho. \tag{7.117}$$

This equation can also help with the design of the GPS-based radiosonde. In the lowest few kilometers of the atmosphere, Z and z are numerically very close, but can have an approximate difference of 136 meters at the geometric height $z = 30$ [km].

(iii) In the troposphere, the scale height $(1/F)$ typically varies smoothly and monotonically with altitude z as shown in Fig. 7.3, and this fact, plus the hydrostatic equation and the ideal gas law, show that pressure decreases exponentially with altitude:

$$p_2 = p_1 \exp(-\bar{F}(z_2 - z_1)), \tag{7.118}$$

where

$$\bar{F} = 2\frac{g_1R_2T_2 + g_2R_1T_1}{R_1R_2T_1T_2}. \tag{7.119}$$

The inverse function of this equation gives the general hypsometric equation

$$z_2 = z_1 + \frac{2R_1R_2T_1T_2}{g_1R_2T_2 + g_2R_1T_1}\ln\left(\frac{p_1}{p_2}\right). \tag{7.120}$$

The hypsometric equation, also known as the thickness equation, relates an atmospheric pressure ratio to the equivalent thickness of an atmospheric layer.

(iv) The general hypsometric equation can accurately predict elevation, as in the example of Mount Mitchell. The accuracy may be affected by the accuracy of the temperature and pressure measurements, and also by the local weather conditions which may determine whether the hydrostatic assumption is justified.

(v) When temperature is changed from T_1 to T_2 for a system, the change of the system's internal energy with a fixed volume is

$$\Delta u = \int_{T_1}^{T_2} c_v(T)dT, \tag{7.121}$$

and that with a fixed pressure is

$$\Delta h = \int_{T_1}^{T_2} c_p(T)dT. \tag{7.122}$$

The fixed pressure condition provides a more efficient approach to increasing the internal energy: $c_p \geq c_v$. More specifically,

$$R = c_p - c_v. \tag{7.123}$$

(vi) Planck's law states that a body having a temperature above absolute zero always radiates energy and the maximum emission of the radiation is at a wavelength determined by the body's temperature. The mathematical expression of Planck's law is

$$E(\lambda, T) = \frac{2hc^2}{\lambda^5} \times \frac{1}{\exp[hc/(k_b \lambda T)] - 1}. \tag{7.124}$$

The SI unit of $E(\lambda, T)$ is W sr^{-1} m^{-3}.

The integration of this equation with respect to the wavelength leads to the Stefan–Boltzmann law of blackbody radiation:

$$E_{bb} = \sigma T^4 \text{ [unit: W m}^{-2}\text{]}. \tag{7.125}$$

The unit [W m^{-2}] is power flux.

The examples of this chapter show the use of integrals in two ways: (a) using an integral to define or express a new quantity, such as using an integral to define the geopotential and internal energy; and (b) using an integral to solve a differential equation, such as the derivation of the hypsometric equation. The fundamental concept of integration is to cut a domain of integration into small pieces, approximate the problem on a small piece, and summarize all the approximations defined on all the small pieces. This method is quite universal and is a critical mathematical skill for climate science. Textbooks of advanced climate science routinely use calculus and differential equations. Readers are referred to the textbooks listed below.

References and Further Readings

[1] Curry, J. A. and P. J. Webster, 1999: *Thermodynamics of Atmospheres and Oceans*. Academic Press, New York.

A standard textbook for an entire generation of students.

[2] Feynman, R. P., R. B. Leighton, and M. Sands, 2013: *The Feynman Lectures on Physics*. HTML edition, California Institute of Technology, URL: www.feynmanlectures.caltech.edu.

A masterful treatment by one of the greatest physicists of the twentieth century. Warning: It is not easy.

[3] Free, M., D. J. Seidel, J. K. Angell, J. Lanzante, I. Durre, and T. C. Peterson, 2005: Radiosonde atmospheric temperature products for assessing climate (RATPAC): A new data set of large-area anomaly time series. *Journal of Geophysical Research*, 110, D22101.

This paper clearly describes the relation between the atmospheric temperature profile and elevation or pressure, based on radiosonde data.

[4] Jacobson, M. Z., 1999: *Fundamentals of Atmospheric Modeling*. Cambridge University Press, New York.

An exceptional account of atmospheric modeling that will repay careful study.

[5] Lewis, H., 2007: *Geodesy Calculations in ROPP*. GRAS SAF Report 02, 24 October 2007 URL: www.romsaf.org.

The Radio Occultation Meteorology Satellite Application Facilities (ROM SAF) is a decentralized facility under EUMETSAT. This particular report describes a mathematical theory that converts the geometric height z to the geopotential height Z and vice versa.

[6] Leung, L.-Y. and G. R. North, 1990: Information theory and climate prediction. *Journal of Climate*, 3, 5–14.

This beautifully written paper briefly introduces the concepts of entropy and transinformation in quantifying climate predicability. A stochastic EBM is used as an example.

[7] Pierrehumbert, R. T., 2010: *Principles of Planetary Climate*. Cambridge University Press, New York.

A highly original and stimulating treatment beginning with the fundamental aspects of climate science and extending to the current research frontier. Highly recommended to all, including those interested in the climates of other planets as well as that of our Earth.

[8] Vallis, G. K, 2017: *Atmospheric and Oceanic Fluid Dynamics: Fundamentals and Large-Scale Circulation*. Cambridge University Press, New York.

This widely praised standard text is a magisterial treatment of geophysical fluid dynamics. Geoffrey K. Vallis is a mathematics professor and an expert in climate dynamics, the circulation of planetary atmospheres, and dynamical meteorology and oceanography.

[9] Wallace, J. M. and P. V. Hobbs, 2006: *Atmospheric Science: An Introductory Survey*. Academic Press, 2nd Edition, New York.

A classic standard textbook, much loved by generations of atmospheric science students.

Exercises

Exercises 7.2, 7.4–7.7, 7.9, and 7.11 require the use of a computer.

7.1 Suppose that the greenhouse effect results in a net energy gain of the Earth's surface by 1.0 [W m^{-2}]. If the gained heat is all used to heat the Earth's atmosphere, how many years will be needed to warm the Earth's entire atmosphere by 1.0°C? If the heat is all used to heat the Earth's surface water, including the water in the oceans, lakes, and rivers, how many years will it take to warm the water by 1.0°C? Make comments about this study's implications for global warming.

Hint: You can find the relevant information about the Earth's atmosphere and water in this book or from the Internet.

7.2 Find the data of geopotential height of 500 mb for a specific location and for a period of time, and write an R code to plot the time series of the data.

7.3 A piece of material with mass m_1, specific heat c_1, and temperature T_1 is put in contact with another piece of material with mass m_2, specific heat c_2, and temperature T_2. Without any loss of energy, the temperatures of the two pieces eventually become the same, T, due to heat conduction. (a) Find T from the given conditions: T_1, T_2, c_1, c_2, m_1, and m_2. (b) Use weighted average to explain your result. (c) Discuss some special cases, such as two very different masses, two the same mass, two very different specific heats, and two the same specific heat.

Hint: Use the energy conservation law: the energy before the pieces of material are in contact is equal to the energy after they have been in contact for a long time, i.e.,

$$c_1 m_1 T_1 + c_2 m_2 T_2 = c_1 m_1 T + c_2 m_2 T. \tag{7.126}$$

7.4 (a) Under the assumptions of hydrostatic equilibrium and ideal gas, derive the hypsometric equation (7.46) using the calculus method: cut a small piece of air of thickness equal to dz and base area equal to A, as shown in Fig. 7.1, and then integrate all the small pieces together. Start with the balance of forces on this small piece of air and derive the final equation (7.46).

(b) Suppose that (i) a layer of atmosphere has an average temperature 253 K, (ii) the layer's bottom is at sea level with $z_1 = 0$ and $p_1 = 1,000$ [mb], and (iii) the layer's top pressure is $p_2 = 500$ [mb] . Calculate the layer top's elevation z_2 using the formula derived in (a) and using appropriate approximations.

Hint: Pay attention to the units of the universal gas constant. Search the Internet and find out how many grams of air are equal to one [mol] of air. If certain conditions are attached to the units conversion, then discuss the conditions and the numerical results.

7.5 Repeat Elisha Mitchell's calculation with the current data you can find online, such as the North Carolina Climate Office NC ECONet, and make your own estimation of the elevation of Mount Mitchell using the hypsometric equation.

7.6 Calculate the elevation of a high mountain peak near your location, or in another place with which you are familiar, using observed data and the hypsometric equation.

7.7 Use the hypsometric equation to calculate the thickness of the dry air layer from atmospheric pressure 1,000 [mb] to 850 [mb] with an average temperature of 5°C. Here, the average temperature is regarded as the isothermal temperature of the layer.

7.8 From

$$\frac{d}{d\lambda}E(\lambda,T) = 0, \tag{7.127}$$

defined by formula (7.88) in Section 7.5, derive that

$$(u-5)e^u + 5 = 0, \tag{7.128}$$

where

$$u = \frac{hc}{k_b T \lambda}. \tag{7.129}$$

7.9 (a) Use R to solve Eq. (7.128).

(b) Use the result from (a) to derive that

$$\lambda_{max} \approx \frac{2,898}{T}, \tag{7.130}$$

where T [K], λ_{max} [μm], and 2,898 [K μm] is Wien's displacement constant. This formula is Eq. (7.91) and is also called Wien's law, or Wien's displacement law since it quantifies the shift of the wavelength of maximum emission as temperature varies. Wilhelm Wien (1864–1928) was a German physicist who won the Nobel Prize in physics in 1911.

7.10 Use integration by parts to evaluate the following integral used in the derivation of the Stefan–Boltzmann law

$$\int_0^\infty x^3 e^{-nx} dx = \frac{6}{n^4} \tag{7.131}$$

for any positive integer n.

7.11 Write an R code to verify the sum of the following infinite series used in the derivation of the Stefan–Boltzmann law

$$\sum_{n=1}^{\infty} \frac{6}{n^4} = \frac{\pi^4}{15} \tag{7.132}$$

7.12 Evaluate the following solid angle integral also used in the derivation of the Stefan–Boltzmann law

$$\int_0^{\pi/2} d\theta \int_0^{2\pi} d\phi \ \sin\theta \cos\theta = \pi. \tag{7.133}$$

7.13 Show that if a gas system is reversible in its cyclic process, then

$$\oint \frac{dq}{T} = 0, \tag{7.134}$$

where T is the temperature of the system, q is the heat per unit mass, and $\oint$ denotes a closed path integral in the 3D gas state space (p, v, T). *Hint: Here, the specific heat capacity c_p was defined by enthalpy:*

$$dh = c_p dT \tag{7.135}$$

per unit mass. The first law of thermodynamics for energy conservation is

$$c_p dT = dq + v dp \tag{7.136}$$

per unit mass. Rewrite this equation as

$$dq = c_p dT - v dp, \tag{7.137}$$

divide both sides by T, use the ideal gas law $pv = RT$ per unit mass, and then make use of a closed path integral using an exact differential.

7.14 This is an information theoretic problem to describe climate prediction: How much information on climate can be gained from a given initial condition?

Let us set up the information theoretic background of this problem. In the field of information theory, entropy is defined in another way:

$$S = -\sum_{i=1}^{N} p_i \ln p_i \tag{7.138}$$

where $\{p_i, i = 1, 2, \ldots, N\}$ is a probability distribution with $\sum_{i=1}^{N} p_i = 1$. This definition is also referred to as the Shannon entropy. Claude Shannon (1916–2001) was an American mathematician and electrical engineer, and is described as "the father of information theory." If the Shannon entropy is reduced, then the system has gained information; and if it is increased, information has been lost. Thus, the Shannon entropy can help quantify the information gain or loss. Leung and North (1990) used this method to quantify the information gained about an unknown temperature anomaly at a given time t with knowledge of the initial temperature anomaly data. They assumed that the temperature anomaly has a normal distribution

$$f(T) = \frac{1}{\sqrt{2\pi\sigma_0^2}} \exp\left[-T^2/(2\sigma_0^2)\right] \tag{7.139}$$

where the expected value of the temperature anomaly is zero and the variance is σ_0^2. The probability distribution of temperature anomaly with the given initial temperature anomaly T_0 at the initial time t_0 is also assumed to be normal

$$f(T|T_0) = \frac{1}{\sqrt{2\pi\sigma^2}} \exp\left[-(T-\bar{T})^2/(2\sigma^2)\right]. \tag{7.140}$$

Here, we assume an exponential decay of the mean $\bar{T}$, and variance σ^2 defined as follows:

$$\bar{T} = T_0 \exp(-t/t_0), \tag{7.141}$$

$$\begin{aligned} \sigma^2(t) &= E[(T(t)-\bar{T})^2] \\ &= \sigma_0^2\left[1-\exp(-2t/\tau_0)\right], \end{aligned} \tag{7.142}$$

where τ_0 is the climate system's time scale.

The exercise problem is to derive the information gained from the given initial condition by carrying out the derivations of the following three steps:

Step (a)

$$\begin{aligned} H(T) &= -\int_{-\infty}^{\infty} dT\, f(T)\ln f(T) \\ &= \frac{1}{2} + \ln\sqrt{2\pi\sigma_0^2}. \end{aligned} \tag{7.143}$$

Step (b)

$$\begin{aligned} H(T|T_0) &= -\int_{-\infty}^{\infty} dT \int_{-\infty}^{\infty} dT_0\, f(T|T_0)\ln f(T|T_0) \\ &= \frac{1}{2} + \ln\sqrt{2\pi\sigma^2}. \end{aligned} \tag{7.144}$$

Step (c)

$$\begin{aligned} I(T,T_0) &= H(T|T_0) - H(T) \\ &= \frac{1}{2}\ln\left[1-\exp(-2t/\tau_0)\right] < 0. \end{aligned} \tag{7.145}$$

This is the amount of Shannon entropy reduced due to the given initial condition. Therefore, the climate information at time t has increased from the initial condition at time t_0.

8 Conservation Laws in Climate Dynamics

Global climate models, sometimes known as general circulation models (both are called GCMs) are a relatively recent development. Until the 1960s, they did not exist, nor did the powerful supercomputers that they require. It is no exaggeration to say that the development of GCMs has revolutionized climate science. GCMs provide us with a virtual Earth, on which we can do controlled experiments, which would (fortunately) be impossible to do on the real Earth. GCMs provide a worthy complement to other research approaches, such as theory and observations. An extraordinarily rapid development of GCMs has occurred since the 1960s. A major aspect of this development has been to extend models of the atmosphere to include models of the ocean circulation and other components of the climate system. For an introductory historical survey of GCMs, see Donner et al. (2011).

At the foundation of GCMs are a set of equations that describe fundamental conservation laws in physics such as the conservation of mass, momentum, and energy. Other conservation laws or principles have proven useful for providing deep physical insight into the dynamics of atmospheres and oceans, such as the conservation of potential vorticity. This chapter introduces examples of both types of conservation laws. The energy balance models (EBMs) described in Chapter 5 are the simplest in a hierarchy of climate models and are based only on a simple form of the conservation of energy: outgoing energy emitted as radiation by the Earth is balanced by incoming energy in the form of solar radiation absorbed by the Earth. Today coupled ocean–atmosphere GCMs, often involving detailed treatments of many aspects of the climate system, from cloud physics to ocean chemistry, occupy the comprehensive end of the spectrum of climate models. Numerical methods are invariably necessary to solve complex climate models. This chapter provides the basic mathematics needed to describe the conservation laws that are the foundation of quantitative climate science.

8.1 Conservation of Mass

At the heart of modern climate models is a set of mathematical equations that describes the natural conservation laws for continuous media, such as the atmosphere, the ocean, and the land. We start by clarifying the idea of continuous media.

8.1.1 Basic Elements of the Continuum Mechanics Method for Climate Modeling

Continuum mechanics is an area of science that deals with the behavior of materials modeled as continuous matter rather than as discrete particles. The basic assumption of continuum mechanics is that in analyzing the large-scale behavior of such matter as the atmosphere or the ocean, we do not have to take account of the fact that the matter is ultimately composed of discrete atoms and molecules with space between them. Instead, we may regard the matter as being continuously distributed and filling the entire space that it occupies. A continuum is assumed to be capable of being continually sub-divided into arbitrarily small elements while still retaining the properties of the bulk material. The prolific French mathematician and physicist Augustin-Louis Cauchy (1789–1857) deserves much of the credit for the early development of continuum mechanics.

This approach to analyzing the continuum of gases, liquids, and solids has been used to develop the fundamental mathematical equations of climate models. This development process involves a comprehensive application of the methods of calculus to derive the conservation laws of mass, momentum, and energy appropriate to a small volume of atmospheric or oceanic mass. The small volume is infinitesimal, meaning a differential volume denoted by $dV = dxdydz$, which is often approximated by a finite spatial volume in numerical climate modeling. Differentials and derivatives will be involved, and the conservation properties will thus lead to a set of equations involving derivatives or differentials, called differential equations (DEs). An equation that involves only ordinary derivatives is called an ordinary differential equation (ODE), such as the ODE used in the simplest EBM in Chapter 5. If more than one independent variable is involved, such as the dependence of air temperature T on the three-dimensional (3D) location (x, y, z) and time t, i.e., $T = T(x, y, z, t)$, then the following partial derivatives

$$\frac{\partial T}{\partial t}, \frac{\partial T}{\partial x}, \frac{\partial T}{\partial y}, \frac{\partial T}{\partial z}$$

will be involved. Equations involving partial derivatives are called partial differential equations (PDEs). Many aspects of modern climate science research are either based on or are evolving toward the consideration of stochastic processes, which involves stochastic PDEs (SPDEs). Here the term stochastic is a general one that means involving chance or randomness or probability. PDEs are often more complex than ODEs. As a generalization, we may safely say that climate model equations are typically extremely complicated, compared to the equations usually studied by mathematicians engaged in proving theorems.

Because of the calculus method of treating a small volume, the numerical solutions of a climate model deal with average values of a climate variable in this small volume and do not represent the value at a particular point (x, y, z). In global climate models (GCMs) the partial differential equations may be solved numerically on finite volumes such as three-dimensional finite-difference grid elements or "boxes." For example, these boxes might have horizontal areas defined by a regular latitude–longitude grid and a vertical dimension defined by finite increments of altitude or pressure. Finer grids require faster computers. Affordable global grids with present-day supercomputers might have grid volumes or boxes with horizontal dimensions of tens or hundreds of kilometers and with tens of layers in the vertical. The numerical GCM solutions on such a grid physically represent the averages of climate parameters in these three-dimensional grid boxes, although the numerical solution output may be written at grid points. Therefore, when comparing climate model output with observational data, both should be analyzed on consistent or identical space–time grids in order to produce a meaningful and objective comparison. This aspect of numerical solutions can easily be misunderstood and incorrectly used by the general public and even by inexperienced scientists who attempt to compare climate model results with observations made at a single location or only a few locations, such as with instruments at an individual weather station.

8.1.2 Lagrangian and Eulerian Observers, and Mass Conservation in the Lagrangian Framework

In analyzing the motions of the ocean and the atmosphere, we typically regard the ocean as a liquid and the atmosphere as a gas, so both ocean and atmosphere are fluids. There are two commonly used means of describing fluid motion mathematically. In each of these methods, we shall define an infinitesimal volume of fluid known as the *control volume*. The Eulerian method uses a control volume that is fixed with respect to the coordinate axes (x, y, z), so that fluid flows into and out of this volume. The Lagrangian method uses a control volume that moves with the fluid in such a way as to always contain the same fluid. Thus, the Eulerian control volume retains its shape and its location. If it is initially a rectangular box with all corners having 90-degree angles, it will always be that same box at the same location. By contrast, the Lagrangian control volume will move in space and distort in shape as time goes on, because no fluid can pass through its "walls." The above description uses the term *method* for the two alternative descriptions, but many equivalent terms are often used, such as Eulerian and Lagrangian *frameworks, frames of reference, systems, observers, and viewpoints*.

In the Lagrangian framework, the fluid mass contained inside a small control volume or "box" with volume V is conserved, meaning that the mass does not change with time. The average density of this small box of fluid is ρ. Thus, the mass in the small volume is $m = \rho V$. The conservation of mass means that the total mass in this box does not change with time:

$$\frac{dm}{dt} = 0, \tag{8.1}$$

or

$$\frac{d}{dt}(\rho V) = 0. \tag{8.2}$$

This formula means that the Lagrangian observer follows the box and sees no "leaking" of fluid into or out of the box, even though the box can deform since it is a fluid body. Because of the box deformation, the volume and the average density can change with time. However, the box is closed: no mass goes out and no mass comes in. It thus somewhat resembles an imaginary sealed flexible tube full of fluid which can change its size and shape but cannot gain or lose any fluid.

The observer following such a fluid volume is said to be in a Lagrangian frame: the fluid flow properties, such as density and velocity, are functions of time and the initial position, i.e., $\rho(x_0, y_0, z_0; t)$. For a given initial position, the fluid box is only a function of time. As a vivid analogy or metaphor, because the observer "follows" the control volume, the Lagrangian observer may be thought of as a kind of imaginary or virtual dog walker, while the small control volume of fluid may be regarded as the dog being walked. We often denote the derivative in a Lagrangian frame as

$$\frac{D}{Dt}, \tag{8.3}$$

rather than $\frac{d}{dt}$. Thus, the mass conservation equation is

$$\frac{D}{Dt}(\rho V) = 0, \tag{8.4}$$

or

$$\frac{Dm}{Dt} = 0. \tag{8.5}$$

However, when considering the atmosphere or the ocean from a Lagrangian viewpoint, in practice it would be extremely difficult, if not impossible, to follow a small fluid control volume. In most cases, we would normally observe the fluid flow at a fixed place, or we would install an instrument in the flow of atmospheric air or ocean water at a fixed place. When we experience wind, for example, we do not follow individual air parcels, rather we remain in one place, and at that place, we feel the wind, and we sense the wind speed and direction. Thus, we resemble an instrument measuring the flow of the atmosphere as it passes by us. We are an Eulerian observer.

Another example of an Eulerian observer is our observation of water flowing in a river. We may stand still on a river bank and observe the waves on the surface of the river and the motion of water within the range of our vision. Our eyes do not follow the individual water parcels or small volumes of water.

Such a fixed-position observer is in an Eulerian framework, so this observer experiences air flowing by, feels the motion of the air at the fixed point of observation, and watches the water flowing down the river.

Still another example might be the observation of ocean waves. In deep water, a surface wave with a wavelength of 6 meters might move at a speed of about 3 meters per second. In 60 seconds, it will have moved 180 meters. However, a small ocean water parcel or control volume located near the ocean surface in a field of such waves does not move 180 meters away from its initial position in one minute. Instead, it moves up and down and circulates in regular oscillatory motion, remaining in the neighborhood of its original position. We can see this by watching a floating object in the water as the wave passes by. The object bobs up and down but experiences no net displacement. Thus, in the same 60 seconds, the Lagrangian observer notes the water parcel and its oscillatory movements around its initial position, while the Eulerian observer sees the wave moving 180 meters. They are observing the same physical phenomenon, but from different viewpoints.

8.1.3 Total Derivative

Consider a fluid density change $\Delta\rho$ in a Lagrangian control volume or box of fluid observed by a Lagrangian observer during an interval of time from time t to $t+\Delta t$. The Lagrangian box's position is given by coordinates (x,y,z), which are Eulerian coordinates and are functions of time t. Thus, $(x(t),y(t),z(t))$ from t_1 to t_2 describes the trajectory of the small Lagrangian fluid box in the time interval $[t_1,t_2]$. The trajectory is observed by the Eulerian framework fixed on the rotating Earth.

When t is increased to $t+\Delta t$, the coordinates of the small Lagrangian box are changed to $(x+\Delta x, y+\Delta y, z+\Delta z)$. The change of the density from $(x(t),y(t),z(t))$, to $(x(t)+\Delta x, y(t)+\Delta y, z(t)+\Delta z)$ due to the time change from t to $t+\Delta t$ is

$$\Delta\rho = \rho(x+\Delta x, y+\Delta y, z+\Delta z, t+\Delta t) - \rho(x,y,z,t). \tag{8.6}$$

The linear approximation of this expression around (x,y,z,t) yields

$$\Delta\rho \approx \frac{\partial\rho}{\partial t}\Delta t + \frac{\partial\rho}{\partial x}\Delta x + \frac{\partial\rho}{\partial y}\Delta y + \frac{\partial\rho}{\partial z}\Delta z. \tag{8.7}$$

Dividing both sides of the above by the time increment Δt leads to

$$\frac{\Delta\rho}{\Delta t} \approx \frac{\partial\rho}{\partial t} + \frac{\partial\rho}{\partial x}\frac{\Delta x}{\Delta t} + \frac{\partial\rho}{\partial y}\frac{\Delta y}{\Delta t} + \frac{\partial\rho}{\partial z}\frac{\Delta z}{\Delta t} \tag{8.8}$$

Here, the partial derivatives are taken at point (x,y,z,t). An Eulerian observer sees

$$\frac{\Delta x}{\Delta t} \tag{8.9}$$

as the x-direction velocity component u of the small Lagrangian box at time t when Δt approaches zero. Similar statements can be made about $\frac{\Delta y}{\Delta t}$ and $\frac{\Delta z}{\Delta t}$.

Thus, as Δt approaches zero,

$$\frac{\Delta\rho}{\Delta t} \tag{8.10}$$

represents the total change rate of density at t and is regarded as the total derivative

$$\frac{D\rho}{Dt}. \tag{8.11}$$

When Δt approaches zero, Eq. (8.8) thus will become

$$\frac{D\rho}{Dt} = \frac{\partial\rho}{\partial t} + \frac{\partial\rho}{\partial x}u + \frac{\partial\rho}{\partial y}v + \frac{\partial\rho}{\partial z}w. \tag{8.12}$$

This means that a total derivative with respect to time is made up of two parts. The first part is

$$\frac{\partial\rho}{\partial t}$$

representing the density change rate directly due to time as if the fluid is at rest. This part is called the local derivative. The second part is

$$\frac{\partial\rho}{\partial x}u + \frac{\partial\rho}{\partial y}v + \frac{\partial\rho}{\partial z}w$$

representing the density change rate due to the motion of the fluid. This part is called the advection part of the total derivative, or it may be called density advection.

In the above total derivative expression, the velocity field (u, v, w) is with respect to the Eulerian observers. Thus, the total derivative provides a mathematical link between what is seen by the two observers, Eulerian and Lagrangian, in terms of the rate of change.

8.1.4 Mass Conservation in the Eulerian Framework

The Eulerian observer thus has his own small control volume or box for fluid. This box is a fixed observation device that allows fluid to freely come in and to freely go out. It is a virtual box, or a conceptual box. The Eulerian observer records the amount of fluid going in and out. This is in contrast to the Lagrangian small box, which is conceptually a container or bag moving with the fluid. This imaginary bag does not allow fluid to go in or out, but the bag can freely deform, by stretching, compression, or distortion. The Eulerian control volume or small box, by contrast, does not change its shape and remains at a fixed position relative to a coordinate system, and the fluid can flow through it.

In the Eulerian framework, the above Lagrangian derivative of density, or mass per unit volume, can be thought of as the sum of two terms. One term is the mass increase with respect to time in a virtual fixed control volume or box. An increase in mass must come from the mass flux from outside the box to inside the box. This can be mathematically described as follows:

$$\frac{\partial(\rho V)}{\partial t} = -\oint_{\partial V} \rho \mathbf{u} \cdot \mathbf{n} dS. \tag{8.13}$$

Here, ∂V stands for the entire surface of the Eulerian fluid box V, and dS is a small surface area of box V and has a unit vector $\mathbf{n}$ perpendicular to the tangent plane of dS and pointing toward the outside of the Eulerian box. This unit vector is called a normal vector of dS. Vector $\mathbf{u}$ is the fluid velocity observed by a fixed instrument. The negative sign of the right-hand side is present because $\mathbf{u} \cdot \mathbf{n} dS$ indicates the flow from the inside to the outside of the box through the area dS. The reverse flow from the outside to the inside will therefore require a negative sign. Thus,

$$-\rho \mathbf{u} \cdot \mathbf{n} dS$$

can be understood as the mass influx into the Eulerian box over a small surface region dS. The integral is a summation of the influx from all the small areas and yields the total mass influx into the Eulerian box. This procedure of starting from a summation and ending with an integral is a standard way of using the integral calculus to develop a mathematical model with derivatives and/or integrals.

The term

$$\frac{\partial(\rho V)}{\partial t}$$

is the mass increase rate of the Eulerian box. Equation (8.13) thus means that the mass increase rate is balanced by the mass influx. This is the integral form of the mass conservation law, a law which is commonly used in the differential form to be derived below.

The divergence theorem (see Appendix A for this theorem) for a 3D surface integral transforms the right-hand side of the equation to yield

$$-\oint_{\partial V} \rho \mathbf{u} \cdot \mathbf{n} dS = -\iiint_V \nabla \cdot (\rho \mathbf{u}) dV, \tag{8.14}$$

where

$$\nabla \cdot (\rho \mathbf{u}) = \frac{\partial(\rho u)}{\partial x} + \frac{\partial(\rho v)}{\partial y} + \frac{\partial(\rho w)}{\partial z} \tag{8.15}$$

is called the divergence of vector $\rho \mathbf{u}$, dV means the volume of a small 3D domain taken from the Eulerian fluid box V, and u is the velocity vector $\mathbf{u}$'s projection component in the direction of the x-axis, v the same for the y-axis, and w the same for the z-axis, i.e., $\mathbf{u} = (u, v, w)$.

When V is small,

$$\iiint_V \nabla \cdot (\rho \mathbf{u}) dV \approx V \nabla \cdot (\rho \mathbf{u}) \tag{8.16}$$

where $\nabla \cdot \mathbf{u}$ is the divergence at any point inside V. This is a result from the mean value theorem of calculus in the integral form.[1] This approximation based on the mean value theorem is valid under the condition that $\rho \mathbf{u}(x, y, z, t)$ is a continuous function in the small volume V for a given time t. In atmospheric and oceanic flows, possible discontinuities include shocks. A shock is an abrupt discontinuity in the flow field. It occurs in flows when the local flow speed exceeds the local sound speed.

Thus, in the mass conservation derivation above, the volume V cannot contain a shock. Shocks in the atmosphere can be produced by supersonic airplanes or by the firing of a powerful cannon, for example. Fortunately, such local phenomena may be regarded as unimportant for climate purposes. In climatically significant motions in the atmosphere or the ocean, shocks do not occur, and our mass conservation law is generally valid for climate models.

Therefore, the law of conservation of mass for a fixed observer, also called the law in Eulerian coordinates, or an Eulerian framework, is

$$\frac{\partial \rho}{\partial t} + \nabla \cdot (\rho \mathbf{u}) = 0. \tag{8.17}$$

[1] The mean value theorem for a single variable is described in Appendix D on calculus. For n variables in the n-dimensional space $\mathbf{R}^n$, the integral form of the mean value theorem can be expressed as follows:

$$\int \cdots \int_D f(x_1, \ldots, x_n)\, dx_1 \ldots dx_n = V f(c_1, \ldots, c_n),$$

where the left-hand side stands for an integral in the domain D in $\mathbf{R}^n$, the integrand $f(x_1, \ldots, x_n)$ is a continuous function in this domain, V is the volume of the domain, and $(c_1, \ldots, c_n)$ is a point in D. The mean value theorem asserts that this point must exist. Thus, the mean value theorem states that for a continuous function in a domain, there exists at least one point inside the domain such that the integral of the function over the domain is equal to the volume of the domain times the function value at the point. Although we use "volume" to describe the domain size in the n-dimensional space $\mathbf{R}^n$, we understand that the measure of the domain size is conventionally called area (not volume) if $n = 2$, and called length if $n = 1$. The continuity condition is usually used in proving this theorem. Examples of discontinuous functions can be found to show that the mean value theorem does not hold for these functions.

This equation is called the continuity equation of fluid mechanics, in the Eulerian framework. Physically, it states that mass is conserved if there are no sources or sinks of mass in the system.

For numerical computing, we write this equation in the *xyz*-coordinate form

$$\rho_t + u\rho_x + v\rho_y + w\rho_z + \rho(u_x + v_y + w_z) = 0, \tag{8.18}$$

where the subscripts x, y, z, t denote partial derivatives with respect to the subscripted variable. This is an equation in the Eulerian framework and the *xyz*-coordinate system.

In this way, the Lagrangian derivative $\frac{D(V\rho)}{Dt}$ is changed to partial derivatives in an Eulerian framework. An Eulerian "observer" places an instrument at the point (x, y, z) and observes at time t. Thus, a climate parameter in the Eulerian coordinates is a function of space and time, but it is a function only of time for a Lagrangian "observer," who begins with a particular small box or control volume. To employ once more the metaphor we introduced earlier, the Lagrangian coordinate frame is moving with the box, like a dog (i.e., the fluid box) and its walker (i.e., the Lagrangian observer), moving together.

In practice, mathematical models of fluid dynamics in climate science are almost always developed in an Eulerian framework.

Thus, the Lagrangian coordinates are metaphorically a dog and its walker, moving together, and the Eulerian coordinates are metaphorically a stationary onlooker watching the dog and its walker as they pass by.

It may be helpful to watch the "classical" film of the late Professor John Lumley's lecture on "Eulerian and Lagrangian Descriptions in Fluid Mechanics," produced in 1968 and supported by the National Science Foundation. It is on YouTube at

www.youtube.com/watch?v=mdN8OOkx2ko

and can also be found in libraries.

In the 1D case, consider two people holding an idealized strip of rubber with one person at each end. The rubber band's length is h. The rubber band's linear density [unit: kg m^{-1}] decreases if the band gets stretched. A way to stretch the band is to have the right end of the band move in the positive x direction (toward the right) faster than the left end. We then have the following balance of mass. The band density decrease rate is $-\partial\rho/\partial t$ where ρ is the linear density of the band and is a function of both location and time: $\rho(x,t)$. The front end (right) of the band stretches, causing the mass decrease at the rate $\rho(x+h,t)u(x+h,t)$ in $(x, x+h)$, while the rear (left) end's compression causes the mass increase in $(x, x+h)$ at the rate $\rho(x,t)u(x,t)$. The mass conservation is

$$\begin{aligned} &[\text{mass decrease}]\ (-\partial(h\rho)/\partial t) \\ &\quad = [\text{mass decrease from the front end}]\ (\rho(x+h,t)u(x+h,t)) \\ &\quad\quad - [\text{mass increase/from the rear end}]\ (\rho(x,t)u(x,t)). \end{aligned} \tag{8.19}$$

This can be simplified as

$$\frac{\partial\rho}{\partial t} + \frac{\rho(x+h,t)u(x+h,t) - \rho(x,t)u(x,t)}{h} = 0. \tag{8.20}$$

The linear approximations of $\rho(x+h,t) = \rho(x,t) + \rho_x(x,t)h$ and $u(x+h,t) = u(x,t) + u_x(x,t)h$ for a small h lead to the 1D mass conservation equation, when we linearize the equation by neglecting the h^2 term:

$$\frac{\partial \rho}{\partial t} + \rho \frac{\partial u}{\partial x} + u \frac{\partial \rho}{\partial x} = 0, \tag{8.21}$$

or

$$\frac{\partial \rho}{\partial t} + \frac{\partial (\rho u)}{\partial x} = 0. \tag{8.22}$$

In the above, $u_x = \partial u/\partial x$ is another notation for partial derivatives.

Extensions of this equation to 2D in the xy-plane and to 3D in xyz-space are below

$$\frac{\partial \rho}{\partial t} + \frac{\partial (\rho u)}{\partial x} + \frac{\partial (\rho v)}{\partial y} = 0 \tag{8.23}$$

$$\frac{\partial \rho}{\partial t} + \frac{\partial (\rho u)}{\partial x} + \frac{\partial (\rho v)}{\partial y} + \frac{(\rho w)}{\partial z} = 0. \tag{8.24}$$

The three equations above are the mass conservation equations in Eulerian coordinates. They have a compact form

$$\frac{\partial \rho}{\partial t} + \nabla \cdot (\rho \mathbf{u}) = 0, \tag{8.25}$$

which is the mass conservation equation or continuity equation (A.20) found in Appendix A.

The two terms in the mass conservation equation (8.25) each have their physical meanings. The first term $\partial \rho/\partial t$ means the density variation in time, due to expansion or compression occurring at the location of the Eulerian observer. In the case of traffic on a road with a traffic light, an Eulerian observer will see an increase of traffic density (the number of cars in a given space) when the light turns red and traffic must stop, and a decrease of density when the red light turns green and traffic starts to move again. This is called the local derivative $\partial \rho/\partial t$.

The second term in equation (8.25) is caused by the speed differences in a fluid flow. If the flow ahead is slower than the flow behind, then the traffic on the road or the fluid mass accumulates and increases the density. In the opposite case, the density decreases. This part of mass change is accomplished by fluid motion and is called mass advection.

In general, a Lagrangian derivative is equal to the local derivative plus the advection. The Lagrangian derivative

$$\frac{D\rho}{Dt}, \tag{8.26}$$

is called the material derivative, or substantial derivative. It is a derivative in the Lagrangian framework.

A material derivative in a flow always has two terms: local derivative and advection. Thus, the following formula is universal:

$$\frac{D\rho}{Dt} = \frac{\partial \rho}{\partial t} + (\mathbf{u} \cdot \nabla)\rho. \tag{8.27}$$

The mass conservation equation (8.25), also called the continuity equation, (A.20), can be rewritten as

$$\begin{aligned}&\frac{\partial \rho}{\partial t}+\nabla\cdot(\rho\mathbf{u})\\&=\left[\frac{\partial \rho}{\partial t}+(\mathbf{u}\cdot\nabla)\rho\right]+\rho\nabla\cdot\mathbf{u}\\&=\frac{D\rho}{Dt}+\rho\nabla\cdot\mathbf{u}=0.\end{aligned} \tag{8.28}$$

The part in the square brackets is the Lagrangian derivative of ρ, and the part following the square-bracketed expression is density times the divergence of velocity.

When a fluid is incompressible, the density of the fluid in the control volume or box does not change with respect to time in the entire flow process. Thus,

$$\frac{D\rho}{Dt}=0. \tag{8.29}$$

Equivalently, for an incompressible fluid, from Eq. (8.28),

$$\nabla\cdot\mathbf{u}=0, \tag{8.30}$$

i.e., the divergence of velocity is zero, or the fluid flow is non-divergent. The divergence theorem for an incompressible fluid flow implies that the flow flux (not the mass flux) through any closed surface must be zero:

$$\oint_{\partial\Omega}\mathbf{u}\cdot\mathbf{n}\,dS=0. \tag{8.31}$$

8.2 Conservation of Momentum Over a Grid Box: F = *ma*

Newton's second law, $\mathbf{F}=m\mathbf{a}$, can be applied to a small Lagrangian control volume or fluid box of volume V and average density ρ:

$$m\frac{D\mathbf{u}}{Dt}=\mathbf{F}_c+\mathbf{F}_p+\mathbf{F}_g+\mathbf{F}_f, \tag{8.32}$$

where

$$m=\rho V \tag{8.33}$$

is the mass of the fluid in the Lagrangian box,

$$\frac{D\mathbf{u}}{Dt}=\mathbf{a} \tag{8.34}$$

is the Lagrangian box's acceleration, and $\mathbf{F}_c,\mathbf{F}_p,\mathbf{F}_g,\mathbf{F}_f$ are the Coriolis force (CF), the pressure gradient force (PGF), the Earth's gravitational force (EGF), and the friction force (FF).

Newton's laws of motion are valid in an inertial coordinate system, that is, a coordinate system in which the vector momentum of a particle is conserved in the absence of external forces. For our purposes, an absolute coordinate system with its origin on the Earth's center and fixed with respect to the stars would be such a system. For convenience, however, we use a coordinate system from the perspective of an Eulerian observer: The system is fixed only with respect to the Earth, not the stars, and therefore rotates in the same way as the Earth rotates. Newton's second law, $\mathbf{F} = m\mathbf{a}$, can be written as a new set of equations in such a coordinate system for the Eulerian observer, provided that we include an apparent force, the centrifugal force, and also a second apparent force, the Coriolis force. We can then use the new set of equations for Newton's second law of motion to study the motion of an object relative to the rotating Earth, such as atmospheric and oceanic fluid flows (Goose 2015).

The Coriolis force is named for Gaspard-Gustave Coriolis (1792–1843), a French mathematician, engineer, and scientist. However, in his paper on this subject, published in 1835, Coriolis dealt with rotating hydraulic machinery, and he was not at all concerned with the atmosphere or the ocean or even with the rotation of the Earth. In addition, he was not the first person to derive the equations of motion in a reference frame rotating with the Earth. The scientist who deserves credit for that is a much more famous French scholar, Pierre-Simon Laplace (1749–1827), who made major contributions in physics, astronomy, mathematics, and statistics. Laplace, who is sometimes called "the Newton of France," derived the mathematical formula for the "Coriolis force" due to the Earth's rotation some four decades before Coriolis.

Furthermore, the equations in a coordinate system rotating with the Earth were used by the American theoretical meteorologist William Ferrel (1817–1891) in an 1859 paper. Ferrel, who was largely self-taught and who appears to have been completely unaware of Coriolis, may have based his work directly on Laplace's great five-volume publication, *Mécanique Céleste (Celestial Mechanics)*. It was Ferrel who first scientifically investigated the implications of the Earth's rotation for atmospheric and oceanic motions. Ferrel made the first attempt to include the consequences of these factors in a theory for the general circulation of the atmosphere and ocean. Ferrel was also perhaps the first to understand the importance for meteorology and oceanography of what we would refer to today as the principle of conservation of angular momentum. Nonetheless, despite the unquestioned priority of Laplace and the great importance of Ferrel's research for our subject, the apparent force that appears in the equations we use to describe motions of an object on the rotating Earth is today always called the Coriolis force. The history of this topic is not simple, and a readable introduction to it is found in the paper by Persson (1998).

The Coriolis force is an apparent force in a rotating coordinate system, which is the Earth in this case. The Coriolis force deflects the direction of fluid flow in this system. The force is perpendicular to both the angular velocity of the rotating Earth and the flow velocity. This force can be expressed by the cross product of two vectors[2]

$$\mathbf{F}_c = -2m\boldsymbol{\Omega} \times \mathbf{u}, \tag{8.35}$$

where m is the mass of the fluid in the control volume, $\mathbf{u}$ is the velocity of the control volume, and $\boldsymbol{\Omega}$ is the angular velocity of the Eulerian frame at the location of interest

[2] See Appendix B for the definition of a cross product.

and is latitude-dependent. In the Northern Hemisphere, the Coriolis force deflects a flow to the right because $\mathbf{F}_c$ points to the right of the direction of the velocity vector **u** based on the right-hand rule of the cross product. From the viewpoint of the Eulerian observer, the Coriolis force $\mathbf{F}_c$ deflects the velocity of the control volume toward the right. For the same reason, in the Southern Hemisphere, the deflection of the velocity of the control volume is to the left. For further discussion of the Coriolis force, see the next subsection and a standard textbook such as Holton and Hakim (2012), Vallis (2017), or Wallace and Hobbs (2006).

The pressure gradient force was discussed earlier and is directed from a region of higher pressure toward a region of lower pressure:

$$\mathbf{F}_p = -V\nabla p, \tag{8.36}$$

where p is the pressure field, and V is the volume of the control box.

The gravitational force is due to the Earth's gravity and is equal to

$$\mathbf{F}_g = m\mathbf{g}, \tag{8.37}$$

where **g** is the gravitational acceleration and points toward the center of the Earth and has a magnitude that is approximately $9.8[\mathrm{m\ s^{-2}}]$.

The friction force can be due to contact between the flow and a boundary, such as the Earth's surface. It can also be due to internal friction (viscosity). Both the atmosphere and the ocean clearly have viscosity, although theoretical treatments and mathematical models in climate science do not always include this force.

If we ignore the friction force $\mathbf{F}_f$, Newton's second law of motion now becomes

$$m\frac{D\mathbf{u}}{Dt} = -2m\boldsymbol{\Omega}\times\mathbf{u} - V\nabla p + m\mathbf{g}. \tag{8.38}$$

Dividing both sides by m, we have the acceleration equation

$$\frac{D\mathbf{u}}{Dt} = -2\boldsymbol{\Omega}\times\mathbf{u} - \frac{\nabla p}{\rho} + \mathbf{g}. \tag{8.39}$$

Here, we have used the relationship that density is equal to mass divided by volume for the control box or volume

$$\rho = \frac{m}{V}. \tag{8.40}$$

When we expand the material derivative $\frac{D\mathbf{u}}{Dt}$ and express the above relationship in the Eulerian framework, we have the following momentum equation for Newton's second law of motion in the form of $\mathbf{a} = \mathbf{F}/m$ $[dimension: LT^{-2}]$:

$$\frac{\partial\mathbf{u}}{\partial t} = -(\mathbf{u}\cdot\nabla)\mathbf{u} - 2\boldsymbol{\Omega}\times\mathbf{u} - \frac{\nabla p}{\rho} + \mathbf{g}. \tag{8.41}$$

The first term on the right-hand side is called the advective acceleration, or velocity advection, which behaves as an acceleration.

8.3 The Equations of Momentum Conservation in x, y, z, t Coordinates

In the Eulerian framework, we often use Cartesian coordinates with the vertical axis z pointing directly upwards (see Fig. 8.1). The z direction is thus normal to the Earth's local surface, being perpendicular to the tangent plane of the spherical Earth at the location of an Eulerian observer. Thus, a z-directional vector is in the direction pointing from the Earth's center outward to the given location. The x-coordinate is a horizontal coordinate in the zonal direction, pointing from west to east. The y-coordinate is a horizontal coordinate in the meridional direction pointing from south to north.

In this xyz-coordinate system, the Earth's angular velocity $\mathbf{\Omega}$ in the momentum equation (8.41) has three components, which can be calculated from the projections of $\mathbf{\Omega}$ on the three axes as shown in Fig. 8.1.

$$\mathbf{\Omega} = (0, \Omega \cos \phi, \Omega \sin \phi), \tag{8.42}$$

where $\Omega = ||\mathbf{\Omega}||$ is the magnitude of the vector $\mathbf{\Omega}$, and ϕ is the latitude of the given location. Note here that the Earth's angular velocity $\mathbf{\Omega}$ is a vector in the yz-plane and hence has a zero projection on the x-axis.

Another way to express a vector in a three-dimensional space is to use three orthogonal unit vectors

- $\mathbf{i}$ a unit vector in the x direction,
- $\mathbf{j}$ a unit vector in the y direction, and
- $\mathbf{k}$ a unit vector in the z direction.

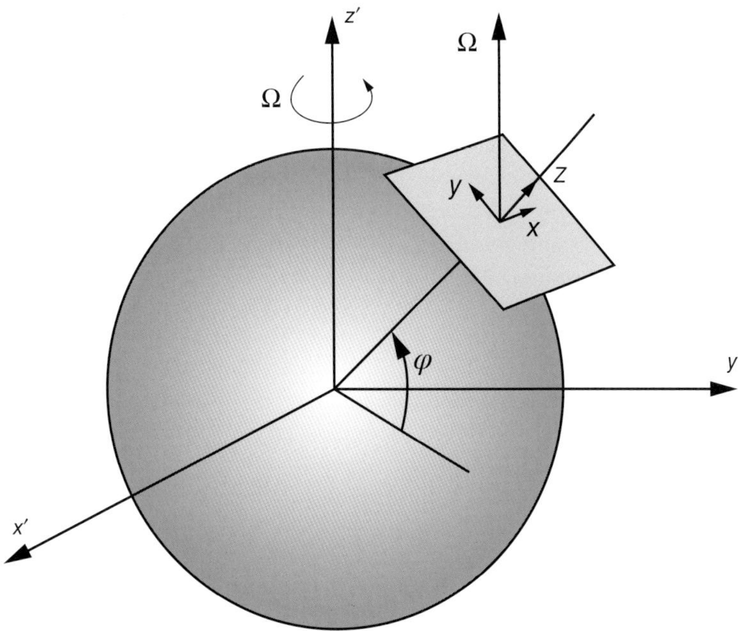

Fig. 8.1 The Cartesian coordinate x, y, z-system on the Northern Hemisphere.

Then, $\boldsymbol{\Omega}$ can be written as

$$\boldsymbol{\Omega} = \Omega\cos\phi\,\mathbf{j} + \Omega\sin\phi\,\mathbf{k}. \tag{8.43}$$

This expression treats a vector explicitly as a summation of three component vectors, while expression (8.42) emphasizes the three projections.

The Earth's rotation speed is a constant with

$$\begin{aligned}\Omega &= 7.292115090\times 10^{-5}\ \text{[radians/second]}\\ &= 72.92115090\ \text{[micro-radians/second]}.\end{aligned} \tag{8.44}$$

This is approximately one revolution each day, or 2π radian per day, because

$$\begin{aligned}&7.292115090\times 10^{-5}\ \text{[radians/second]}\times 24\ \text{[hours/day]}\times 3{,}600\ \text{[seconds/hour]}\\ =\ &6.300387438\ \text{[radians]}\\ =\ &360.9820095^\circ.\end{aligned} \tag{8.45}$$

This is about 361°, a revolution plus one degree. The Earth revolves around the Sun one cycle per year, thus moving about one degree per day around the Sun. Therefore, the Sun is overhead at approximately noon every day for a given location.

The Coriolis force can be calculated from its definition as $-2m$ times the cross product of the Earth's angular velocity $\boldsymbol{\Omega}$ and the flow velocity $\mathbf{u}$:

$$\mathbf{F}_c = -2m\boldsymbol{\Omega}\times\mathbf{u}. \tag{8.46}$$

This cross product yields the following x, y, z-component expression of the Coriolis force per unit mass:

$$\begin{aligned}\frac{\mathbf{F}_c}{m} &= -2\det\begin{bmatrix}\mathbf{i} & \mathbf{j} & \mathbf{k}\\ 0 & \Omega\cos\phi & \Omega\sin\phi\\ u & v & w\end{bmatrix}\\ &= 2\Omega\left[(v\sin\phi - w\cos\phi)\,\mathbf{i} - u\sin\phi\,\mathbf{j} + u\cos\phi\,\mathbf{k}\right].\end{aligned} \tag{8.47}$$

The notation *det* stands for determinant. See Appendix B for cross product and its representation by a determinant.

The pressure gradient force (PGF) per unit mass can also be written explicitly in terms of x, y, z-components:

$$\frac{\mathbf{F}_p}{m} = -\frac{1}{\rho}\left[p_x\mathbf{i} + p_y\mathbf{j} + p_z\mathbf{k}\right]. \tag{8.48}$$

The gravitational force per unit mass is

$$\frac{\mathbf{F}_g}{m} = -g\mathbf{k}. \tag{8.49}$$

The friction force is complicated and can be due to boundary friction or internal friction (viscosity). In this book, we do not discuss it in detail. Instead, we simply retain the generic notation:

$$\frac{\mathbf{F}_f}{m} = f_{(rx)}\mathbf{i} + f_{(ry)}\mathbf{j} + f_{(rz)}\mathbf{k}, \tag{8.50}$$

where $f_{(rx)}$ is the x-component of the friction force per unit mass, and $f_{(ry)}$ and $f_{(rz)}$ the y- and z-components, respectively.

Finally, the momentum equation in an Eulerian framework and in xyz-components may be written as follows:

$$u_t = -(uu_x + vu_y + wu_z) + 2\Omega(v\sin\phi - w\cos\phi) - \frac{p_x}{\rho} + f_{(rx)}, \tag{8.51}$$

$$v_t = -(uv_x + vv_y + wv_z) - 2\Omega u\sin\phi - \frac{p_y}{\rho} + f_{(ry)}, \tag{8.52}$$

$$w_t = -(uw_x + vw_y + ww_z) + 2\Omega u\cos\phi - \frac{p_z}{\rho} - g + f_{(rz)}, \tag{8.53}$$

where the subscripts x, y, z, t denote the partial derivative with respect to the given variable.

Idealized Eulerian observers could in principle measure every quantity in the above equations, except the friction force. The nonlinearity of the equations arises from the velocity advection and is described by the first three terms in the curved brackets on the right-hand side of equations (8.51)–(8.53).

These momentum equations and the continuity equation (8.18) form the four basic equations of conservation laws to describe atmospheric and oceanic flows. We see immediately that these four equations involve five dependent variables, if the friction force is ignored for the present: the three components of the velocity vector, plus pressure and density. Thus, additional conditions or assumptions will be required.

Due to the numerous varieties of atmospheric and oceanic flows on different spatial and temporal scales, solving these four partial differential equations (PDEs) is a very challenging task. However, there are useful special cases with given conditions.

Next we will briefly discuss two special cases that provide physical insight and can help us understand many important atmospheric and oceanic phenomena:

(a) geostrophic flow, characterized by a balance between the pressure gradient force and the Coriolis force, and
(b) the conservation of potential vorticity.

The following sections will describe these special results from the above four equations derived from conservation laws.

8.4 Geostrophic Approximation of the Momentum Equations

The geostrophic approximation is a simplification of the momentum equations under the following assumptions

(i) The acceleration $D\mathbf{u}/Dt$ is small and can be ignored. This simplification is appropriate for a relatively stationary system that is not changing rapidly.
(ii) The vertical velocity w is much smaller than the horizontal velocity and can be ignored.

(iii) The friction force is small and can be ignored.
(iv) The gravitational force is balanced by the vertical pressure gradient force (PGF), and the entire equation (8.53) for the vertical acceleration in the momentum equations (8.51)–(8.53) can be ignored.

8.4.1 Mathematical Description of the Geostrophic Approximation

Under these assumptions, the three momentum equations (8.51)–(8.53) are reduced to the following two linear equations

$$2\Omega v \sin\phi - \frac{p_x}{\rho} = 0, \tag{8.54}$$

$$-2\Omega u \sin\phi - \frac{p_y}{\rho} = 0. \tag{8.55}$$

These equations are called the geostrophic approximation. The terms in these two equations are defined on the xy-plane and involve only the Coriolis force and the pressure gradient force, also on the xy-plane.

According to Eq. (8.47), the Coriolis force per unit mass on the xy-plane is

$$\frac{\mathbf{F}_{c(xy)}}{m} = 2\Omega v \sin\phi \mathbf{i} - 2\Omega u \sin\phi \mathbf{j}. \tag{8.56}$$

Here, the vertical velocity w is ignored.

According to Eq. (8.48), the pressure gradient force per unit mass on the xy-plane is

$$\frac{\mathbf{F}_{p(xy)}}{m} = -\frac{p_x}{\rho}\mathbf{i} - \frac{p_y}{\rho}\mathbf{j}. \tag{8.57}$$

Thus, the geostrophic approximation equations (8.54) and (8.55) can be written as

$$\mathbf{F}_{c(xy)} + \mathbf{F}_{p(xy)} = 0, \tag{8.58}$$

which explicitly shows the exact balance between the Coriolis force and the PGF on the xy-plane.

8.4.2 Flow Direction Perpendicular to the PGF under the Geostrophic Approximation

The quantity

$$f = 2\Omega \sin\phi \approx 146 \sin\phi \ \text{[microradian/sec]}, \tag{8.59}$$

is called the Coriolis frequency, which is twice the projection of the angular velocity of the rotating Earth on the local vertical coordinate z at latitude ϕ. We will now show that the flow which results from the geostrophic assumption has a direction which is not parallel to the direction of the PGF but is instead perpendicular to the direction of the PGF.

The geostrophic approximation equations (8.54) and (8.55) let us explicitly represent the horizontal velocity in terms of pressure gradient and Coriolis forces.

$$u = -\frac{p_y}{\rho f}, \tag{8.60}$$

$$v = \frac{p_x}{\rho f}. \tag{8.61}$$

Substitution of the above two equations for (u, v) into the following dot product

$$(u, v) \cdot (p_x, p_y) = \frac{1}{\rho f}(-p_y, p_x) \cdot (p_x, p_y) = \frac{1}{\rho f}(p_x p_y - p_y p_x) = 0 \tag{8.62}$$

yields zero, which implies that the two vectors (u, v) and (p_x, p_y) are perpendicular, i.e., the flow velocity is orthogonal to PGF. Thus, the flow is parallel to the isobars, which are curves on the xy-plane on which pressure is constant. The PGF, however, is perpendicular to the isobars.

The large-scale flows in much of both the atmosphere and the ocean are approximately geostrophic. We illustrate geostrophic flow here using an atmospheric example. See Fig. 8.2 for an example of the isohypses and wind velocities on a map showing a forecast of the atmosphere at a pressure altitude of 500 mb over the North Atlantic at a specific time on March 27, 2018. Here, isohypses, lines of equal geopotential height, replace isobars.

The figure shows that the wind flows approximately parallel to the isohypses in most locations. There the geostrophic approximation is largely valid. However, there also exist many locations where the wind does not flow along the isohypses. In these locations, the geostrophic approximation is less valid or not valid, because at least one of the four assumptions for the geostrophic approximation is violated, such as the nonlinear advective terms not being negligibly small.

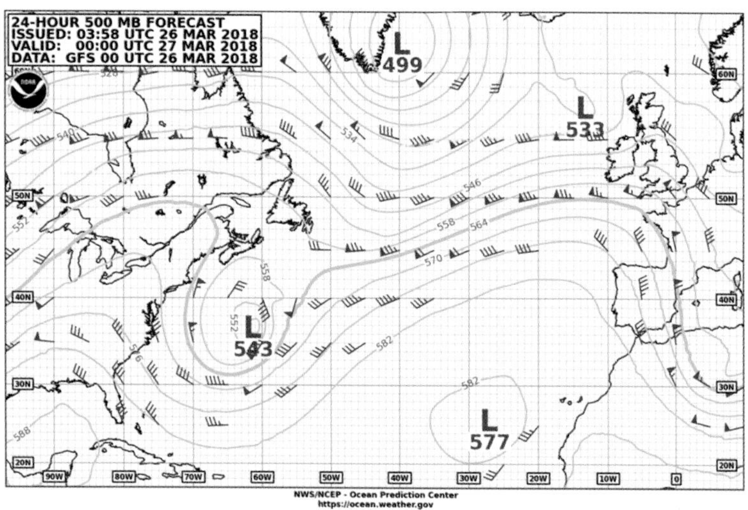

Fig. 8.2 A forecast map of geopotential height of 500 mb, March 27, 2018: isohypses, wind barbs, and centers of low pressures. The wind barb's handle points in the direction of the wind flow vector. Each full tick of the wind barb indicates 10 knots (1 knot = 1.852 km hour^{-1}), and each half tick indicates 5 knots. Each triangle flag indicates 50 knots. *Credit: NOAA Ocean Prediction Center. Source:* https://ocean.weather.gov

8.5 The Potential Vorticity Conservation Equation

We now very briefly introduce the concept of vorticity, which is of critical importance in discussing the motions of both the ocean and the atmosphere. Vorticity is defined as the curl of velocity, and so is mathematically expressed as

$$\zeta = \nabla \times \mathbf{u}, \tag{8.63}$$

where $\mathbf{u} = (u, v, w)$ is the velocity vector.

8.5.1 Absolute Vorticity and Relative Vorticity

The definition of the cross product (see Appendix B)

$$\nabla \times \mathbf{u} = \det \begin{bmatrix} \mathbf{i} & \mathbf{j} & \mathbf{k} \\ \partial/\partial x & \partial/\partial y & \partial/\partial z \\ u & v & w \end{bmatrix} \tag{8.64}$$

implies that the vorticity vector is equal to

$$\zeta = \nabla \times \mathbf{u} = (w_y - v_z)\mathbf{i} - (w_x - u_z)\mathbf{j} + (v_x - u_y)\mathbf{k}. \tag{8.65}$$

Thus, the local vorticity component in the vertical z-direction (i.e., $\mathbf{k}$ direction) that points from the Earth's center to the point of interest is

$$\zeta^{(z)} = v_x - u_y. \tag{8.66}$$

This quantity is defined in terms of the components of the velocity vector in the coordinate system rotating with the Earth. This velocity is known as the relative velocity, and the vorticity associated with this velocity is called the relative vorticity, because these definitions are relative to the rotating coordinate system.

However, the rotation of the planet Earth adds an additional component to the velocity and vorticity defined in this relative way. This additional component is called the planetary vorticity, and it is simply the local vertical component of the vorticity of the Earth, which is the Coriolis frequency defined by (8.59). Taking account of this additional component gives the absolute vorticity, which is the vorticity measured in an inertial coordinate system. Thus, the absolute vorticity is the sum of the relative vorticity defined by (8.66) and the Coriolis frequency defined by (8.59):

$$\zeta_a = f + \zeta^{(z)}. \tag{8.67}$$

8.5.2 Potential Vorticity and Its Conservation

A very important concept in the dynamics of the ocean and the atmosphere is the potential vorticity, which has different definitions depending on the specific characteristics of the system being considered. For a homogeneous incompressible system, which is a system in which the density is a constant that does not vary in space or time, the potential vorticity is defined as

$$\zeta_p = \frac{f+\zeta}{H}. \tag{8.68}$$

Here, H is the depth of the fluid, which we know must be finite, because the density is constant; and ζ replaces the vertical component $\zeta^{(z)}$ of the vorticity vector for simplicity. Note that the numerator has the dimension of vorticity, and the denominator H has the dimension of length, so potential vorticity defined in this way does not have the dimension of vorticity. Under certain simplifying assumptions, we will prove that the potential vorticity of an ocean water column defined in this way is a constant during the motion of the column. Namely, if a water column moves from location P_1 to another location P_2, then we have $\zeta_{p1} = \zeta_{p2}$:

$$\frac{f_1+\zeta_1}{H_1} = \frac{f_2+\zeta_2}{H_2}. \tag{8.69}$$

8.5.3 Mathematical Derivations of the Conservation of Potential Vorticity

Here we present a simple and beautiful derivation of a potential vorticity conservation equation due to Carl-Gustaf Rossby (1898–1957).

The assumptions for potential vorticity conservation are as follows:

(i) The friction force is small and ignored.
(ii) The gravity force is balanced by the vertical pressure gradient, the entire equation for the vertical acceleration Dw/Dt can be ignored, and thus the momentum equation will consist of only the component on the horizontal plane.
(iii) The vertical velocity w is small compared with u and v and can be ignored when it appears in a formula together with u and v (but w cannot be ignored when it is not used together with u and v).
(iv) However, in the continuity equation, the term $\partial w/\partial z$ cannot be ignored.

These assumptions for potential vorticity conservations are different from those of the geostrophic approximation in the following two aspects:

(a) the horizontal acceleration must be included, and
(b) the three-dimensional continuity equation must be used.

The above assumptions imply the following momentum equations

$$u_t = -(uu_x + vu_y) - fv - p_x/\rho, \tag{8.70}$$

$$v_t = -(uv_x + vv_y) + fu - p_y/\rho. \tag{8.71}$$

These can be written as

$$u_t + (uu_x + vu_y) + fv = -p_x/\rho, \tag{8.72}$$

$$v_t + (uv_x + vv_y) - fu = -p_y/\rho. \tag{8.73}$$

From these two equations, we can compute vorticity: $\zeta^{(z)} = v_x - u_y$. Taking $\partial/\partial x$ for Eq. (8.73) minus $\partial/\partial y$ for Eq. (8.72) leads to the following

$$(v_x - u_y)_t + u(v_x - u_y)_x + v(v_x - u_y)_y + v_x(u_x + v_y) - u_y(u_x + v_y) + (fu)_x + (fu)_y = 0. \tag{8.74}$$

This simplifies to

$$\zeta_t + u(\zeta + f)_x + v(\zeta + f)_y + (\zeta + f)(u_x + v_y) = 0. \tag{8.75}$$

Here, the vertical local vorticity component $\zeta^{(z)}$ has been simplified to ζ, since the two horizontal vorticity components $\zeta^{(x)}$ and $\zeta^{(y)}$ are sufficiently small to be neglected.

The continuity equation for an incompressible fluid

$$u_x + v_y + w_z = 0 \tag{8.76}$$

leads to

$$(\zeta + f)_t + u(\zeta + f)_x + v(\zeta + f)_y - (\zeta + f)w_z = 0, \tag{8.77}$$

or

$$\frac{D}{Dt}(\zeta + f) - (\zeta + f)w_z = 0. \tag{8.78}$$

The z derivative gives us a hint that we should integrate this equation vertically in the fluid domain shown in Fig. 8.3 from the bottom $z = b(x,y)$ to the surface $z = b(x,y) + H(x,y,t)$. This integration yields

$$\frac{D}{Dt}(\zeta + f)H + (\zeta + f)\left[w(x,y,b,t) - w(x,y,z,b + H,t)\right] = 0. \tag{8.79}$$

In the integration, we have assumed that the absolute vorticity $\zeta + f$ does not depend on z and is only a function of x, y, t.

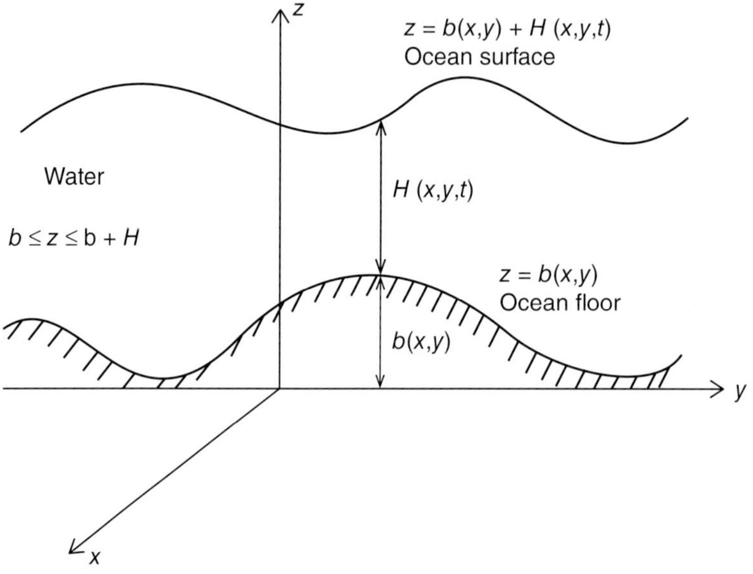

Fig. 8.3 The ocean water domain for the derivation of the vorticity conservation equation: from the three-dimensional ocean bottom to the three-dimensional free surface at the top.

If we think of this fluid column as being seawater, the kinematic boundary conditions for the bottom and the free surface at the top of the column are that a small fluid parcel at the boundary remains at the boundary:

$$\frac{D}{Dt}(z-b)=0 \quad \text{(at ocean bottom)} \tag{8.80}$$

$$\frac{D}{Dt}[z-(b+H)]=0 \quad \text{(on sea surface)}. \tag{8.81}$$

From the definition of a material derivative, we have

$$\frac{\partial}{\partial t}(z-b)+u\frac{\partial}{\partial x}(z-b)+v\frac{\partial}{\partial y}(z-b)=0 \tag{8.82}$$

$$\frac{\partial}{\partial t}[z-(b+H)]+u\frac{\partial}{\partial x}[z-(b+H)]+v\frac{\partial}{\partial y}[z-(b+H)]=0. \tag{8.83}$$

These can be simplified to

$$w(x,y,b,t)=ub_x+vb_y \tag{8.84}$$

$$w(x,y,b+H,t)=H_t+u(b+H)_x+v(b+H)_y, \tag{8.85}$$

where $\partial z/\partial t = w, \partial b/\partial t = 0, \partial z/\partial x = 0, \partial z/\partial y = 0$. The first equation above minus the second yields

$$\begin{aligned} & w(x,y,b+H,t)-w(x,y,b,t) \\ &= H_t+u(b+H)_x+v(b+H)_y-(ub_x+vb_y) \\ &= -(H_t+uH_x+vH_y) \\ &= -\frac{DH}{Dt}. \end{aligned} \tag{8.86}$$

It was defined earlier that

$$\zeta_a=\zeta+f \tag{8.87}$$

is the absolute vorticity, sometimes called the total vorticity. Using this notation ζ_a and from the above two formulas, Eq. (8.79) becomes

$$\frac{D\zeta^{(a)}}{Dt}H-\zeta^{(a)}\frac{DH}{Dt}=0. \tag{8.88}$$

The quotient rule of differentiation

$$\left(\frac{f}{g}\right)_x=\frac{f_xg-g_xf}{g^2} \tag{8.89}$$

is applicable to the material derivative D/Dt. Dividing equation (8.88) by H^2 leads to

$$\frac{\frac{D\zeta_a}{Dt}H-\zeta_a\frac{DH}{Dt}}{H^2}=\frac{D}{Dt}\left(\frac{\zeta_a}{H}\right)=0. \tag{8.90}$$

This completes the derivation of the potential vorticity conservation principle:

$$\frac{D\zeta_p}{Dt}=0 \tag{8.91}$$

where $\zeta_p = \zeta_a/H$ is the potential vorticity. The zero material derivative implies that the potential vorticity of a water column is a constant along the parcel's path of motion.

If a water column of the ocean moves from a deeper region to the shallower coastal region, the depth H becomes smaller, which forces the total vorticity $\zeta_a = \zeta + f$ to increase. If the move is approximately at the same latitude, then the local vorticity $\zeta^{(z)} = v_x - u_y$ must increase.

Here we have barely scratched the surface of the significance of the concept of potential vorticity and its conservation. For an excellent discussion at an advanced level, see Vallis (2017). Rossby was the first to use the term "potential vorticity." He did so in an atmospheric context. Vallis (2017) quotes from the 1940 paper by Rossby who wrote, "This quantity, which may be called the potential vorticity, represents the vorticity the air column would have if it were brought, isopycnally or isentropically, to a standard latitude and stretched or shrunk vertically to a standard depth." Also see the standard textbooks by Talley et al. (2011) and Wallace and Hobbs (2006).

8.6 Chapter Summary

This chapter describes some basic equations in climate science from the perspective of conservation laws: the conservation of mass, momentum, energy, potential vorticity, and others. Some of these laws are foundational in developing climate models for simulating the past climate and predicting the future climate under different scenarios of human–Earth interactions. Others have proven to be valuable in developing physical insight into how the climate system works. The development of these conservation laws uses calculus extensively. In climate modeling, the atmosphere or ocean is divided into small volumes, called grid cells in climate modeling language. When a proper coordinate framework is introduced, the conservation of mass, momentum, and energy in a small volume of the atmosphere or ocean can yield differential equations. The commonly used coordinate framework is the Eulerian framework which is fixed on the rigid rotating Earth, has its z-axis in the Earth's radial direction, and describes atmospheric and oceanic flows on the rotating Earth. In this framework, the conservation laws are expressed by PDEs with the independent variables as x, y, z, t and the dependent variables as the unknown climate parameters, such as atmospheric and oceanic velocity, pressure, and temperature. Because of the Earth's rotation, the Coriolis force is an essential part of the equations expressing the conservation laws. A summary of the conservation laws described in this chapter is as follows.

(i) The law of mass conservation is that mass cannot be created or destroyed. Based on this law and the assumption of continuous media, the application of the divergence theorem and the mean value theorem of calculus to a small volume of fluid leads to the continuity equation for fluid dynamics (8.28):

$$\frac{D\rho}{Dt} + \rho \nabla \cdot \mathbf{u} = 0. \tag{8.92}$$

(ii) Ocean water is often considered incompressible. The density of water in a small parcel then does not change during the parcel's motion. This means that the density's material derivative, often called total derivative, is zero:

$$\frac{D\rho}{Dt} = 0. \tag{8.93}$$

Based upon the general continuity equation, this is equivalent to

$$\nabla \cdot \mathbf{u} = 0. \tag{8.94}$$

This means that an incompressible fluid flow must have a divergence-free velocity field. The atmosphere is a compressible fluid and is not divergence-free.

(iii) The law of momentum conservation over a grid cell in an Eulerian coordinate framework yields the following momentum equations for fluid dynamics (i.e., equations (8.51)–(8.53))

$$u_t = -(uu_x + vu_y + wu_z) + 2\Omega(v\sin\phi - w\cos\phi) - \frac{p_x}{\rho} + f_{(rx)}, \tag{8.95}$$

$$v_t = -(uv_x + vv_y + wv_z) - 2\Omega u\sin\phi - \frac{p_y}{\rho} + f_{(ry)}, \tag{8.96}$$

$$w_t = -(uw_x + vw_y + ww_z) + 2\Omega u\cos\phi - \frac{p_z}{\rho} - g + f_{(rz)}. \tag{8.97}$$

Here, the gravitational force, Coriolis force, and friction force are included. The continuity equation and these three equations form a set of nonlinear PDEs with unknowns as u, v, w, p, and ρ. These nonlinear PDEs are extremely difficult to solve in general. Numerical methods and additional assumptions are needed to solve these equations, except for some simplified special cases.

(iv) The geostrophic approximation results from a special assumption that the momentum equation retains only the balance of the Coriolis force and the pressure gradient force on the horizontal xy-plane, and all the other terms are small enough to be ignored. From the point of view of mathematics, this simplification is severe and reduces the three nonlinear momentum PDEs into only two linear PDEs (i.e., Eqs. (8.54) and (8.55)):

$$2\Omega v\sin\phi - \frac{p_x}{\rho} = 0, \tag{8.98}$$

$$-2\Omega u\sin\phi - \frac{p_y}{\rho} = 0. \tag{8.99}$$

These two linear equations provide insight and can serve as adequate descriptions of many of the large-scale features of the motions of both the ocean and the atmosphere. Geostrophic winds are parallel to isobars, so pressure fields can often serve as good proxies for the wind fields. On a map of the geopotential height of a constant pressure surface, such as Fig. 8.2, isohypses play the role of isobars.

(v) Vorticity measures the strength of the fluid rotational motion and is defined as the curl of the velocity field, i.e., Eq. (8.63):

$$\zeta = \nabla \times \mathbf{u}. \tag{8.100}$$

The conservation of potential vorticity states that the potential vorticity ζ_p for an ocean water column under idealized approximations does not change in its path of motion, i.e.,

$$\zeta_{p1} = \zeta_{p2}, \tag{8.101}$$

or

$$\frac{f_1 + \zeta_1}{H_1} = \frac{f_2 + \zeta_2}{H_2}, \tag{8.102}$$

for any locations P_1 and P_2 on the path of the column motion. Here, $f_i \approx 146 \times 10^{-6} \sin\phi_i$ [1/s] is the Coriolis frequency at latitude ϕ_i $(i = 1,2)$. This law of potential vorticity conservation can be invaluable in explaining many atmospheric and physical oceanographic phenomena. This law is a result derived from the laws of both conservation of mass and momentum.

The four equations of fluid dynamics derived from the conservation of mass and momentum can govern infinitely many kinds of fluid motion existing in nature. When we examine a certain specific circulation feature, an appropriate simplification of these four equations can lead to fruitful and insightful results. The well-known geostrophic approximation is such a simplification. Many other idealizations and simplifications have been explored in the literature of geophysical fluid dynamics.

References and Further Readings

[1] Donner, L., W. Schubert, and R. C. J. Somerville (eds.), 2011: *The Development of Atmospheric General Circulation Models: Complexity, Synthesis and Computation.* Cambridge University Press, New York.

> The book is a multi-authored treatment, and many of the authors are leading pioneers in the field of climate modeling.

[2] Goose, H., 2015: *Climate System Dynamics and Modeling,* Cambridge University Press, New York.

> This is a textbook, a reference book, and an excellent and very readable introduction to climate dynamics and climate modeling. It is aimed at graduate students and includes a glossary and exercises.

[3] Holton, J. R. and G. J. Hakim, 2012: *An Introduction to Dynamic Meteorology,* 5th Edition. Academic Press, New York.

> The standard text, known to an entire generation of atmospheric science students.

[4] Persson, A. 1998: How do we understand the Coriolis force? *Bulletin of the American Meteorological Society*, 79, 1373–1385.

An interesting summary of the history of the concept of the Coriolis force.

[5] Talley, L. D., G. L. Pickard, W. J. Emery, and J. H. Swift, 2011: *Descriptive Physical Oceanography: An Introduction.* Academic Press, 6th Edition, New York.

A popular standard text in physical oceanography with many creative and informative figures.

[6] Vallis, G. K, 2017: *Atmospheric and Oceanic Fluid Dynamics: Fundamentals and Large-Scale Circulation.* Cambridge University Press, New York.

Geoffrey K. Vallis is a professor of mathematics and an expert in climate dynamics, the circulation of planetary atmospheres, and dynamical meteorology and oceanography.

[7] Wallace, J. M. and P. V. Hobbs, 2006: *Atmospheric Science: An Introductory Survey.* Academic Press, 2nd Edition, New York.

A classic standard textbook, much loved by generations of atmospheric science students.

Exercises

Exercises 8.3 and 8.4 require the use of a computer.

8.1 Following the derivation approach for the 1D mass conservation law shown in Equations (8.19)–(8.22), derive the 3D mass conservation law (8.25). Draw a schematic 3D cubic mass parcel to assist the derivation.

8.2 (a) Describe the four assumptions of the geostrophic approximation in Section 8.4 in terms of mathematical formulas in the three momentum equations (8.51)–(8.53).

(b) Following (a), derive the geostrophic approximation equations (8.54) and (8.55) from (8.51)–(8.53).

8.3 Calculate the Coriolis force of the Gulf Stream current at 100 km east of Cape Hatteras (35.3°N, 75.5°W), North Carolina. The current velocity there is assumed to be 1.0 m s^{-1}.

8.4 For the coastal region of the Cape of Good Hope, the ocean current speed can change from 5–8 m s^{-1} to near zero m s^{-1} in a short range of 30–60 km near the Cape of Good Hope (34.3581°S, 18.4719°E).

(a) Estimate the range of local vorticities. *Hint: The maximum local vorticity may be estimated to be* $\zeta^{(z)} = 8/(30 \times 1000) = 27 \times 10^{-5}$[radian s^{-1}].

(b) Search the Internet to find local vorticity data at other regions, such as the Gulf Stream east of Cape Hatteras, and compare the magnitudes of these local vorticities with that of the coastal region of the Cape of Good Hope.

8.5 State and prove the quotient rule of the total derivative. *Hint: A special case of the rule is shown as follows and was used in the derivation of the potential vorticity conservation in the text.*

$$\begin{aligned}\frac{D}{Dt}\left(\frac{\zeta_a}{H}\right) &= \left(\frac{\zeta_a}{H}\right)_t + u\left(\frac{\zeta_a}{H}\right)_x + v\left(\frac{\zeta_a}{H}\right)_y \\ &= \frac{(\zeta_a)_t H - H_t\zeta_a}{H^2} + u\frac{(\zeta_a)_x H - H_x\zeta_a}{H^2} + v\frac{(\zeta_a)_y H - H_y\zeta_a}{H^2} \\ &= \frac{1}{H^2}\left[((\zeta_a)_t + u(\zeta_a)_x + v(\zeta_a)_y)H - (H_t + uH_x + vH_y)\zeta_a\right] \\ &= \frac{\left[H\frac{D\zeta_a}{Dt} - \zeta_a\frac{DH}{Dt}\right]}{H^2}.\end{aligned} \tag{8.103}$$

8.6 Following the materials of Section 8.5.3, write down the detailed steps to prove the principle of the conservation of potential vorticity (8.91).

8.7 Compute the potential vorticity $\zeta_p = (f + \zeta^{(z)})/H$ for the coastal current of the Cape of Good Hope. Here, f is the planetary vorticity, $\zeta^{(z)}$ is the local vorticity, and H is the depth of the ocean water. However, the data of H are not given. You may assume a depth profile to estimate the potential vorticity. Discuss the conservation of the potential vorticity when a column of ocean water with vorticity moves toward the coastline.

9 R Graphics for Climate Science

Research very often includes the need for graphic displays of scientific data. Such graphics are necessary both to carry out the research and to present it to others in formats such as publications and talks. If your objective is to become a master of good graphical design, presenting quantitative scientific information in an artistic way that is both aesthetically appealing and informative, then we commend you for this worthy goal, and we suggest you might begin with careful study of the visualization books by Edward R. Tufte.

Our goal in this book is much more modest. In the final three chapters, beginning with this one, we want to show you how versatile and useful the R language can be in both graphics and data analysis. In this chapter, we demonstrate that R is well suited for producing a wide variety of graphics including simple line plots and color contour maps. We illustrate these capabilities of R using a wide variety of climate data.

This chapter is an introduction to the basic skills needed to use R graphics for climate science. These skills are sufficient to meet most needs for climate science research, teaching, and publications. We have divided these skills into the following categories:

(i) plotting multiple data time series in the same figure, including multiple panels in a figure, adjusting margins, and using proper fonts for text, labels, and axes;
(ii) creating color maps of a climate parameter, such as the surface air temperature on the globe or over a given region; and
(iii) animating plots.

9.1 Two-Dimensional Line Plots and Setups of Margins and Labels

R can generate almost all the two-dimensional (2D) line plots for climate science applications. The *R Graphics Cookbook* by Chang (2012) provides details of simple R graphics for statistical analysis of data. This section describes two skills of 2D line plotting that are commonly used in climate science: (a) Putting several time series of two different units on the same figure, and (b) adjusting the margins and labels to meet various kinds of application demands.

9.1.1 Plot Two Different Time Series on the Same Plot

In Chapter 3, we showed how to plot a simple time series using `plot(xtime, ydata)`. Climate science often requires one to plot two different quantities, such as two time series,

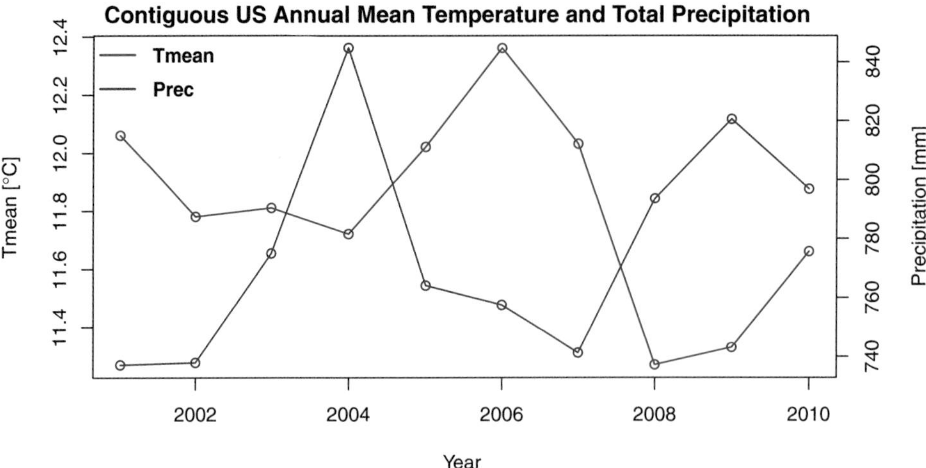

Fig. 9.1 Contiguous United States annual mean temperature and annual total precipitation.

on the same plot so that direct comparisons can be made. For example, to see whether a hot year is also a dry year, one may plot the temperature data on the same figure as the precipitation data. The left side of the y-axis shows temperature and the right side shows precipitation. The following code plots a figure containing the contiguous United States (CONUS) annual mean temperature and annual total precipitation from 2001–2010 (see Fig. 9.1).

```
setEPS()
postscript("fig0901.eps", height=4, width=8)
par(mar=c(4.0,4.2,1.8,4.1))
Time = 2001:2010
Tmean = c(12.06,11.78,11.81,11.72,12.02,12.36,12.03,11.27,11.33,11.66)
Prec = c(737.11,737.87,774.95,844.55,764.03,757.43,741.17,793.50,
       820.42,796.80)
plot(Time,Tmean,type="o",col="red", lwd=1.5, xlab="Year",
     ylab=expression(paste("Tmean [", degree,"C]")),
main="Contiguous U.S. Annual Mean Temperature and Total Precipitation")
legend(2000.5,12.42, col=c("red"),lty=1,lwd=2.0,
       legend=c("Tmean"),bty="n",text.font=2,cex=1.0)
#Allows a figure to be overlaid on the first plot
par(new=TRUE)
plot(Time, Prec,type="o",col="blue",lwd=1.5,axes=FALSE,xlab="",ylab="")
legend(2000.5,839, col=c("blue"),lty=1,lwd=2.0,
       legend=c("Prec"),bty="n",text.font=2,cex=1.0)
#Suppress the axes and assign the y-axis to side 4
axis(4)
mtext("Precipitation [mm]",side=4,line=3)
#legend("topleft",col=c("red","blue"),lty=1,legend=c("Tmean","Prec"),
#        cex=0.6)
#Plot two legends at the same time make it difficult to adjust
#the font size because of different scale
dev.off()
```

Figure 9.1 shows that during the ten years from 2001 to 2010, the CONUS precipitation and temperature are in opposite phase: higher temperature tends to occur in dry years

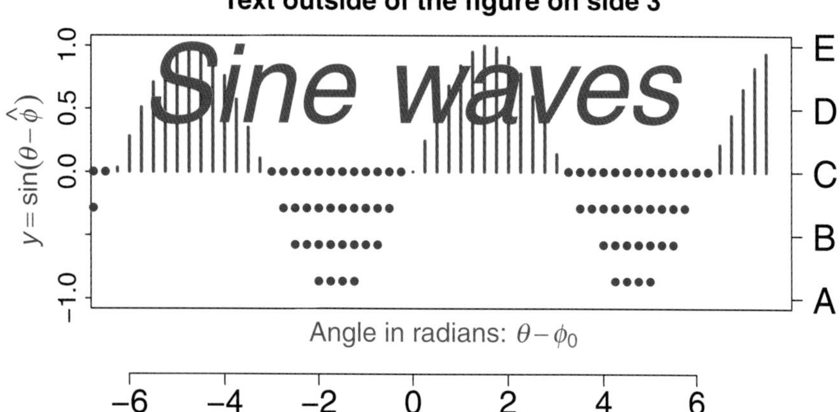

Fig. 9.2 Illustrating how to set margins, insert mathematical symbols, and write text outside a figure.

with less precipitation, and lower temperature tends to occur in wet years with more precipitation.

9.1.2 Figure Setups: Margins, Fonts, Mathematical Symbols, and More

R has the flexibility to create plots with specific margins, mathematical symbols for text and labels, text fonts, text size, and more. R also allows one to merge multiple figures. These capabilities are often useful in producing a high-quality figure for presentations or publication.

`par(mar=c(2,5,3,1))` specifies the four margins of a figure. The first margin 2 (i.e., two line space) is the x-axis, the second 5 is for the y-axis, 3 is for the top, and 1 is for the right. One can change the numbers in `par(mar=c(2,5,3,1))` to adjust the margins. A simple example is shown in Fig. 9.2, which may be generated by the following R program.

```
#Margins, math symbol, and figure setups
setEPS()
postscript("fig0902.eps", height=4, width=8)
#Margins, math symbol, and figure setups
par(mar=c(5,4.5,2.5,2.5))
x<-0.25*(-30:30)
y<-sin(x)
x1<-x[which(sin(x) >=0)]
y1<-sin(x1)
x2<-x[which(sin(x) < 0)]
y2<-sin(x2)
plot(x1,y1,xaxt="n", xlab="",ylab="",lty=1,type="h",
     lwd=3, tck=-0.02, ylim=c(-1,1), col="red",
     col.lab="purple",cex.axis=1.4)
lines(x2,y2,xaxt="n", xlab="",ylab="",lty=3,type="h",
      col="blue",lwd=8, tck=-0.02)
axis(1, at=seq(-6,6,2),line=3, cex.axis=1.8)
axis(4, at=seq(-1,1,0.5), lab=c("A", "B", "C", "D","E"),
     cex.axis=2,las=2)
text(0,0.7,font=3,cex=6, "Sine waves", col="darkgreen") #Itatlic font
mtext(side=2,line=2, expression(y==sin(theta-hat(phi))),
      cex=1.5, col="blue")
```

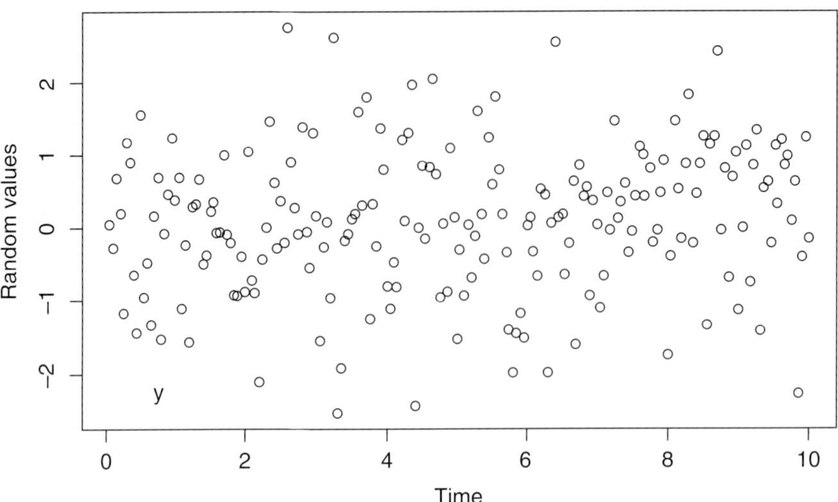

Fig. 9.3 Illustrating how to adjust font size, axis labels space, and margins.

```
mtext(font=2,"Text outside of the figure on side 3",
      side=3,line=1, cex=1.5)#Bold font
mtext(font=1, side=1,line=1,
      expression(paste("Angle in radians: ",
                       theta-phi[0])),cex=1.5, col="red")
dev.off()
```

Similar to using `cex.axis=1.8` to change the font size of the tick values, one can use `cex.lab=1.5, cex.main=1.5, cex.sub=1.5` to change the font sizes for axis labels, the main title, and the subtitle. An example is shown in Fig. 9.3 generated by the R code below.

```
par(mar=c(8,6,3,2))
par(mgp=c(2.5,1,0))
plot(1:200/20, rnorm(200),sub="Sub-title: 200 random values",
     xlab= "Time", ylab="Random values", main="Normal random values",
     cex.lab=1.5, cex.axis=2, cex.main=2.5, cex.sub=2.0)
```

Here `par(mgp=c(2.5,1,0))` is used to adjust the positions of axis labels, tick values, and tick bars, where 2.5 means the xlab is two and a half lines away from the figure's lower and left borders, 1 means the x-axis tick values are one line away from the borders, 0 means the tick bars are on the border lines. The default mgp values are 3,1,0. Another simple example is below.

```
par(mgp=c(2,1,0))
plot(sin,xlim=c(10,20))
```

The above R code used many R plot functions. An actual climate science line plot is often simpler than this. One can simply remove the redundant functions in the above R code to produce the desired figure.

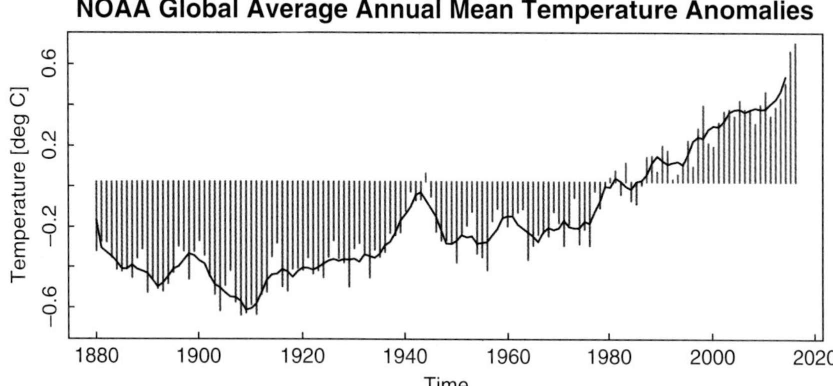

Fig. 9.4 Global average annual mean SAT based on the NOAAGlobalTemp data.

Let us plot the global average annual mean surface air temperature (SAT) from 1880–2016 using the above plot functions (see Fig. 9.4). The data is from the NOAAGlobalTemp dataset

www.ncdc.noaa.gov/data-access/marineocean-data/noaa-global-surface-temperature-noaaglobaltemp

We write the data in two columns in a file named `NOAATemp`. The first column is the years, and the second is the temperature anomalies.

Figure 9.4 can be generated by the following R code.

```
#A fancy plot of the NOAAGlobalTemp time series
setwd("/Users/sshen/climmath")
NOAATemp =
  read.table("data/aravg.ann.land_ocean.90S.90N.v4.0.1.2016.txt",
             header=F)
par(mar=c(4,4,3,1))
x<-NOAATemp[,1]
y<-NOAATemp[,2]
z<-rep(-99,length(x))
for (i in 3:length(x)-2) z[i]=mean(c(y[i-2],y[i-1],y[i],y[i+1],y[i+2]))
n1<-which(y>=0)
x1<-x[n1]
y1<-y[n1]
n2<-which(y<0)
x2<-x[n2]
y2<-y[n2]
x3<-x[2:length(x)-2]
y3<-z[2:length(x)-2]
plot(x1,y1,type="h",xlim=c(1880,2016),lwd=3,
     tck=0.02, ylim=c(-0.7,0.7), #tck>0 makes ticks inside the plot
     ylab="Temperature [deg C]",
     xlab="Time",col="red",
     main="NOAA Global Average Annual Mean Temperature Anomalies")
lines(x2,y2,type="h",
      lwd=3, tck=-0.02,  col="blue")
lines(x3,y3,lwd=2)
```

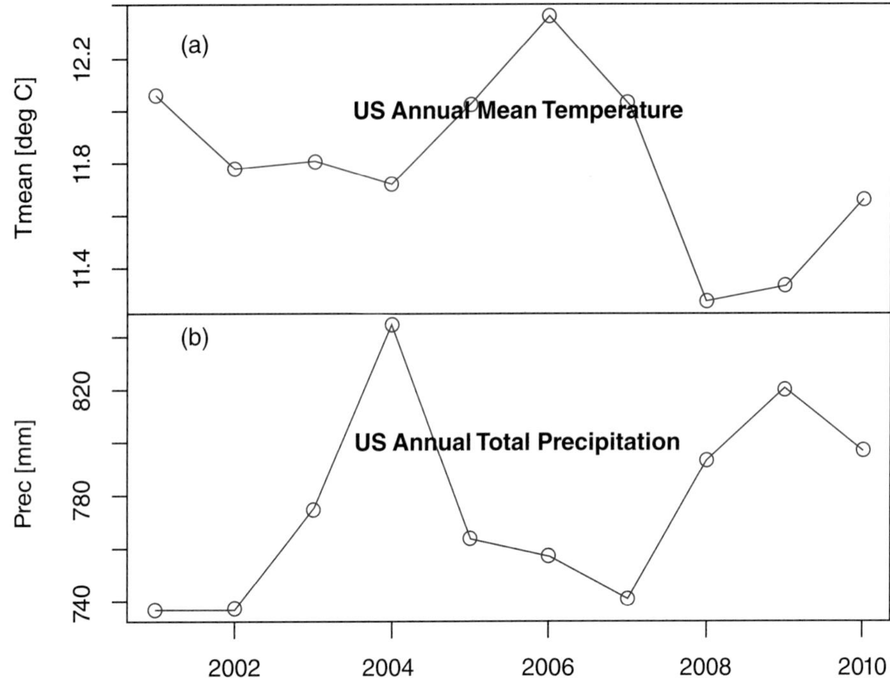

Fig. 9.5 (a) Contiguous United States annual mean temperature; and (b) annual total precipitation.

9.1.3 Plot Two or More Panels on the Same Figure

Another way to compare the temperature and precipitation time series is to plot them in different panels and display them in one figure, as shown in Fig. 9.5.

Figure 9.5 can be generated by the following R code. This figure's arrangement has used the setups described in the previous subsection.

```
#Plot US temp and prec times series on the same figure
par(mfrow=c(2,1))
par(mar=c(0,5,3,1)) #Zero space between (a) and (b)
Time = 2001:2010
Tmean = c(12.06,11.78,11.81,11.72,12.02,12.36,12.03,11.27,11.33,11.66)
Prec = c(737.11,737.87,774.95,844.55,764.03,757.43,741.17,793.50,
         820.42,796.80)
plot(Time,Tmean,type="o",col="red",xaxt="n",
     xlab="",ylab="Tmean [deg C]")
text(2006, 12,font=2,"US Annual Mean Temperature", cex=1.5)
text(2001.5,12.25,"(a)")
#Plot the panel on row 2
par(mar=c(3,5,0,1))
plot(Time, Prec,type="o",col="blue",xlab="Time",ylab="Prec [mm]")
text(2006, 800, font=2, "US Annual Total Precipitation", cex=1.5)
text(2001.5,840,"(b)")
```

After completing this figure, the R console may "remember" the setup. When you plot the next figure expecting the default setup, R may still use the previous setup. One can remove the R "memory" by

```
rm(list=ls())
plot.new()
```

A more flexible way to stack multiple panels together as a single figure is to use the `layout` matrix. The following example has three panels on a 2-by-2 matrix space. The first panel occupies the first row's two positions. Panels 2 and 3 occupy the second row's two positions.

```
layout(matrix(c(1,1,2,3), 2, 2, byrow = TRUE),
       widths=c(3,3), heights=c(2,2))
plot(sin,type="l", xlim=c(0,20))
plot(sin,xlim=c(0,10))
plot(sin,xlim=c(10,20))
```

This layout setup does not work for the plot function `filled.contour` described in the next section, since it has already used a layout and overwrites any other layout.

9.2 Color Contour Maps

Modern color contour maps, instead of the traditional black–white line contours, are routinely used in displaying weather forecasting products and many other kinds of data. Colors can effectively represent values of a meteorological parameter, such as temperature, pressure, and precipitation. R is able to use colors in many kinds of plots, including color contour maps.

9.2.1 Basic Principles for an R Contour Plot

The basic principles for an R contour plot are below.

(i) The main purpose of a contour plot is to show a 3D surface with contours or filled contours, or simply a color map for a climate parameter;
(ii) (x,y,z) coordinates data or a function $z = f(x,y)$ should be given; and
(iii) a color scheme should be defined, such as `color.palette = heat.colors`.

A few simple examples are below.

```
x <- y <- seq(-1, 1, len=25)
z <- matrix(rnorm(25*25),nrow=25)
contour(x,y,z, main="Contour Plot of Normal Random Values")
filled.contour(x,y,z, main="Filled Contours of Normal Random Values")
filled.contour(x,y,z, color.palette = heat.colors)
filled.contour(x,y,z,
  color.palette = colorRampPalette(c("red", "white", "blue")))
```

9.2.2 Plot Contour Color Maps for Random Values on a Map

For climate applications, a contour plot is often overlaid on a geographic map, such as a world map or a map of a country or a region. Our first example illustrates a very simple

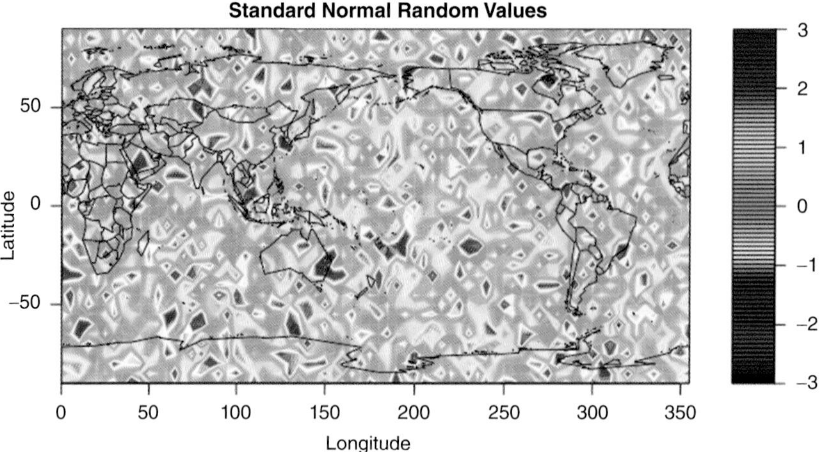

Fig. 9.6 Color map of standard normal random values on a $5^\circ \times 5^\circ$ grid over the globe.

color plot over the world: plotting standard normal random values on a $5^\circ \times 5^\circ$ grid over the globe. See Fig. 9.6.

```
#Plot a 5-by-5 grid global map of standard normal random values
library(maps)
plot.new()
#Step 1: Generate a 5-by-5 grid (pole-to-pole, lon 0 to 355)
Lat<-seq(-90,90,length=37) #Must be increasing
Lon<-seq(0,355,length=72) #Must be increasing
#Step 2: Generate the random values
mapdat<-matrix(rnorm(72*37),nrow=72)
#The matrix uses lon as row going and lat as column
#Each row includes data from south to north
#Define color
int=seq(-3,3,length.out=81)
rgb.palette=colorRampPalette(c('black','purple','blue','white',
'green', 'yellow','pink','red','maroon'), interpolate='spline')
#Step 3: Plot the values on the world map
filled.contour(Lon, Lat, mapdat, color.palette=rgb.palette, levels=int,
               plot.title=title(xlab="Longitude", ylab="Latitude",
main="Standard Normal Random Values on a World Map: 5 Lat-Lon Grid"),
plot.axes={ axis(1); axis(2);map('world2', add=TRUE);grid()}
)
#filled.contour() is a contour plot on an x-y grid.
#Background maps are added later in plot.axes={}
#axis(1) means ticks on the lower side
#axis(2) means ticks on the left side
```

Similarly one can plot a regional map. See Fig. 9.7.

```
#Plot a 5-by-5 grid regional map to cover USA and Canada
Lat3<-seq(10,70,length=13)
Lon3<-seq(230,295,length=14)
mapdat<-matrix(rnorm(13*14),nrow=14)
int=seq(-3,3,length.out=81)
rgb.palette=colorRampPalette(c('black','purple','blue','white',
'green', 'yellow','pink','red','maroon'), interpolate='spline')
```

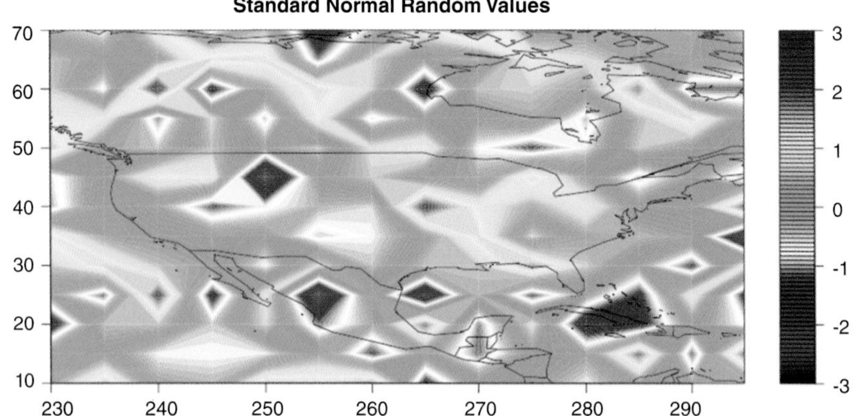

Fig. 9.7 Color map of standard normal random values on a $5^\circ \times 5^\circ$ grid over Canada and USA.

```
filled.contour(Lon3, Lat3, mapdat, color.palette=rgb.palette,
               levels=int, plot.title=title(
main="Standard Normal Random Values", xlab="Lon", ylab="Lat"),
      plot.axes={axis(1); axis(2);map('world2', add=TRUE);grid()})
```

9.2.3 Plot Contour Maps from Climate Model Data in NetCDF Files

Here we show how to plot a downloaded netCDF NCEP/NCAR Reanalysis Monthly Means dataset of surface air temperature from the data site of the NOAA Earth System Research Laboratory.

www.esrl.noaa.gov/psd/data/gridded/data.ncep.reanalysis.derived.surface.html

The reanalysis data are generated by climate models that have "assimilated" (i.e., been constrained by) observed data. The reanalysis output is the complete space–time gridded data. Reanalysis data in a sense is still model data, although some scientists prefer to regard the reanalysis data as dynamically interpolated observational data because the assimilation of observational data has taken place. Gridded observational data in this context may thus be the interpolated results from observational data which have been adjusted in a physically consistent way with the assistance of climate models. The data assimilation system is a tool to accomplish such a data adjustment process correctly.

9.2.3.1 Read `.nc` File

We first download the Reanalysis data, which gives a `.nc` data file: `air.mon.mean.nc`. The R package `ncdf4` can read the data into R.

```
#R plot of NCEP/NCAR Reanalysis PSD monthly temp data .nc file
#www.esrl.noaa.gov/psd/data/gridded/data.ncep.
#reanalysis.derived.surface.html

rm(list=ls(all=TRUE))
setwd("/Users/sshen/climmath")
```

```
# Download netCDF file
# Library
#install.packages("ncdf4")
library(ncdf4)

# 4 dimensions: lon,lat,level,time
nc=ncdf4::nc_open("data/air.mon.mean.nc")
nc
nc$dim$lon$vals # output values 0.0->357.5
nc$dim$lat$vals #output values 90->-90
nc$dim$time$vals
#nc$dim$time$units
#nc$dim$level$vals
Lon <- ncvar_get(nc, "lon")
Lat1 <- ncvar_get(nc, "lat")
Time<- ncvar_get(nc, "time")
head(Time)
#[1] 65378 65409 65437 65468 65498 65529
library(chron)
month.day.year(1297320/24,c(month = 1, day = 1, year = 1800))
#1948-01-01
precnc<- ncvar_get(nc, "air")
dim(precnc)
#[1] 144  73 826, i.e., 826 months=1948-01 to 2016-10, 68 years 10 mons
#plot a 90S-90N precip along a meridional line at 160E over Pacific
par(mar=c(4.5,5,3,1))
plot seq(90,-90, length-73), precnc[65,,1],
     type="l", xlab="Latitude",
     ylab="Temperature  [deg C]",
     main="90S-90N Temperature [degree C]
     along a meridional line at 160E: January 1948",
     lwd=3, cex.lab=1.5, cex.axis=1.5)
```

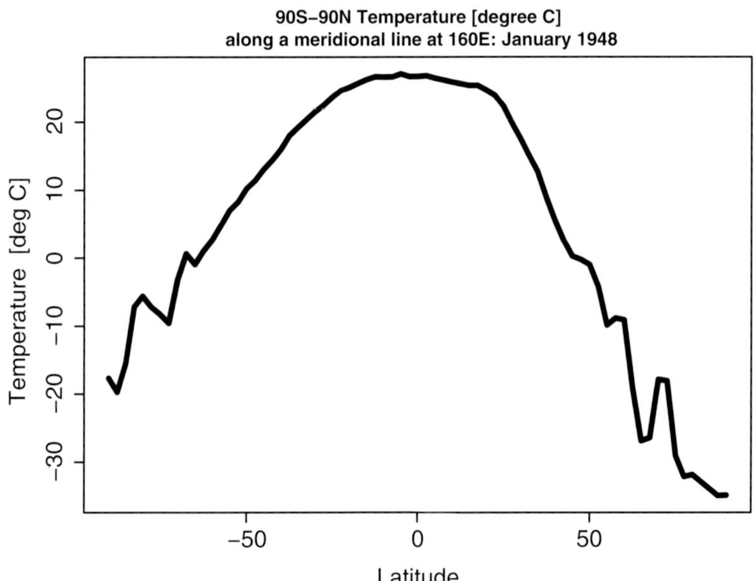

Fig. 9.8 The surface air temperature along a meridional line at 160°E over the Pacific.

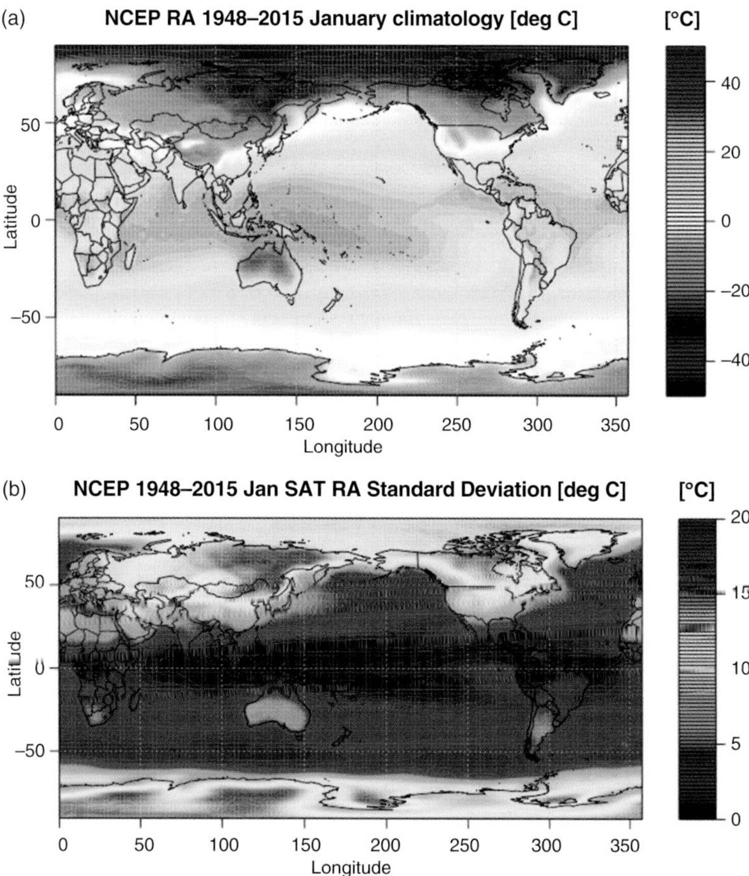

Fig. 9.9 NCEP Reanalysis January climatology (a) computed as the January temperature mean from 1948–2015. (b) shows the standard deviation of the same 1948–2015 January temperature data.

Here, our first example is to plot the temperature variation in the meridional (i.e., north–south) direction from pole to pole, for a given longitude. See Fig. 9.8.

Next we plot the global color contour map showing the January temperature climatology as the average of the January temperature from 1948 to 2015, plus the surface air temperature of January 1983, and finally its anomaly calculated as the difference defined as the January 1983 data minus the January climatology. The R code is below, and the results are shown in Figs. 9.9–9.11.

```
#Compute and plot climatology and standard deviation Jan 1948-Dec 2015
library(maps)
climmat=matrix(0,nrow=144,ncol=73)
sdmat=matrix(0,nrow=144,ncol=73)
Jmon<-12*seq(0,67,1)
for (i in 1:144){
  for (j in 1:73) {climmat[i,j]=mean(precnc[i,j,Jmon]);
  sdmat[i,j]=sd(precnc[i,j,])
  }
}
mapmat=climmat
#R requires coordinates increasing from south to north -90->90
```

```
#and from west to east from 0->360. We must arrange Lat and Lon this way.
#Correspondingly, we have to flip the data matrix left to right.
#The data matrix precnc[i,j,]: 360 (i.e. 180W) lon and from North Pole
#and South Pole, then lon 178.75W, 176.75W, ..., 0E. This puts Greenwich
#at the center, China on the right, and USA on the left. We wish to
#have Pacific at the center, and USA on the right. Thus, we make a flip.
Lat=-Lat1
mapmat= mapmat[,length(mapmat[1,]):1]#Matrix flip around a column
#mapmat= t(apply(t(mapmat),2,rev))
int=seq(-50,50,length.out=81)
rgb.palette=colorRampPalette(c('black','blue','darkgreen','green',
'white','yellow','pink','red','maroon'),interpolate='spline')
par(cex.axis=1.3,cex.lab=1.3)
filled.contour(Lon, Lat, mapmat, color.palette=rgb.palette, levels=int,
 plot.title=title(main="NCEP RA 1948-2015 January climatology [deg C]",
                                   xlab="Longitude",ylab="Latitude"),
 plot.axes={axis(1); axis(2); map('world2', add=TRUE);grid()},
 key.title=title(main="[oC]"))

#plot standard deviation
par(mgp=c(2,1,0))
par(mar=c(3.2,3.3,2.2,0))
par(cex.axis=1.3,cex.lab=1.3)
mapmat= sdmat[,seq(length(sdmat[1,]),1)]
mapmat=pmax(pmin(mapmat,6),0)
int=seq(0,6,length.out=81)
rgb.palette=colorRampPalette(c('black','blue', 'green','yellow',
                      'pink','red','maroon'), interpolate='spline')
filled.contour(Lon, Lat, mapmat, color.palette=rgb.palette, levels=int,
  plot.title=title(
  main="NCEP 1948-2015 Jan SAT Standard Deviation [deg C]",
                 xlab="Longitude", ylab="Latitude"),
  plot.axes={axis(1); axis(2);map('world2', add=TRUE);grid()},
   key.title=title(main=expression(paste("[", degree, "C]"))))
```

9.2.3.2 Plot Data for Displaying Climate Features

The next figure is the January 1983 temperature. The 1982–83 winter is noteworthy because of a strong El Niño event. However, the full temperature field depicted in Fig. 9.10 cannot show the El Niño feature: the warming of the eastern tropical Pacific. The reason is that the full temperature field is dominated by its variation with latitude: hot in the tropics and cold in the polar regions. El Niño is a phenomenon of climate anomalies: the temperature over the eastern tropical Pacific is warmer than normal, sometimes by as much as 6 °C.

```
#Plot the January 1983 temperature using the above setup
mapmat83J=precnc[,,421]
mapmat83J= mapmat83J[,length(mapmat83J[1,]):1]
int=seq(-50,50,length.out=81)
rgb.palette=colorRampPalette(c('black','blue','darkgreen',
'green', 'white','yellow','pink','red','maroon'),interpolate='spline')
filled.contour(Lon, Lat, mapmat83J, color.palette=rgb.palette,
       levels=int,
 plot.title=title(main="January 1983 surface air temperature [deg C]",
                                   xlab="Longitude",ylab="Latitude"),
 plot.axes={axis(1); axis(2);map('world2', add=TRUE);grid()},
 key.title=title(main=expression(paste("[", degree, "C]"))))
```

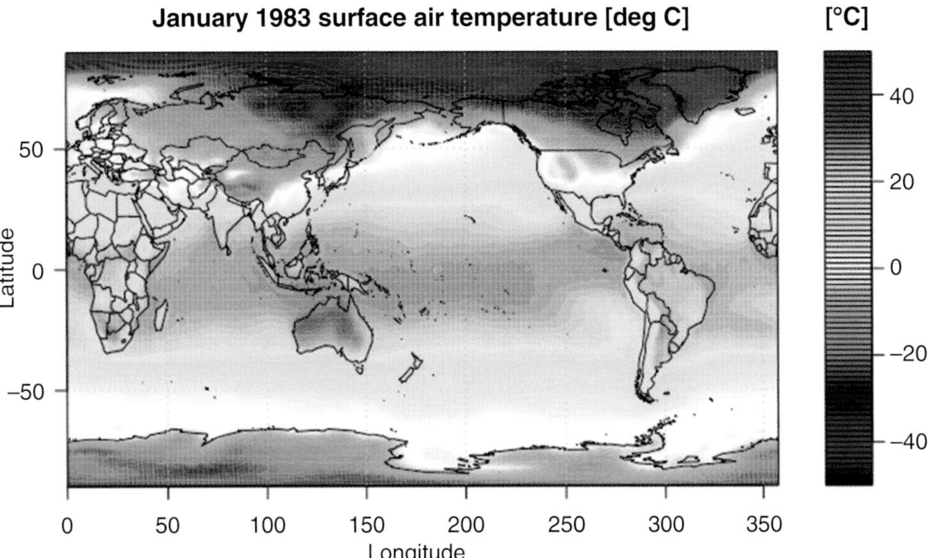

Fig. 9.10 NCEP Reanalysis temperature of January 1983: an El Niño event.

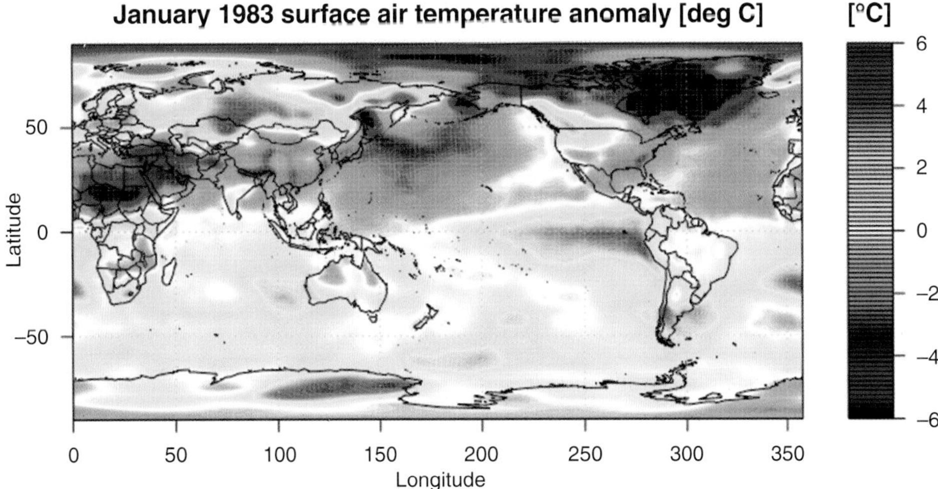

Fig. 9.11 NCEP Reanalysis temperature anomaly of January 1983, showing the eastern tropical Pacific's El Niño warming tongue.

To visualize the El Niño, we compute the temperature anomaly, which is the January 1983 temperature minus the January climatology. A large tongue-shaped region over the eastern tropical Pacific appears with temperatures up to almost 6°C warmer than the climatological average temperatures (Fig. 9.11). This is the typical El Niño signal.

```
#Plot the January 1983 temperature anomaly from NCEP data
plot.new()
anomat=precnc[,,421]-climmat
anomat=pmin(anomat,6)
anomat=pmax(anomat,-6)
anomat= anomat[,seq(length(anomat[1,]),1)]
```

```
int=seq(-6,6,length.out=81)
rgb.palette=colorRampPalette(c('black','blue','darkgreen','green',
 'white','yellow','pink','red','maroon'),interpolate='spline')
filled.contour(Lon, Lat, anomat, color.palette=rgb.palette, levels=int,
plot.title=title(main="January 1983 surface air temperature anomaly
                              [deg C]", xlab="Longitude",ylab="Latitude"),
            plot.axes={axis(1); axis(2);map('world2', add=TRUE);grid()},
       key.title=title(main=expression(paste("[",degree, "C]"))))
```

Sometimes one needs to zoom in to a given latitude–longitude box of the above maps, in order to see the detailed spatial climate pattern over the region. For example, Fig. 9.12 shows the January 1983 SAT anomalies over the Pacific and North America. The El Niño pattern over the Pacific and El Niño's influence over North America are much more clear than in the global map shown in Fig. 9.11.

Figure 9.12(a) for the Pacific region may be generated by the following code, which is a minor change from the global map generation: limiting the `xlim` and `ylim` to the desired region Pacific region (100°E, 60°W) and (50°S, 50°N).

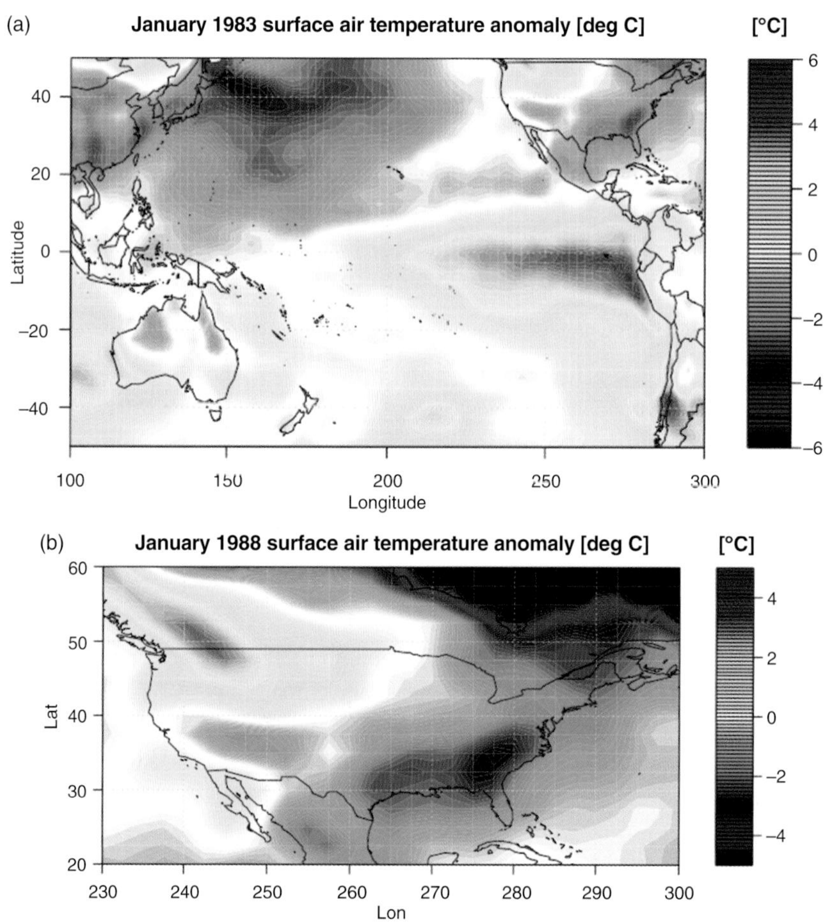

Fig. 9.12 NCEP Reanalysis temperature anomaly of January 1983: the Pacific region (a), and the North America region (b).

```
#Zoom in to a specific lat-lon region: Pacific
int=seq(-6,6,length.out=81)
rgb.palette=colorRampPalette(c('black','blue','darkgreen','green',
'white','yellow','pink','red','maroon'), interpolate='spline')
matdiff = precnc[,,421] -climmat
matdiff= matdiff[,length(matdiff[1,]):1]
filled.contour(Lon, Lat, matdiff,
               xlim=c(100,300), ylim=c(-50,50), zlim=c(-6,6),
               color.palette=rgb.palette, levels=int,
      plot.title=title(
        main="January 1983 surface air temperature anomaly [deg C]",
        xlab="Longitude",ylab="Latitude"),
      plot.axes={axis(1); axis(2);map('world2', add=TRUE);grid()},
      key.title=title(main="[°C]"))
```

Figure 9.12(b) the Northern American region can be generated in a similar way by changing the `xlim` and `ylim`: (130°W, 60°W) and (20°N, 60°N).

9.3 Plot Wind Velocity Field on a Map

Wind velocity is a vector quantity, having both direction and speed. Plotting wind velocity is, therefore, more complex than plotting a scalar such as temperature or precipitation. Fortunately, R can make plotting wind velocity easy.

9.3.1 Plot a Wind Field Using `arrow.plot`

To describe the use of `arrow.plot`, we use the ideal geostrophic wind field as an example to plot a vector field on a map (see Fig. 9.13). The geostrophic wind field is a result of the balance between the pressure gradient force (PGF) and the Coriolis force (CF).

Figure 9.13 can be generated by the following R code.

```
#Wind directions due to the balance between PGF and Coriolis force
#using an arrow plot for vector fields on a map
library(fields)
library(maps)
library(mapproj)

lat<-rep(seq(-75,75,len=6),12)
lon<-rep(seq(-165,165,len=12),each=6)
x<-lon
y<-lat
u<- rep(c(-1,1,-1,-1,1,-1), 12)
v<- rep(c(1,-1,1,-1,1,-1), 12)
wmap<-map(database="world", boundary=TRUE, interior=TRUE)
grid(nx=12,ny=6)
#map.grid(wmap,col=3,nx=12,ny=6,label=TRUE,lty=2)
points(lon, lat,pch=16,cex=0.8)
arrow.plot(lon,lat,u,v, arrow.ex=.08, length=.08, col='blue', lwd=2)
box()
```

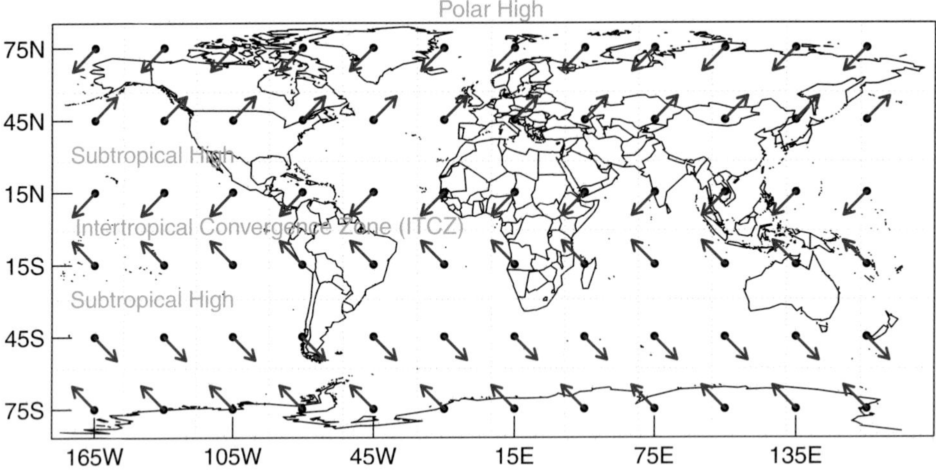

Fig. 9.13 Vector field of the ideal geostrophic wind field.

```
axis(1, at=seq(-165,135,60), lab=c("165W","105W","45W","15E","75E","135E"),
     col.axis="black",tck = -0.05, las=1, line=-0.9,lwd=0)
axis(1, at=seq(-165,135,60),
     col.axis="black",tck = 0.05, las=1, labels = NA)
axis(2, at=seq(-75,75,30),lab=c("75S","45S","15S","15N","45N","75N"),
     col.axis="black", tck = -0.05, las=2, line=-0.9,lwd=0)
axis(2, at=seq(-75,75,30),
     col.axis="black", tck = 0.05, las=1, labels = NA)
text(0, 0, "Intertropical Convergence Zone (ITCZ)", col="green")
text(0, 30, "Subtropical High", col="green")
text(0, -30, "Subtropical High", col="green")
mtext(side=3, "Polar High", col="green", line=0.0)
```

9.3.2 Plot a Surface Wind Field from netCDF Data

This subsection uses `vectorplot` in `rasterVis` to plot the wind velocity field. The surface wind data over the global ocean are used as an example. The procedure is described from the data download to the final product of a wind field. The NOAA wind data were generated from a variety of satellite observations on a global $1/4^\circ \times 1/4^\circ$ grid with a time resolution of 6 hours. See Fig. 9.14.

```
#Plot the wind field over the ocean
#Ref: https://rpubs.com/alobo/vectorplot
#Agustin.Lobo@ictja.csic.es
#20140428

library(ncdf4)
library(chron)
library(RColorBrewer)
library(lattice)

install.packages("rasterVis")
install.packages("latticeExtra")
```

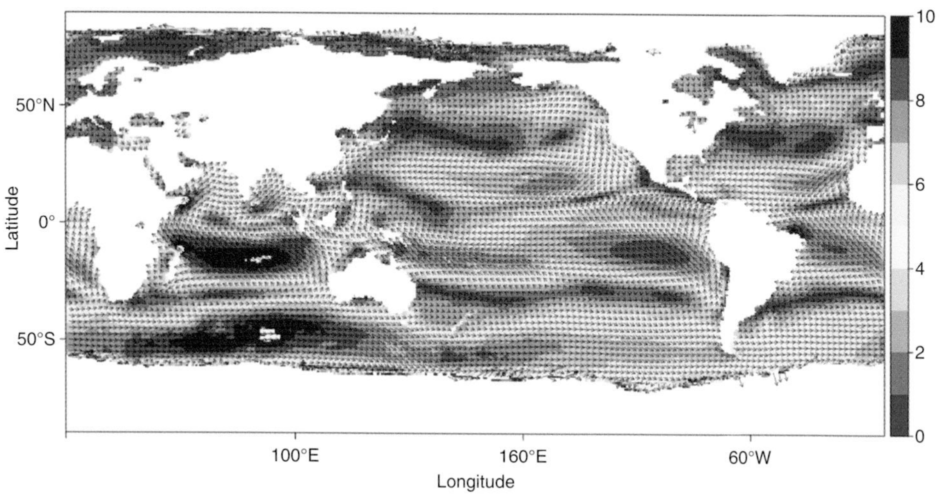

Fig. 9.14 The NOAA sea wind field of January 1, 1995 at time UTC00Z and 1/4° × 1/4° resolution.

```
library(latticeExtra)
library(rasterVis)

install.packages("raster")
library(raster)
library(sp)
library(rgdal)

#download.file(
#"ftp://eclipse.ncdc.noaa.gov/pub/seawinds/SI/uv/clm/uvclm95to05.nc",
#              "uvclm95to05.nc", method = "curl")
mincwind <- nc_open("uvclm95to05.nc")
length(mincwind)
#[1] 14
u <- ncvar_get(mincwind, "u")
dim(u)
#[1] 1440  719   12 #lon, lat, and month
v <- ncvar_get(mincwind, "v")
dim(v)
u9 <- raster(t(u[, , 9])[ncol(u):1, ])
v9 <- raster(t(v[, , 9])[ncol(v):1, ])
filled.contour(u[, , 9])
filled.contour(u[, , 9],color.palette = heat.colors)
filled.contour(u[, , 9],
   color.palette = colorRampPalette(c("red", "white", "blue")))
contourplot(u[, , 9])
u9 <- raster(t(u[, , 9])[ncol(u):1, ])
v9 <- raster(t(v[, , 9])[ncol(v):1, ])
w <- brick(u9, v9)
wlon <- ncvar_get(mincwind, "lon")
wlat <- ncvar_get(mincwind, "lat")
range(wlon)
range(wlat)
```

```
projection(w) <- CRS("+init=epsg:4326")
extent(w) <- c(min(wlon), max(wlon), min(wlat), max(wlat))

plot(w[[1]])
plot(w[[2]])

vectorplot(w * 10, isField = "dXY", region = FALSE,
           margin = FALSE, narrows = 10000)
slope <- sqrt(w[[1]]^2 + w[[2]]^2)
aspect <- atan2(w[[1]], w[[2]])
vectorplot(w*6, isField = "dXY", region = slope,
           margin = FALSE,
           par.settings=BuRdTheme,
           narrows = 10000, at = 0:10)
#vectorplot(stack(slope * 10, aspect), isField = TRUE,
#                  region = FALSE, margin = FALSE)
```

Also see the following websites for more vector field plots:
(a) *Vectorplot in rasterVis* posted on R-Bloggers by Oscar Perpiñán Lamigueiro
www.r-bloggers.com/vectorplot-in-rastervis
(b) *Vectorplot* posted on RPubs by Agustin Lobo Aleu
https://rpubs.com/alobo/vectorplot

9.4 ggplot for Data

`ggplot` is a data-oriented R plot tool developed by Hadley Wickham based on Leland Wilkinson's landmark 1999 book entitled *The Grammar of Graphics* (gg). `ggplot` can generally produce graphic-artist-quality default output and can make plotting complicated data easy with a relatively simple code. For example, ggplot can graphically display multiple columns of data in a `.csv` file after its conversion into a `data.frame`. `ggplot` can save plots as objects, which allows superposition of different layers in a figure and hence enables one to see the evolution of a figure from an initial framework to the final product. The `ggplot2` library was built using a logical mapping between data and graphical elements and includes many maps and datasets that are useful in climate science.

However, `ggplot` syntax is not the same as the syntax of a conventional R plot. There is a learning curve, and a novice may need to spend some time on it before becoming an expert user of `ggplot`.

A simple example is given here for plotting the contiguous "lower 48" states of the United States shown in Fig. 9.15. The figure may be generated by the following `ggplot` code.

```
#ggplot for USA States
library(ggplot2)
states <- map_data("state") #"states" is in a data.frame
p<- ggplot(data = states) +
  geom_polygon(aes(x = long, y = lat, fill = region, group = group),
               color = "white") +
  coord_fixed(1.3)
```

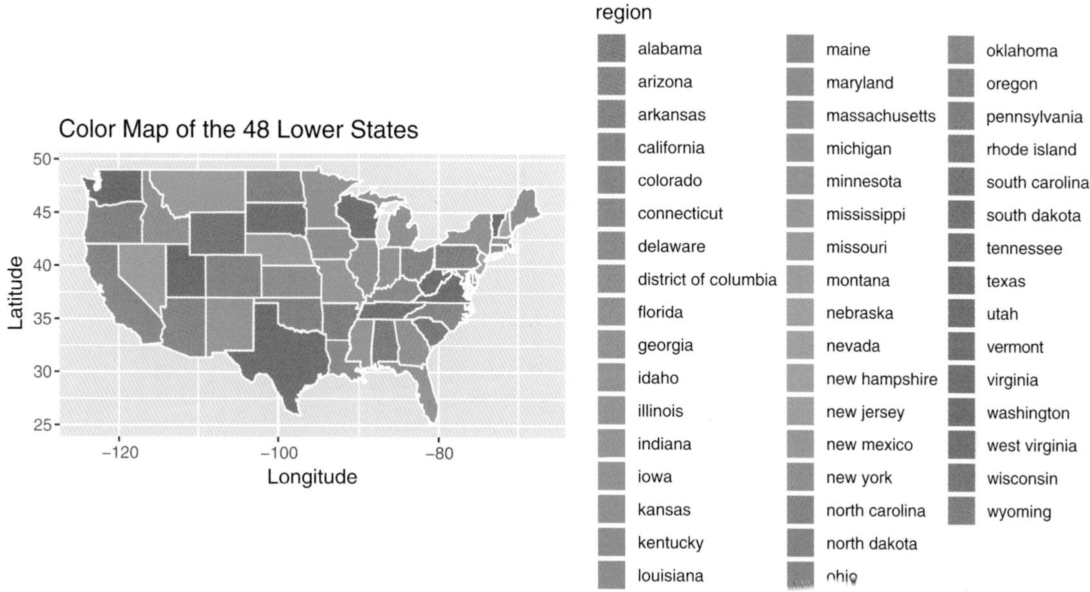

Fig. 9.15 "Lower 48" contiguous states of the United States.

```
#if fill=FALSE, the large color legend on the right is off.
p<- p + xlab("Longitude")+ ylab("Latitude")
p + ggtitle("Color Map of the 48 Lower States")
```

Although some R users strongly advocate the use of `ggplot`, a non-expert in R may remain with the regular R codes to produce plots that might be sufficient for his or her applications. However, `ggplot` is always a good resource if a figure cannot be generated by the usual R plot. Many good `ggplot` tutorial materials and examples are online and can be easily found with a search engine, such as the `ggplot` tutorial by P. Bartlein of the University of Oregon (2018), and that by the Harvard Data Science Service (2018).

9.5 Animation

R has an animation package called `animation` that can animate picture frames in the plot window of RStudio or on an HTML website for you to see and to distribute. The R animation principle is the same as other animation tools: create all the picture frames, and animate them. Let us use the free fall of a round ball as an example. The ball falls from a point that is 490 meters high, under the assumption of no friction force and no wind. Let z be the height of the ball's position at time t. Then

$$z = 490 - \frac{1}{2}gt^2 \tag{9.1}$$

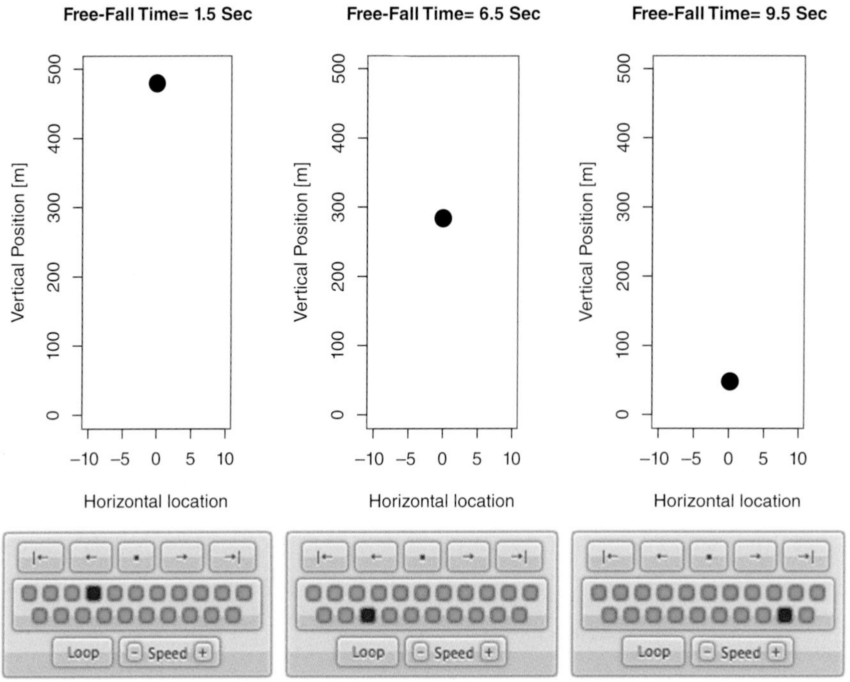

Fig. 9.16 Three frames in a 21-frame animation of a freely falling ball from 490 meters to the ground.

where $g = 9.8$ [m s^{-2}] is the Earth's gravitational acceleration. When $t = 10$ [s], the ball reaches the ground since $z = 0$ [m] if $t = 10$ [s]. Figure 9.16 shows three frames of the animation. The frames and the entire animation can be generated by the following R code.

```
#Free fall animation by 21 frames
g=9.8
n=21
t=seq(0,10,len=n)
#install.packages("animation")
library(animation)
## set up an empty frame, then add points one by one
par(bg = "white") # ensure the background color is white
ani.record(reset = TRUE) # clear history before recording
for (i in 1:n) {
  plot(0, 490-(1/2)*g*(t[i])^2, pch=19, lwd=12, col="black",
       xlab="Horizontal location", xlim=c(-10,10),
       ylim=c(0,500), ylab="Vertical Position [m]",
main=paste("Free Fall Time=", format(t[i],digits = 2, nsmall=1), "sec")
  )
  ani.record() # is: function (reset = FALSE, replay.cur = FALSE)
}
## Now we can replay it, with an appropriate pause between frames:
## Smaller interval means faster animation. Default: interval=1
oopts = ani.options(interval = 0.5,
                    ani.width=200,
                    ani.height=400,
                    title="Free Fall"
                    )
```

```
#Animate the frames in the plot window of RStudio
ani.replay()
## Show the animation on an HTML page
saveHTML(ani.replay(), img.name = "Fall_animation")
```

The last command `saveHTML` generates four items: (i) `index.html` file that animates the picture frames generated, (ii) a folder called `images` that contains all the frames to be animated, (iii) a folder of Java Script (i.e., `.js`) files that support the HTML file, and (iv) a folder of Cascading Style Sheet (i.e., `.css`) files that link to the HTML page. One can go to the `images` folder to check each picture frame. With these generated picture frames, one can of course animate them using other animation tools besides R, such as Adobe Animate.

9.6 Chapter Summary

In the big data era, graphics and data visualization are critical skills for a student to master for her or his education or career. An important difference between our book and some of the more traditional mathematics books encountered in climate science education is that this book provides readers with practical skills for real applications of both mathematical methods and computer data analysis and graphics. Numerous R codes in this book are included for you to modify and use in your own work in climate science applications, education, and research.

This chapter is the beginning of a three-chapter part of our book to show readers how to produce publication-quality figures from climate data. This chapter has described the following basic graphics skills for climate science students.

(i) Two time series on the same plot: On the same figure panel, we plot two series for comparison, such as one for temperature with its scale ticks on the left edge of the graphics box, and another for precipitation with its scale ticks on the right edge of the graphics box. The key R command is

```
par(new=TRUE)
```

which allows a new curve to be superimposed on the previous curve.

(ii) Color contour maps: A given grid box has a temperature datum which is represented by a specific color, such as red for higher temperature and blue for lower temperature. The same type of representation can be arranged for any climate parameter. Thus, the spatial distribution of a climate parameter at any moment in time or in the mean over a time interval can be shown as a color map. This type of color map in climate science replaces the traditional contour lines marked with contour values. The key R command is

```
filled.contour(Lon, Lat, mapdat)
```

with the given longitude grid, latitude grid, and the climate data for the grid. Our sample R codes show the ways to add different features to a color contour map.

(iii) Plot the climate model data: Presenting climate model data in an attractive and informative manner for the general public has become a popular way to display climate science results in a manner that is relevant to society. There is a great demand for more and better data visualization of this type. One example is generating and presenting creative graphics from climate model data, such as the very large Reanalysis dataset of weather data. We have presented an easy-to-use R code for you to read the climate model output as netCDF files and to convert the netCDF data into the color contour maps plotted for a given time.

(iv) Plot wind velocity field on a map: We have presented an R code to plot the velocity field on a map. This technique is very useful in plotting weather maps which often need to superimpose the wind field on a pressure or temperature map.

(v) Graphics by `ggplot`: We have shown a simple example of using `ggplot` to generate figures. The `ggplot` has the advantage of flexibility and the capability of many superpositions, i.e., many layers of figures. It is thus very useful in plotting weather maps with a wind field, for example.

(vi) Animation: We have presented simple R code to generate an animation with a movie output or an html output.

What we have presented here are the basic R graphics skills for climate data visualization. Although these R codes are very basic, it is still unrealistic for you to try to memorize all the key R commands and to expect that you will quickly be able to write excellent R codes for your own application problems. As in every field, practice and experience will lead you to greater proficiency in using R. To use this book effectively and efficiently in a short period of time, you can simply copy and paste the R codes in this book or from the book's website www.cambridge.org/climatemathematics and make modifications based on your own applications. It is normal that you will frequently need to use an Internet search to find the R commands that are not provided in this book. Our book thus provides the basic R codes and key words for you to make efficient Internet searches for generating your own effective graphics.

References and Further Readings

[1] Bartlein, P., 2018: *GeogR: Geographic Data Analysis Using R*, University of Oregon. URL: http://geog.uoregon.edu/GeogR/index.html

> This online material was written for the Geographic Data Analysis course (GEOG 495), University of Oregon. It contains both R codes and basic mathematical methods for geoscience. It features many useful R map-plotting examples, such as http://geog.uoregon.edu/GeogR/topics/maps01.html
> http://geog.uoregon.edu/GeogR/topics/maps02.html
> http://geog.uoregon.edu/GeogR/topics/maps03.html

[2] Chang, W., 2012: *R Graphics Cookbook: Practical Recipes for Visualizing Data.* O'Reilly Media, Inc, Sebastopol, California.

> This basic R-graphics book contains more than 150 sample codes for commonly used graphs based on data.

[3] Harvard Data Science Service, 2018: *R Graphics Tutorial with ggplot2.* URL: http://tutorials.iq.harvard.edu/R/Rgraphics/Rgraphics.html

> This is a good ggplot2 tutorial, starting from the beginning and ending with relatively complex plots.

Exercises

All exercises in this chapter require the use of a computer.

9.1 Use R to plot the surface air temperature (SAT) and sea level pressure (SLP) anomaly time series of Tahiti and Darwin. Put the four time series on the same figure, and explain their behaviors during the El Niño and La Niña periods. You may use the NCEP/NCAR Reanalysis surface data for the Darwin and Tahiti grid boxes.

9.2 (a) Use R to compute the 1971–2000 January climatology of the SAT from the NCEP/NCAR Reanalysis data for each grid box. Plot the climatology map.
(b) Perform the same procedure for June.

9.3 (a) Use R to compute the 1971–2000 January standard deviation of the SAT from the NCEP/NCAR Reanalysis data for each grid box. Plot the climatology map.
(b) Perform the same procedure for June.

9.4 (a) Use R to generate the annual mean SAT data for each grid box from the monthly mean data of NCEP/NCAR Reanalysis.
(b) Use the above result to compute the 1971–2000 annual SAT climatology from the NCEP/NCAR Reanalysis data for each grid box. Plot the climatology map.

9.5 Use R to compute the 1971–2000 standard deviation from the NCEP/NCAR Reanalysis annual SAT data for each grid box. Plot the standard deviation map.

9.6 (a) Use R and NCEP/NCAR Reanalysis data to display the El Niño temperature anomaly for January 2016 with respect to the 1971–2000 climatology.
(b) Find the latitude and longitude of the grid box on which the maximum temperature anomaly of the month occurred. What was the maximum anomaly? Where did it occur?

9.7 (a) Compute the global average monthly mean SAT from January 1948 to December 2015 using the NCEP/NCAR Reanalysis data.
(b) Plot the time series.
(c) Compute the temporal mean of this time series.

9.8 (a) Compute the global average monthly mean SAT *anomalies* from 1948 to 2015 for January with respect to the 1971–2000 January climatology, using the NCEP/NCAR Reanalysis data.

(b) Plot the time series and its linear trend on the same figure.

9.9 Use R to plot the map of North America and plot the December 1997 SAT anomaly data with respect to the 1971–2000 December climatology on this map. Choose your own gridded dataset from the Internet, such as the NOAAGlobalTemp dataset used in this book.

9.10 Use R to generate an HTML animation for a cosine wave

$$w = a\cos(k(x - ct)) \tag{9.2}$$

where $a = 1.5$ [m], $k = 0.2$ [km^{-1}], and $c = 8$ [km hour^{-1}]. The animation time is from 0 to 10 hours.

9.11 Use R to animate the annual SAT anomalies from 1951 to 2000 based on the NCEP/NCAR Reanalysis data.

9.12 (a) Use R to animate the January SAT anomalies from 1948 to 2015 based on the NCEP/NCAR Reanalysis data.

(b) Describe your observation of the El Niño phenomenon over the globe, particularly over the eastern tropical Pacific region.

10 Advanced R Analysis and Plotting: EOFs, Trends, and Global Data

The empirical orthogonal function (EOF) method is a commonly used tool for climate data analysis in research. This chapter provides basic ideas and mathematical theory for the EOF analysis, also known as principal component analysis (PCA) in the statistical literature. This chapter provides recipe-like R codes for analyzing and visualizing space–time climate data using EOFs and PCs. The codes can (i) compute EOFs and principal components (PCs) using the singular value decomposition (SVD) analysis approach, (ii) plot the EOFs on a world map and PC time series, (iii) compute temporal trends, data standardization, and de-trended data over a spatial grid, and (iv) plot the spatial distribution of temporal trends. NCEP/NCAR Reanalysis data are used as examples. As described in Chapter 4, the SVD method helps to reveal the spatial and temporal patterns of any dataset obtained by sampling in space and time. EOFs show spatial patterns of climate data, such as the El Niño warm anomaly pattern of the eastern tropical Pacific. The corresponding temporal patterns are depicted in PCs that can show the times when El Niños occur. The SVD approach can aid in developing physical insight and visualizing climate information, and thus can help lead to an improved understanding of the phenomena under study. Our description makes EOFs and PCs a natural space–time decomposition technique that can be readily carried out by a simple R command: `svd(datamatrix)`. This method is different from the traditional approach of an eigenvalue problem based on a covariance matrix, which focuses only on spatial patterns.

10.1 Ideas of EOF, PC, and Variances Computed from SVD

SVD as described in Chapter 4 decomposes a space–time climate data matrix A into three parts: spatial patterns U, temporal patterns V, and energy D, i.e.,

$$A = UDV^t. \tag{10.1}$$

Both U and V are orthogonal matrices, meaning that each column has length equal to one and is orthogonal to a different column vector. The first column of U determines the spatial pattern of mode 1. The pattern is called an empirical orthogonal function, and the method was introduced into meteorology in the 1950s by Edward Lorenz, who is well known for his contributions to theoretical meteorology, the theory of chaos, and the "butterfly effect." The corresponding first column of V determines the temporal pattern, which is called a principal component. For example, for the SLP data of Darwin and Tahiti analyzed in Chapter 4, the EOFs are the SLP patterns at the two locations, and the PCs are the time series. EOF1 shows that Darwin and Tahiti have opposite SLP anomalies. When Darwin's SLP anomaly is positive and Tahiti's negative, it is an El Niño. The corresponding PC1's positive peaks show the time when El Niño actually occurred. PC1's negative peaks indicate the time of La Niña.

As discussed in Section 4.6.2, EOF1 and PC1 are both referred to as mode 1, and mode 1 has "energy": 31.35^2, which is the variance of $A\mathbf{u}_1$, i.e., the data's projection onto the first EOF pattern. The variance of a mode relative to the total variance is a more useful piece of information. In the Darwin–Tahiti SLP case, as discussed in Section 4.6.2, the relative variance of EOF1 is

$$\frac{d_1^2}{d_1^2+d_2^2} = \frac{31.35^2}{31.35^2+22.25^2} = 67\%. \tag{10.2}$$

EOF1 thus accounts for 2/3 of the total variance. EOF2's variance relative to the total variance is thus 33%, or 1/3.

The elements of U and V are dimensionless and consist of othonormal vectors. The dimension of $d_i (i = 1,2)$ is the same as that of the elements of the data matrix A, and measures the "amplitude" that is proportional to the system's variance or "energy." The dimension of d_1^2 is the square of the dimension of A, i.e., the dimension of the variance. Variance is thus a measure of the "energy" of the El Niño–Southern Oscillation system.

10.2 2Dim Spatial Domain EOFs and 1Dim Temporal PCs

The spatial fields of many climate data applications are two-dimensional, or 2Dim or 2D for short. The corresponding EOFs are over a 2Dim domain on the Earth's surface, and the corresponding PCs are on a time interval. This section describes the basic concepts of using SVD to compute EOFs and PCs and using R graphics to show them. We shall first use a simple synthetic dataset to illustrate the procedures. Afterwards, we shall demonstrate using real climate data.

10.2.1 Generate Synthetic Data by R

The spatial domain is $\Omega = [0,2\pi] \times [0,2\pi]$, and the time interval is $T = [1,10]$. The synthetic data are generated by the following function

$$z(x,y,t) = c_1(t)\psi_1(x,y) + c_2(t)\psi_2(x,y), \tag{10.3}$$

where $\psi_1(x,y)$ and $\psi_1(x,y)$ are two orthonormal basis functions given below

$$\psi_1(x,y) = (1/\pi)\sin x \sin y, \tag{10.4}$$

$$\psi_2(x,y) = (1/\pi)\sin(8x)\sin(8y), \tag{10.5}$$

with

$$\int_{\Omega} d\Omega \psi_k^2(x,y) = 1, \quad k = 1,2 \tag{10.6}$$

$$\int_{\Omega} d\Omega \psi_1(x,y)\psi_2(x,y) = 0. \tag{10.7}$$

The corresponding basis expansion coefficients are

$$c_1(t) = \sin(t), \tag{10.8}$$

$$c_2(t) = \exp(-0.3t). \tag{10.9}$$

These are not orthogonal. Thus, the generating function for $z(x,y,t)$ in Eq. (10.3) is not an SVD decomposition.

The spatial domain Ω is divided into a 100×100 grid. The time grid is $1,2,\ldots,10$. There are 10,000 spatial grid points $(100 \times 100 = 10,000)$ and 10 temporal grid points. The space–time data may be generated by the following R commands.

```
x<-seq(0, 2*pi, len=100)
y<-seq(0, 2*pi, len=100)
mydat<-array(0,dim=c(100,100,10))
for(t in 1:10){
  z<-function(x,y){z=sin(t)*(1/pi)*sin(x)*sin(y)+
    exp(-0.3*t)*(1/pi)*sin(8*x)*sin(8*y)}
  mydat[,,t]=outer(x,y,z)
               }
```

The $z(x,y,t)$ is a superposition of a large-scale spatial wave $\psi_1(x,y)$ (see Fig. 10.2) with a small-scale wave $\psi_2(x,y)$. The first wave's coefficient $c_1(t)$ varies periodically, while the second wave's coefficient $c_2(t)$ decays exponentially. Thus, for a large value of time t, the z pattern will be dominated by the large-scale spatial wave $\psi_1(x,y)$. Figure 10.1 shows $z(x,y,1)$ and $z(x,y,10)$. When time is 1, the figure shows the superposition of a large-scale wave and a small-scale wave. When the time is 10, the figure shows only the large-scale wave, and the small-scale wave's influence is negligible.

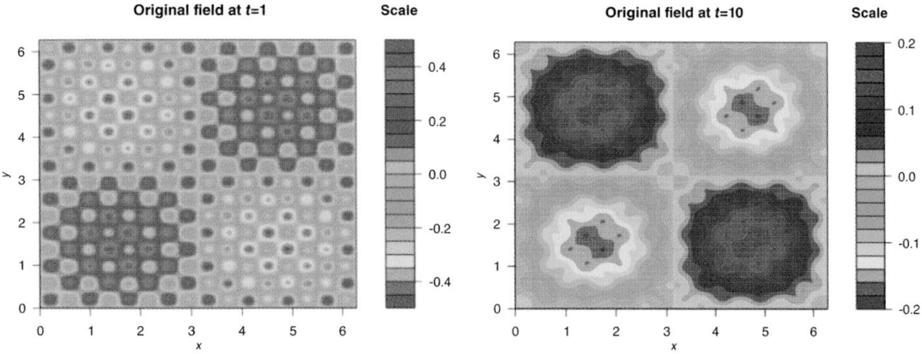

Fig. 10.1 The $z(x,y,t)$ function at $t = 1$ and $t = 10$.

This figure can be generated by the following `filled.contour` command for matrix data.

```
#Plot the original z(x,y,t) waves for a given t
filled.contour(x,y,mydat[,,10], color.palette =rainbow,
               plot.title=title(main="Original field at t=10",
                                xlab="x", ylab="y", cex.lab=1.0),
               key.title = title(main = "Scale"),
               plot.axes =  {axis(1,seq(0,3*pi, by = 1), cex=1.0)
                 axis(2,seq(0, 2*pi, by = 1), cex=1.0)}
               )
```

10.2.2 SVD for the Synthetic Data EOFs, Variances, and PCs

We first convert the synthetic data array into a $10{,}000 \times 10$ matrix of space–time data. Then SVD can be applied to the space–time data to generate EOFs, variances, and PCs.

The following code converts the 3Dim array `mydat(,,,)` into a 2Dim space–time data matrix `da1`.

```
da1<- matrix(0,nrow=length(x)*length(y),ncol=10)
for (i in 1:10) {da1[,i]=c(t(mydat[,,i]))}
```

Applying SVD on this space–time data is shown below.

```
da2<-svd(da1)
uda2<-da2$u
vda2<-da2$v
dda2<-da2$d
dda2
#[1] 3.589047e+01 1.596154e+01 7.764115e-14 6.081008e-14
```

The EOFs shown in Fig. 10.2 can be plotted by the following R code.

```
par(mgp=c(2,1,0))
filled.contour(x,y,matrix(-uda2[,1],nrow=100), color.palette =rainbow,
               plot.title=title(main="SVD Mode 1: EOF1",
               xlab="x", ylab="y", cex.lab=1.0),
               key.title = title(main = "Scale"),
               plot.axes =  {axis(1,seq(0,2*pi, by = 1), cex=1.0)
                 axis(2,seq(0, 2*pi, by = 1), cex=1.0)})
```

Figure 10.2 shows that the EOF patterns from SVD are similar to the original orthonormal basis functions $\psi_1(x,y)$ and $\psi_2(x,y)$. This means that SVD has recovered the original orthonormal basis functions. However, this is not always true, when the variances of the two modes are close to each other. These two SVD eigenvalues will then be close to each other. Consequently, the EOFs, as eigenfunctions, will have large differences from the original true orthonormal basis functions. This is quantified by North's rule-of-thumb, which states that both EOFs will have large errors, which are inversely proportional to the difference between the two eigenvalues. Thus, when the two eigenvalues have a small difference, the two corresponding eigenfunctions will have large errors due to mode mixing. In linear algebra terms, this means that when two eigenvalues are close to each other, the corresponding eigenspaces tend to be close to each other. They form a two-dimensional eigenspace, which has infinitely many eigenvectors due to the mixture of the two eigenvectors. A physically

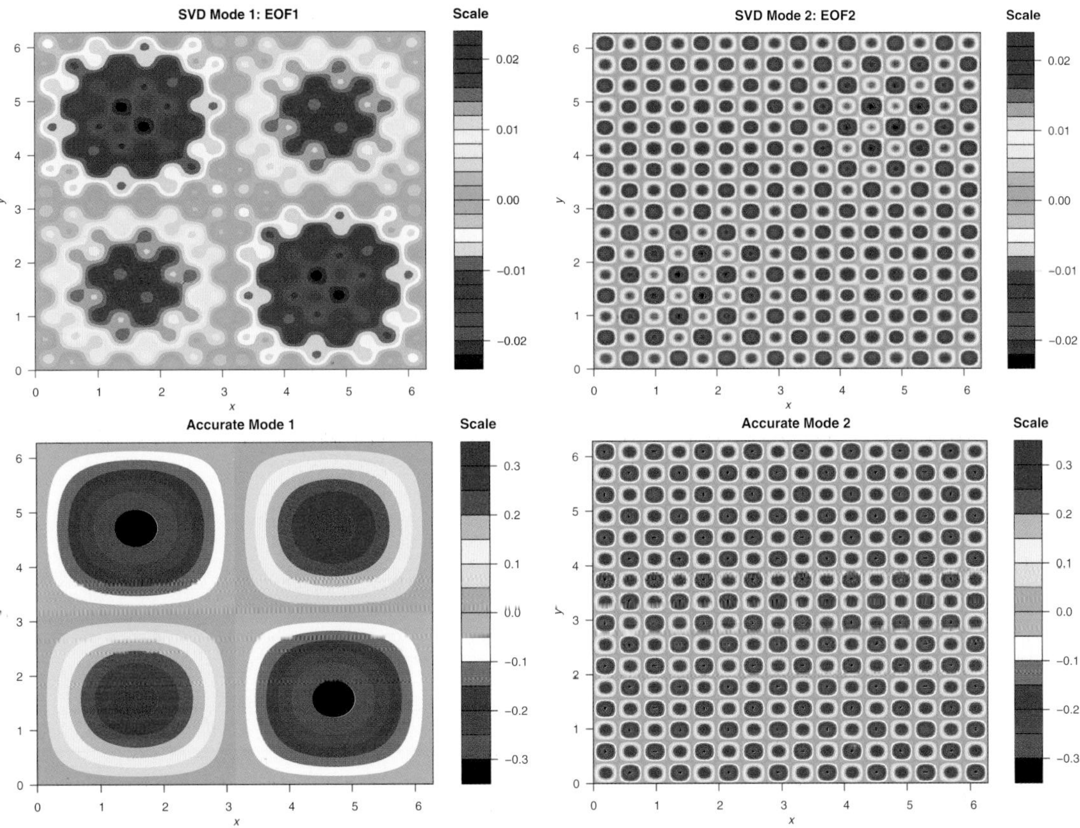

Fig. 10.2 The first row shows two EOFs from SVD, and the second row shows two orthonormal basis functions on the xy-domain: $\phi_1(x,y) = -(1/\pi)\sin x \sin y$, and $\phi_2(x,y) = (1/\pi)\sin 8x \sin 8y$.

meaningful eigenvector should have no ambiguity, and infinitely many eigenvectors imply uncertainties, large errors, and no physical interpretation.

The original orthonormal basis functions can be plotted by the following R codes.

```
#Accurate spatial patterns from functions that generate data
z1 <- function(x,y){(1/pi)*sin(x)*sin(y)}
z2 <- function(x,y){(1/pi)*sin(8*x)*sin(8*y)}
fcn1<-outer(x,y,z1)
fcn2<-outer(x,y,z2)
par(mgp=c(2,1,0))
filled.contour(x,y,fcn1, color.palette =rainbow,
               plot.title=title(main="Accurate Mode 1",
                            xlab="x", ylab="y", cex.lab=1.0),
               key.title = title(main = "Scale"),
               plot.axes = {axis(1,seq(0,3*pi, by = 1), cex=1.0)
                 axis(2,seq(0, 2*pi, by = 1), cex=1.0)}
)
```

The first two principal components (PCs) are shown in Fig. 10.3, which can be generated by the following code.

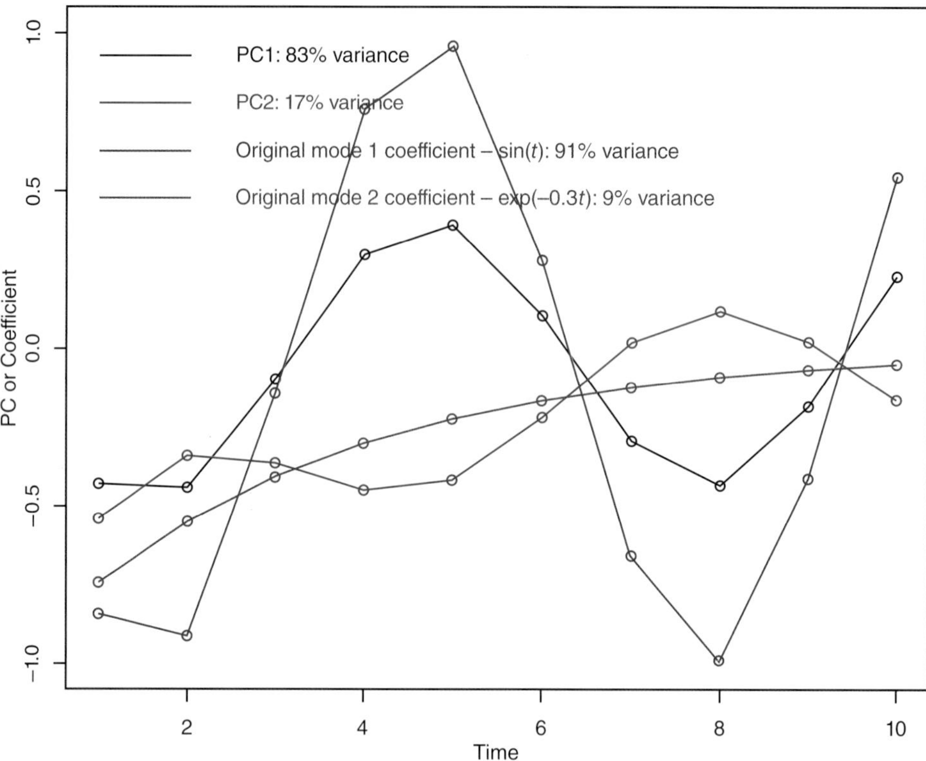

Fig. 10.3 Two principal components (PCs) from SVD approximation and two accurate time coefficients: $-\sin t$ and $\exp(-0.3t)$.

```
#Plot PCs and coefficients of the functional patterns
t=1:10
plot(1:10, vda2[,1],type="o", ylim=c(-1,1), lwd=2,
     ylab="PC or Coefficient", xlab="Time",
     main="SVD PCs vs. Accurate Temporal Coefficients")
legend(0.5,1.15, lty=1, legend=c("PC1: 69% variance"),
  bty="n",col=c("black"))
lines(1:10, vda2[,2],type="o", col="red", lwd=2)
legend(0.5,1.0, lty=1, legend=c("PC2: 31% varance"),
  col="red", bty="n",text.col=c("red"))
lines(t, -sin(t), col="blue", type="o")
legend(0.5,0.85, lty=1, legend=c("Mode 1 coefficient: 80% variance"),
  col="blue", bty="n",text.col="blue")
lines(t, -exp(-0.3*t), type="o",col="purple")
legend(0.5,0.70, lty=1, legend=c("Mode 2 coefficient: 20% variance"),
  col="purple", bty="n",text.col="purple")
```

PC1 demonstrates sinusoidal oscillation, while PC2 shows a wavy increase. These two temporal patterns are similar to the original time coefficients $-\sin(t)$ and $-\exp(-0.3t)$. Here, the negative signs are added to make the EOF patterns have the same sign as the original basis functions, because the EOFs are determined up to the sign, i.e., the plus or minus sign is indeterminate.

PC1 and PC2 are orthogonal, but coefficients $c_1(t)$ and $c_2(t)$ are not orthogonal. This can be verified by the following code.

```
#Verify orthogonality of PCs
t(vda2[,1])%*%vda2[,2]
#  [1,] -5.551115e-17
t=1:10
t(-sin(t))%*%(-exp(-0.3*t))
#[1,] 0.8625048
```

The SVD theory tells us that the original data can be recovered from the EOFs, PCs, and the variances by the following formula

$$z = UDV^t. \tag{10.10}$$

Because the eigenvalues for this problem, except the first two, are close to zero, we can have an accurate reconstruction by using the first two EOFs, PCs, and their corresponding eigenvalues. The R code for both 2-mode approximation and all-mode recovery is below.

```
B<-uda2[,1:2]%*%diag(dda2)[1:2,1:2]%*%t(vda2[,1:2])
B1<-uda2%*%diag(dda2)%*%t(vda2)
```

Figure 10.4(a) shows the recovered z at time $t = 5$ using only two EOF modes. It can be plotted by the following R code.

```
plot.new()
filled.contour(x,y,matrix(B[,5],nrow=100), color.palette =rainbow,
               plot.title=title(main="2-mode SVD reconstructed field t=5",
                        xlab="x", ylab="y", cex.lab=1.0),
               key.title = title(main = "Scale"),
               plot.axes = {axis(1,seq(0,3*pi, by = 1), cex=1.0)
                 axis(2,seq(0, 2*pi, by = 1), cex=1.0)})
```

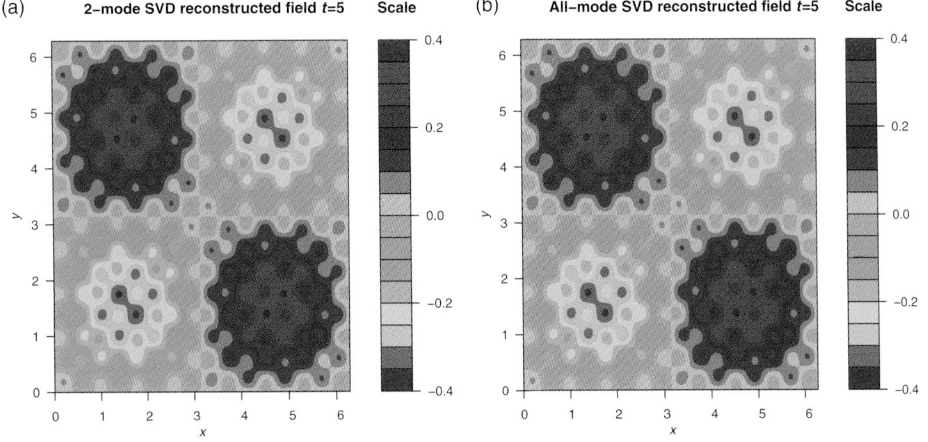

Fig. 10.4 Recovery of the original data using two modes (a) and using all modes (b).

The full recovery or the original field at time $t = 5$, shown in Fig. 10.4(b), is virtually identical to the 2-mode approximation. The difference is less than 10^{-10} at any given point. This high level of accuracy may not always be achieved even with the full recovery UDV^t when high spatial variability appears. To be specific, the full recovery UDV^t may have non-negligible numerical truncation errors (single precision or double precision), which can cause large errors in the recovery results, when high spatial variability is involved.

10.3 From Climate Data Download to EOF and PC Visualization: An NCEP/NCAR Reanalysis Example

This section presents an example of computing EOFs and PCs from a netCDF file downloaded from the Internet. In climate research and teaching, data from netCDF files are often encountered. We shall download the data and make an EOF analysis. The example is the surface temperature data from the NCEP/NCAR Reanalysis I, which outputs 2.5 degree monthly data from January 1948 until at least October 2016 and perhaps much longer. We choose the most frequently used surface air temperature (SAT) field.

10.3.1 Download and Visualize the NCEP Temperature Data

We downloaded the monthly surface air temperature (SAT) data from
www.esrl.noaa.gov/psd/data/gridded/data.ncep.reanalysis.surface.html.
The datafile is called `air.month.mean.nc`, and its size is 26 MB for the data extending from January 1948 to October 2016. The spatial resolution is a 2.5-degree latitude–longitude (lat–lon) grid. The following code reads the `.nc` file into R.

```
# Read netCDF file
# Library
install.packages("ncdf")
library(ncdf4)

# 4 dimensions: lon,lat,level,time
setwd("/Users/sshen/climmath")
nc=ncdf4::nc_open("data/air.mon.mean.nc")
nc
nc$dim$lon$vals #output lon values 0.0->357.5
nc$dim$lat$vals #output lat values 90->-90
nc$dim$time$vals #output time values in GMT hours: 1297320, 1298064
nc$dim$time$units
#[1] "hours since 1800-01-01 00:00:0.0"
#nc$dim$level$vals
Lon <- ncvar_get(nc, "lon")
Lat1 <- ncvar_get(nc, "lat")
Time<- ncvar_get(nc, "time")
#Time is the same as nc$dim$time$vals
head(Time)
#[1] 1297320 1298064 1298760 1299504 1300224 1300968
library(chron)
```

```
Tymd<-month.day.year(Time[1]/24,c(month = 1, day = 1, year = 1800))
#c(month = 1, day = 1, year = 1800) is the reference time
Tymd
#$month
#[1] 1
#$day
#[1] 1
#$year
#[1] 1948
#1948-01-01
precnc<- ncvar_get(nc, "air")
dim(precnc)
#[1] 144  73 826, i.e., 826 months=1948-01 to 2016-10, 68 years 10 mons
```

To check whether our downloaded data appear to be reasonable, we plot the first month's temperature data at longitude 180°E from the South Pole to the North Pole (see Fig. 10.5). The figure shows a reasonable temperature distribution: a high temperature nearly 30°C over the tropics, a lower temperature below −30°C over the Arctic region at the right, and between −20°C and 0°C over the Antarctic region at the left. We are thus reasonably confident that our downloaded data are correct and that the data values correctly correspond to their assigned positions on the latitude–longitude grid.

Figure 10.5 may be generated by the following R code:

```
#plot a 90S 90N temp along a meridional line at 180E
plot(seq(90,-90, length=73),precnc[72,,1], type="o", lwd=2,
```

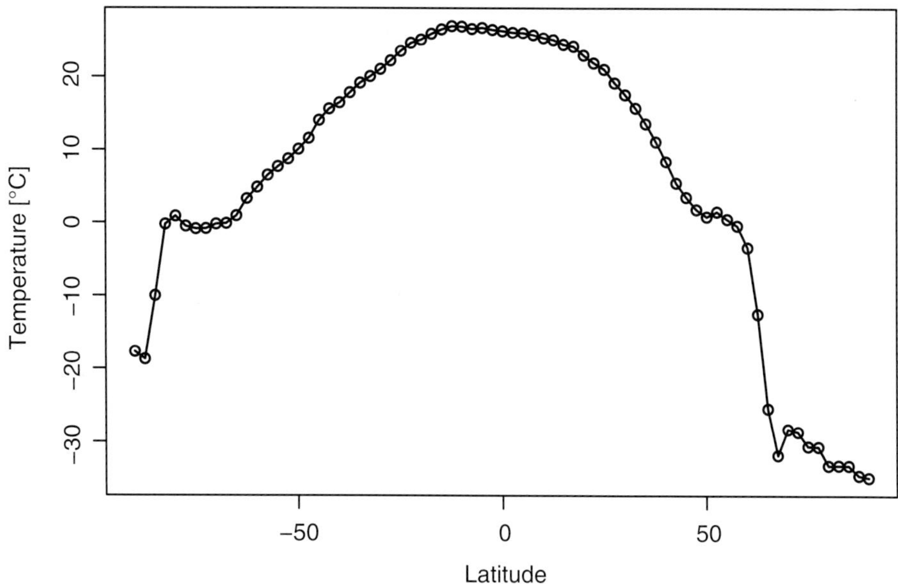

Fig. 10.5 The north–south distribution of SAT along the meridional line at 180°E for January 1948, based on the NCEP/NCAR Reanalysis data.

```
xlab="Latitude", ylab="Temperature [oC]",
main="NCEP/NCAR Reanalysis Surface Air Temperature [deg C]
along a Meridional Line at 180E: Jan 1948")
```

10.3.2 Space–Time Data Matrix and SVD

Having downloaded the data, the next step is to put the data into a convenient format. Anticipating the use of SVD, this is easily done, as we shall now show.

10.3.2.1 Reformat the Data into a Space–Time Data Matrix

We convert the downloaded nc file into a space–time matrix and write it into a csv file (see Section 2.3.3) which is easy for users first to read the dataset and then to use it in their computer programs. The R code for this procedure is below.

```
#Write data as a CSV space-time matrix with a header
precst=matrix(0,nrow=10512,ncol=826)
temp=as.vector(precnc[,,1])
head(temp)
for (i in 1:826) {precst[,i]=as.vector(precnc[ , , i])}
dim(precst)
#[1] 10512   826
#Build lat and lon for 10512 spatial positions usig rep
LAT=rep(Lat1, 144)
LON=rep(Lon[1],each=73)
gpcpst=cbind(LAT, LON, precst)
dim(gpcpst)
#[1] 10512   828
#The first two columns are lat and lon. 826 mons: 1948.01-2016.10
#Convert the Julian day and hour into calendar mons for time
tm=month.day.year(Time/24, c(month = 1, day = 1, year = 1800))
tm1=paste(tm$year,"-",tm$month)
tm2=c("Lat","Lon",tm1) #This is the header
#Assign the header to the space-time data matrix
colnames(gpcpst) <- tm2
setwd("/Users/sshen/climmath")
#setwd routes the desired csv file to a given directory
write.csv(gpcpst,file="ncepJan1948_Oct2016.csv")
```

The resulting csv file's first column is latitude, the second is longitude, and the first row is the time mark for each month from January 1948 to October 2016. The csv file can be read and then used by Excel, R, Matlab, Python, and almost all other commonly encountered software.

10.3.2.2 Climatology and Standard Deviation

Here we show a different way to compute the climatology and standard deviation, which we previously discussed in Section 9.2.3.1. The method described here is to use the space–time data matrix. With this matrix, computing the climatology and standard deviation becomes very easy. Graphically showing the spatial distribution of climatology and standard deviation (see Fig. 9.9) can further verify the correctness of the downloaded data, such as the cold temperature over the Himalayas and Tibetan Plateau regions of Asia, and the Andes region of South America. These regions are at relatively low latitudes but they experience very low temperatures due to their high altitudes, which may be 4,000 or more meters.

The R code for computing the January climatology and standard deviation is below.

```
#Compute the January climatology and standard deviation: 1948-2015
monJ=seq(1,816,12)
gpcpdat=gpcpst[,3:818]
gpcpJ=gpcpdat[,monJ]
climJ<-rowMeans(gpcpJ) #Use all the 68 Januarys from 1948 to 2015
library(matrixStats)# rowSds command is in the matrixStats package
sdJ<-rowSds(gpcpJ)
```

The climatology and standard deviation are shown in Fig. 9.9, which can be generated by the following R code.

```
#Plot Jan climatology
Lat=-Lat1
mapmat=matrix(climJ,nrow=144)
mapmat= mapmat[,seq(length(mapmat[1,]),1)]
plot.new()
int=seq(-50,50,length.out=81)
rgb.palette=colorRampPalette(c('black','blue', 'darkgreen','green',
    'white','yellow','pink','red','maroon'),interpolate='spline')
filled.contour(Lon, Lat, mapmat, color.palette=rgb.palette, levels=int,
 plot.title=title(main="NCEP Jan SAT RA 1948-2015 climatology [deg C]"),
 plot.axes={axis(1); axis(2);map('world2', add=TRUE);grid()},
 key.title=title(main="[oC]"))
#----------------------
#Plot Jan Standard Deviation
Lat=-Lat1
mapmat=matrix(sdJ,nrow=144)
mapmat= mapmat[,seq(length(mapmat[1,]),1)]
plot.new()
int=seq(0,3,length.out=81)
rgb.palette=colorRampPalette(c('black','blue', 'green',
    'yellow','pink','red','maroon'),interpolate='spline')
filled.contour(Lon, Lat, mapmat, color.palette=rgb.palette, levels=int,
  plot.title=title(
      main="NCEP Jan SAT RA 1948-2015 Standard Deviation [deg C]"),
  plot.axes={axis(1); axis(2);map('world2', add=TRUE);grid()},
  key.title=title(main="[oC]"))
```

The very large standard deviation, more than 5°C over the high latitudes of the Northern Hemisphere, shown in Fig. 9.9, may be artificially amplified by the climate model employed to carry out the NCEP/NCAR reanalysis. This error may be due to the model's handling of a physically complex phenomenon, namely the sea ice and albedo feedback. The true standard deviation might thus be smaller over the same region. This type of error highlights a caution that should be kept in mind when using reanalysis datasets. Combining a complex climate model with observational data can improve the realism of datasets, but it can also introduce a source of error that would not exist if no model were used.

10.3.2.3 Plot EOFs, PCs, and Variances

The EOFs, PCs, and variances for a month can be easily calculated by SVD for the space–time data matrix for the given month, by the following R code:

```
#Compute the Jan EOFs
monJ=seq(1,816,12)
gpcpdat=gpcpst[,3:818]
gpcpJ=gpcpdat[,monJ]
climJ<-rowMeans(gpcpJ)
library(matrixStats)
sdJ<-rowSds(gpcpJ)
anomJ=(gpcpdat[,monJ]-climJ)/sdJ #standardized anomalies
anomAW=sqrt(cos(gpcpst[,1]*pi/180))*anomJ #Area weighted anomalies
svdJ=svd(anomAW)   #execute SVD
```

The eigenvalues of a covariance matrix are variances, and the SVD eigenvalues from a data matrix correspond to standard deviations. Climate science often uses variance for measuring a signal strength that may be attributed to a particular mode or modes. Further, what is often most important is the relative variance, i.e., the percentage of the variance attributable to a specific mode, relative to the sum of the variances in all the modes. We thus plot the percentage of the square of each SVD eigenvalue, and the cumulative percentage from the first mode to the last mode. See Fig. 10.6 for the plot. This figure can be plotted by the following R code.

```
#plot eigenvalues
par(mar=c(3,4,2,4))
plot(100*(svdJ$d)^2/sum((svdJ$d)^2), type="o",
     ylab="Percentage of variance [%]",
     xlab="Mode number", main="Eigenvalues of covariance matrix")
legend(20,5, col=c("black"),lty=1,lwd=2.0,
       legend=c("Percentange variance"),bty="n",
       text.font=2,cex=1.0, text.col="black")
```

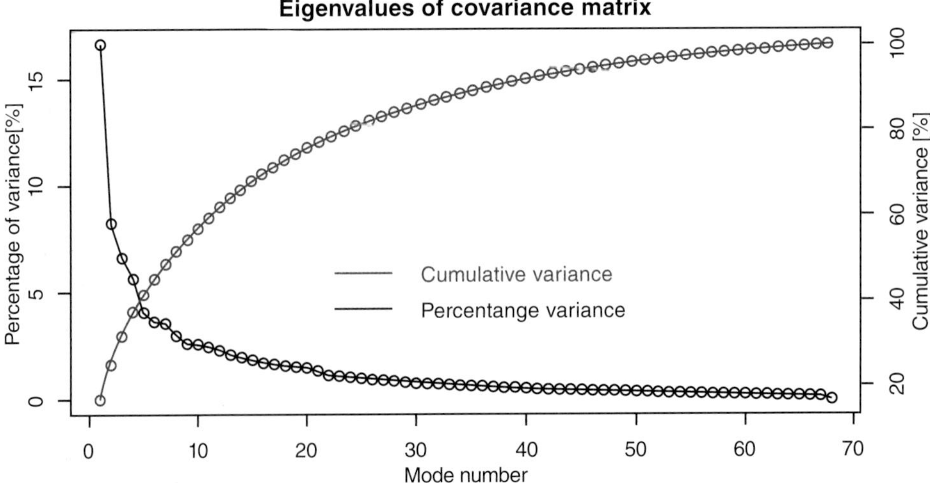

Fig. 10.6 Percentage variance, and the cumulative variance of the covariance matrix of the January SAT from 1948–2015.

```
par(new=TRUE)
plot(cumsum(100*(svdJ$d)^2/sum((svdJ$d)^2)),type="o",
col="blue",lwd=1.5,axes=FALSE,xlab="",ylab="")
legend(20,50, col=c("blue"),lty=1,lwd=2.0,
       legend=c("Cumulative variance"),bty="n",
       text.font=2,cex=1.0, text.col="blue")
axis(4)
mtext("Cumulative variance [%]",side=4,line=2)
```

The EOFs are from the column vectors of the SVD's U matrix and the PCs are the SVD's V columns. The first three EOFs and PCs are shown in Figs. 10.7–10.9, which may be generated by the following R code.

```
#plot EOF1: The physical EOF= eigenvector divided by area factor
mapmat=matrix(svdJ$u[,1]/sqrt(cos(gpcpst[,1]*pi/180)),nrow=144)
rgb.palette=colorRampPalette(c('blue','green','white',
                'yellow','red'),interpolate='spline')
int=seq(-0.04,0.04,length.out=61)
mapmat=mapmat[, seq(length(mapmat[1,]),1)]
filled.contour(Lon, Lat, -mapmat, color.palette=rgb.palette, levels=int,
```

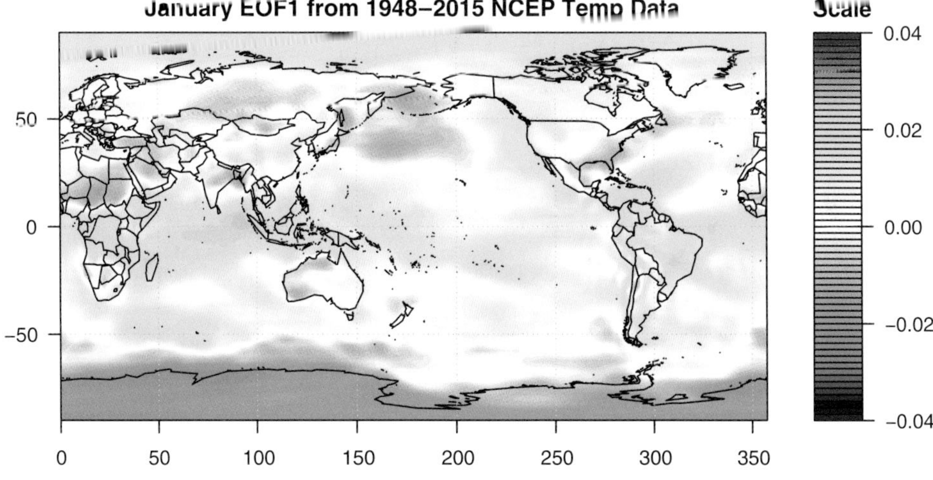

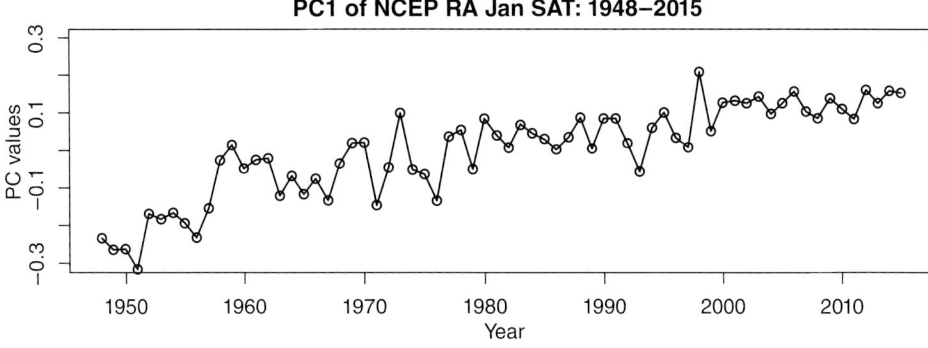

Fig. 10.7 The first EOF and PC from the January SAT standardized area-weighted anomalies.

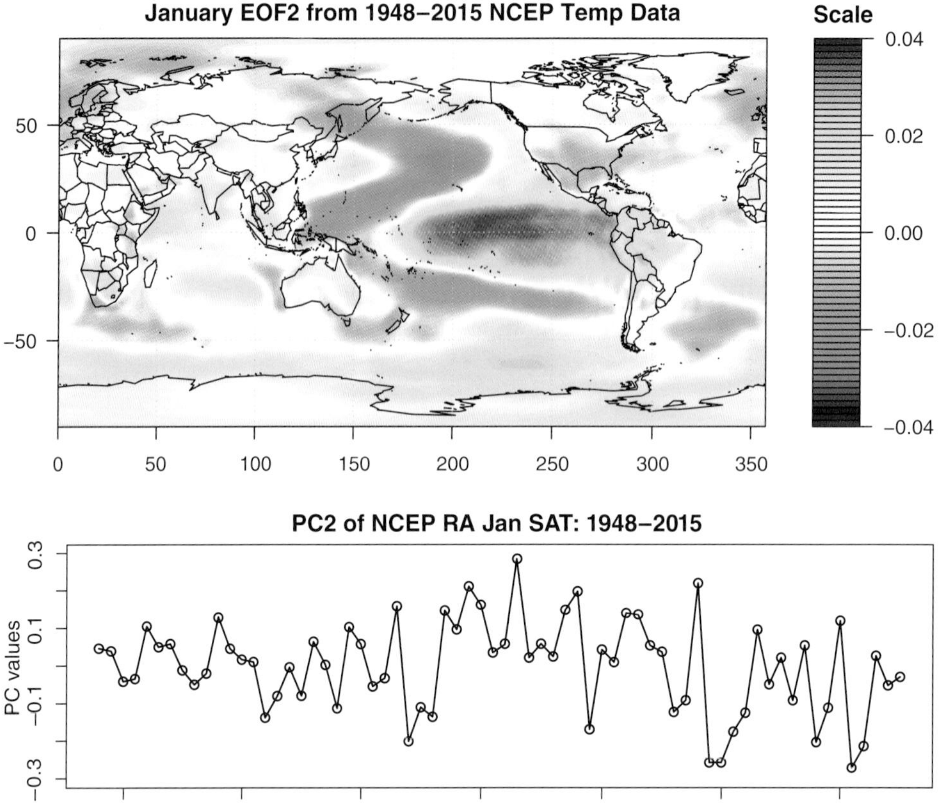

Fig. 10.8 The second EOF and PC from the January SAT standardized area-weighted anomalies.

```
  plot.title=title(main="January EOF1 from 1948-2015 NCEP Temp Data"),
                 plot.axes={axis(1); axis(2);map('world2', add=TRUE);grid()},
                 key.title=title(main="Scale"))
#
#plot PC1
pcdat<-svdJ$v[,1]
Time<-seq(1948,2015)
plot(Time, -pcdat, type="o", main="PC1 of NCEP RA Jan SAT: 1948-2015",
xlab="Year",ylab="PC values",
      lwd=2, ylim=c(-0.3,0.3))
```

Often, only the first two or three EOFs and PCs will have some physical interpretations, because the higher modes' eigenvalues are too close to one another, and hence have a large amount of uncertainty.

In the case of the January SAT Reanalysis data here, PC1 shows an increasing trend. The corresponding EOF1 shows the spatially non-uniform pattern of temperature increasing.

EOF2 shows an El Niño pattern, with a warm tongue over the eastern tropical Pacific. PC2 shows the temporal variation of the El Niño signal. For example, the January 1983 and 1998 peaks correspond to two strong El Niños.

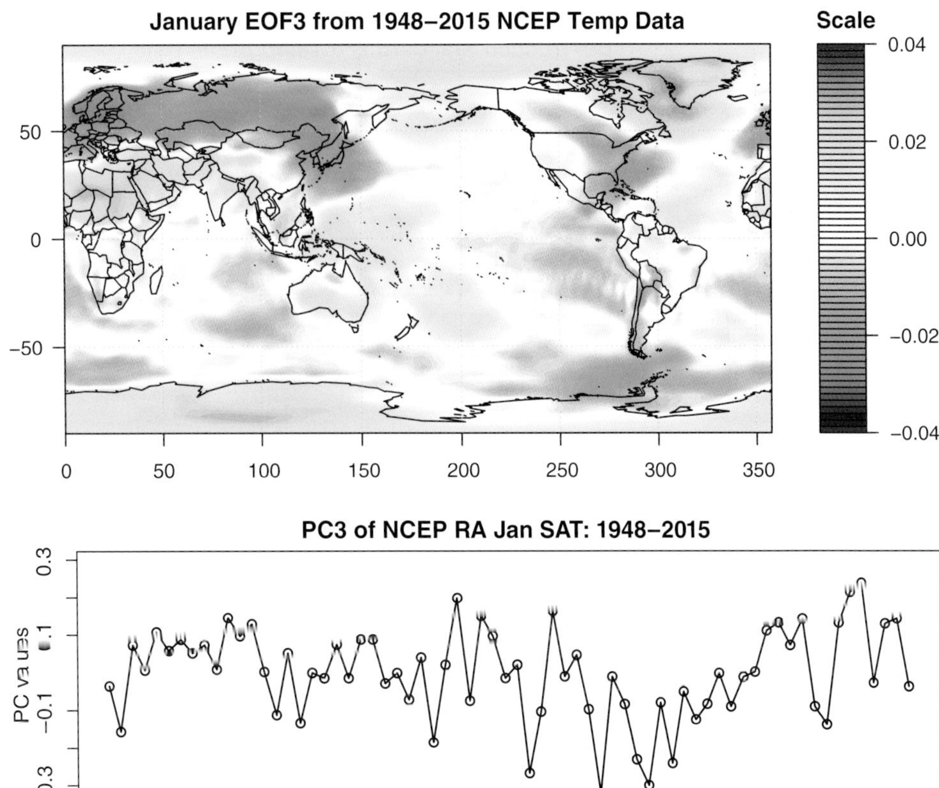

Fig. 10.9 The third EOF and PC from the January SAT's standardized area-weighted anomalies.

EOF3 appears to correspond to a mode known as the Pacific Decadal Oscillation. It shows a dipole pattern over the Northern Pacific. PC3 shows quasi-periodicity in this mode with a period of about 20 to 30 years.

10.3.2.4 EOFs from the De-trended Standardized Data

We can also de-trend the standardized anomaly data first and then compute the EOFs and PCs. As expected, the new EOF1 then will no longer be the trend pattern, rather it is the El Niño mode, i.e. the EOF2 of the non-detrended anomalies (see Figs. 10.10 and 10.11). This implies that the de-trending process has removed the EOF1 mode in the original non-de-trended data.

The de-trending and SVD procedures can be carried out by the following R code.

```
#EOF from de-trended data
monJ=seq(1,816,12)
gpcpdat=gpcpst[,3:818]
gpcpJ=gpcpdat[,monJ]
climJ<-rowMeans(gpcpJ)
```

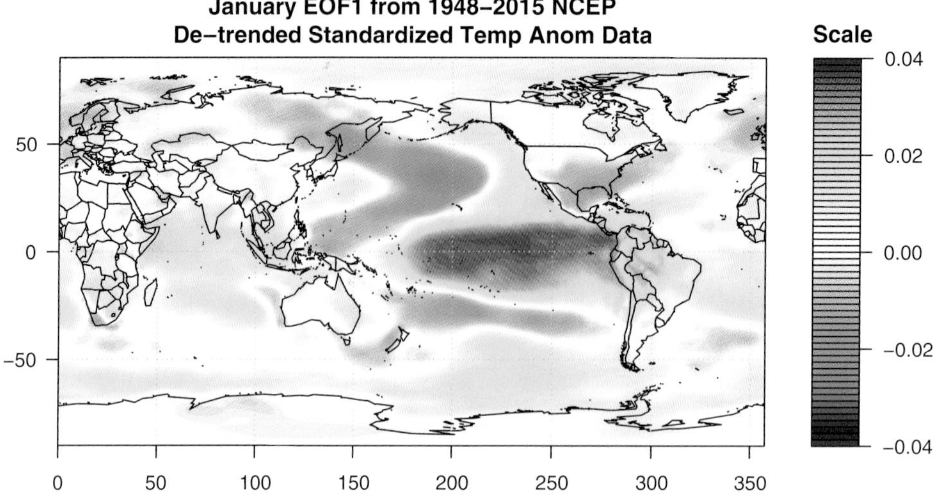

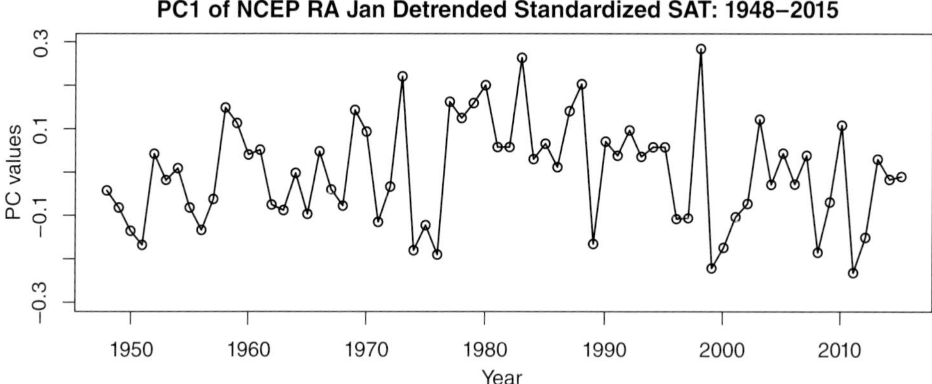

Fig. 10.10 The first EOF and PC from the January SAT de-trended standardized area-weighted anomalies.

```
library(matrixStats)
sdJ<-rowSds(gpcpJ)
anomJ=(gpcpdat[,monJ]-climJ)/sdJ
trendM<-matrix(0,nrow=10512, ncol=68)#trend field matrix
trendV<-rep(0,len=10512)#trend for each grid box: a vector
for (i in 1:10512) {
 trendM[i,] = (lm(anomJ[i,] ~ Time))$fitted.values
 trendV[i]<-lm(anomJ[i,] ~ Time)$coefficients[2]
}
dtanomJ = anomJ - trendM
dim(dtanomJ)
dtanomAW=sqrt(cos(gpcpst[,1]*pi/180))*dtanomJ
svdJ=svd(dtanomAW)
```

One can then use the EOF plotting code described in Section 10.3.2.3 to make the plots of eigenvalues, EOFs, and PCs. Comparing with the EOFs and PCs of Section 10.3.2.3, it is clear that the de-trended EOF1 here is similar to the non-de-trended EOF2. However,

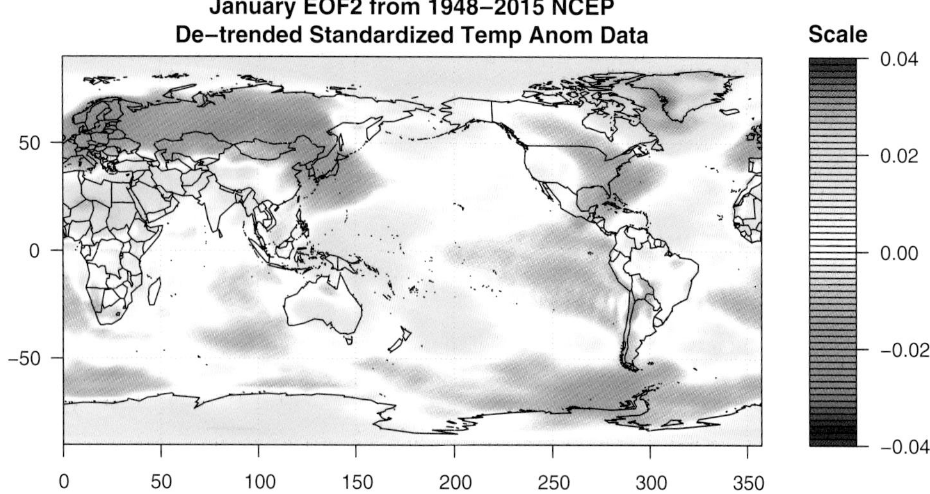

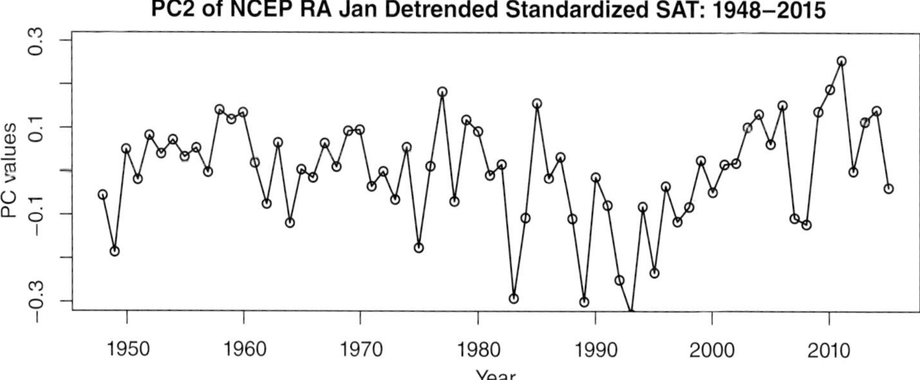

Fig. 10.11 The second EOF and PC from the January SAT de-trended standardized area-weighted anomalies.

they are not exactly the same. Thus, the de-trending process is approximately similar to the EOF1 filtering, although not exactly the same. Similar statements can be made for the other EOFs and PCs.

The first eigenvalue of the de-trended anomalies explains about 10% of the total variance, approximately equivalent to the 8% of the total variance explained by EOF2 of the non-de-trended anomalies (see Fig. 10.6). This can be derived from the non-de-trended SVD results. Let

$$c_i = 100 \frac{d_i^2}{\sum_{i=1}^{K} d_i^2}, \quad i = 1, 2, \ldots, K \tag{10.11}$$

be the percentage of variance explained by the ith mode, where K is the total number of modes available. In our case of January temperature from 1948–2015, $K = 68$. The SVD calculation of Section 10.3.2.3 found that

$$c_1 = 16.63[\%], c_2 = 8.25[\%]. \tag{10.12}$$

Figure 10.6 shows these values. From the following formula

$$c_2 = 100\frac{d_2^2}{\sum_{i=1}^K d_i^2} = 100\frac{d_2^2}{d_1^2 + \sum_{i=2}^K d_i^2}, \quad (10.13)$$

one can derive that

$$\begin{aligned} 100\frac{d_2^2}{\sum_{i=2}^K d_i^2} &= 100\frac{1}{1/c_2 - d_1^2/d_2^2} \\ &= 100\frac{c_2}{1-c_1} = 100\frac{0.0825}{1-0.1663} = 9.9[\%] \end{aligned} \quad (10.14)$$

which is approximately equal to the percentage 10% of the variance explained by the first EOF mode of the de-trended standardized data.

10.4 Area-Weighted Average and Spatial Distribution of Trend

10.4.1 Global Average and PC1

To verify that PC1 of the non-de-trended anomalies represents the trend, we compute and plot the area-weighted SAT (see Fig. 10.12) from the NCEP/NCAR RA1 data using the following R code (see p. 271).

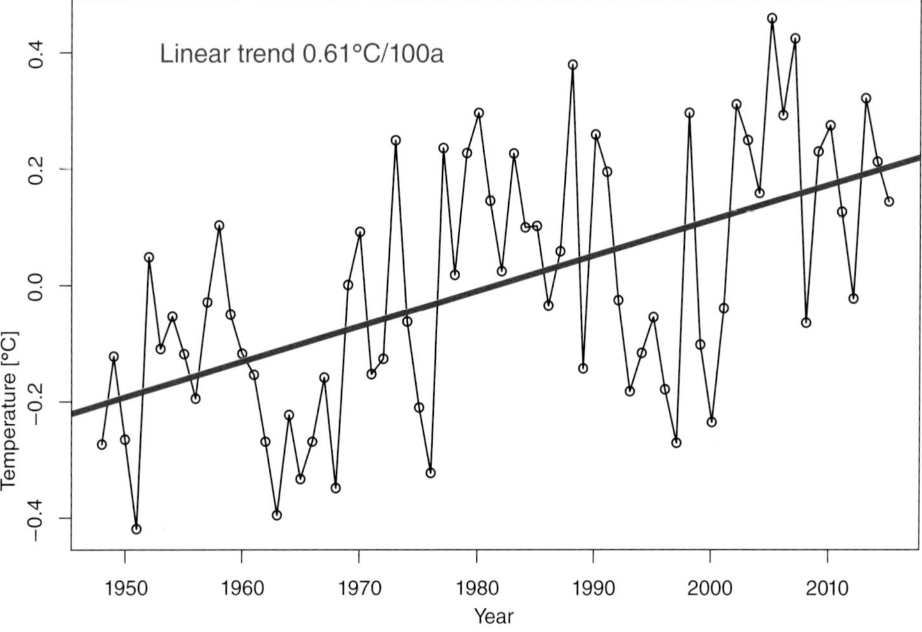

Fig. 10.12 The global area-weighted January SAT anomalies from 1948–2015 based on the NCEP/NCAR RA1 data.

```
#Plot the area-weighted global average Jan temp from 1948-2015
#Begin from the space-time data matrix gpcpst[,1]
vArea=cos(gpcpst[,1]*pi/180)
anomA=vArea*anomJ
dim(anomA)
JanSAT<-colSums(anomA)/sum(vArea)
plot(Time, JanSAT, type="o", lwd=2,
     main="Global Average Jan SAT Anomalies from NCEP RA",
     xlab="Year",ylab="Temperature [oC]")
regSAT<-lm(JanSAT ~ Time)
#0.48oC/100a trend
abline(regSAT, col="red", lwd=4)
text(1965,0.35,"Linear trend 0.48oC/100a", col="red", cex=1.3)
```

Figures 10.12 and 10.7 clearly show that this global average is similar to PC1 of the non-de-trended data. Their trends are very close: 0.53°C/100 a (i.e., 0.53 °C per century) for the global average, and 0.48°C/100 a for PC1. Their correlation is 0.61, not as large as one might expect, because the global average has more extremes and large temporal variances. PC1 may be understood as the nonlinear trend of large-scale spatial patterns. In other words, PC1 and EOF1 may be regarded as a low frequency and small wave number filter of the temperature anomaly field.

10.4.2 Spatial Pattern of Linear Trends

Next we compute and plot the temperature trend from 1948 to 2015 based on the NCEP/NCAR Reanalysis January temperature data. Each grid box has a time series of 68 years of January SAT anomalies from 1948 to 2015, and each grid box has a linear trend. These trends form a spatial pattern (see Fig. 10.13), which is similar to that of EOF1 for the non-de-trended SAT anomaly data. The trend value for each grid box is computed by R's linear model command:
`lm(anomJ[i,]~Time)$coefficients[2]`.
The anomaly data are assumed to be written in the space–time data matrix `gpcpst` with the first two columns as latitude and longitude. The following R codes make the trend calculation and plot.

```
#plot the trend of Jan SAT non-standardized anomaly data
#Begin with the space-time data matrix
monJ=seq(1,816,12)
gpcpdat=gpcpst[,3:818]
gpcpJ=gpcpdat[,monJ]
plot(gpcpJ[,23])
climJ<-rowMeans(gpcpJ)
anomJ=(gpcpdat[,monJ]-climJ)
trendV<-rep(0,len=10512)#trend for each grid box: a vector
for (i in 1:10512) {
  trendV[i]<-lm(anomJ[i,] ~ Time)$coefficients[2]
}
mapmat1=matrix(10*trendV,nrow=144)
mapv1=pmin(mapmat1,1) #Compress the values >5 to 5
mapmat=pmax(mapv1,-1) #compress the values <-5 t -5
rgb.palette=colorRampPalette(c('blue','green','white', 'yellow','red'),
```

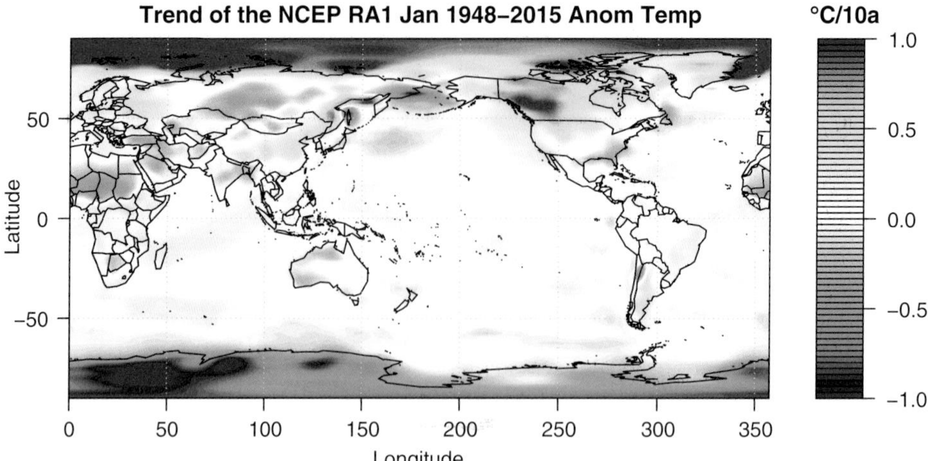

Fig. 10.13 Linear trend of NCEP Reanalysis January SAT from 1948–2015.

```
interpolate='spline')
int=seq(-1,1,length.out=61)
mapmat=mapmat[, seq(length(mapmat[1,]),1)]
filled.contour(Lon, Lat, mapmat, color.palette=rgb.palette, levels=int,
               plot.title=title(
               main="Trend of the NCEP RA1 Jan 1948-2015 Anom Temp",
               xlab="Latitude",ylab="Longitude"),
       plot.axes={axis(1); axis(2);map('world2', add=TRUE);grid()},
       key.title=title(main=expression(paste(degree, "C/10a"))))
```

Figure 10.13 shows that the trends are non-uniform. The trend magnitudes over land are larger than those over ocean. The largest positive trends are over the Arctic region, while the Antarctic region has negative trends. These trends may not be reliable since the reanalysis climate model has an amplified variance over the polar regions, another possible example of model deficiencies leading to erroneous information in the reanalysis produced by using that model.

Indeed, the spatial distribution of the trends appears similar to EOF1 of the non-detrended data (see Fig. 10.7). The spatial correlation between the trend map of Fig. 10.13 and the EOF1 map in Fig. 10.7 is very high and in fact is 0.97. This result implies that temperature's spatial patterns are more coherent than the temporal patterns, which is consistent with the existence of large spatial correlation scales for the monthly mean temperature field.

10.5 GPCP Precipitation Data: Analysis and Visualization by R

The preceding two sections used the gridded temperature data as examples to describe the R analysis and visualization. This section will use the gridded precipitation data to present some additional R analysis and visualization methods.

The Global Precipitation Climatology Project (GPCP) is a NASA project and contains several gridded global precipitation data products beginning in 1979 (Huffman et al. 1997). The GPCP data have incorporated both satellite remote sensing observations and the ground-based precipitation gauge measurements. The data time scales include daily, 5-day, and monthly values. GPCP datasets are often used as a convenient resource for climate model validation.

10.5.1 Read and Write GPCP Data

GPCP's native URL is https://precip.gsfc.nasa.gov. Several climate centers have provided GPCP data, such as the NOAA Earth System Research Laboratory (ESRL)
www.esrl.noaa.gov/psd/data/gridded/data.gpcp.html
ESRL has converted the GPCP data and many other popular climate datasets into the unified netCDF format to facilitate various kinds of applications. This section will use the monthly GPCP data downloaded from the ESRL. The download can be made using the following procedures

```
#Download the GPCP data from the ESRL website
#www.esrl.noaa.gov/psd/data/gridded/data.gpcp.html
#Select the monthly precipitation data by clicking
#the Download File: precip.mon.mean.nc
#Move the data file to your assigned working directory
#/Users/sshen/climmath/data
```

Read the netCDF data into R and check some basic parameters.

```
setwd("/Users/sshen/climmath")
#install.packages("ncdf")
library(ncdf4)
# 4 dimensions: lon,lat,level,time
nc=ncdf4::nc_open("data/precip.mon.mean.nc")
nc
nc$dim$lon$vals
nc$dim$lat$vals
nc$dim$time$vals
#nc$dim$time$units
#nc$dim$level$vals
Lon <- ncvar_get(nc, "lon")
Lat <- ncvar_get(nc, "lat")
Time<- ncvar_get(nc, "time")
head(Time)
#[1] 65378 65409 65437 65468 65498 65529
library(chron)
month.day.year(65378,c(month = 1, day = 1, year = 1800))
#1979-01-01
precnc<- ncvar_get(nc, "precip")
dim(precnc)
#[1] 144  72 451 #2.5-by-2.5, 451 months from Jan 1979-July 2016
```

This dataset is from January 1979 to July 2016 and has a 2.5-degree spatial resolution.

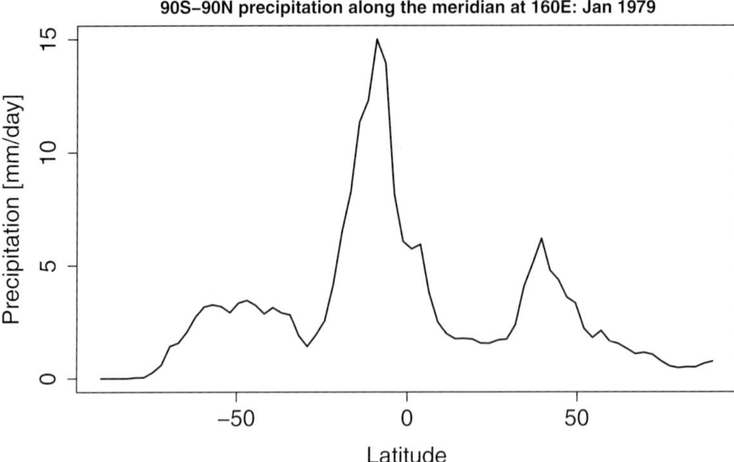

Fig. 10.14 The precipitation distribution along the meridian at 160°E for January 1979.

To see if the dataset is climatologically reasonable before proceeding for further analysis, one can make a simple plot of the precipitation along a meridional line over the Pacific. The tropical zone should have more precipitation than the polar regions (see Fig. 10.14). This can be done by the following R code for January 1979 on the meridional line at 160°E.

```
#plot a 90S-90N precip along a meridional line at 160E
par(mar=c(4.5,4.5,2,0.5))
plot(seq(-90,90,length=72), precnc[64,,1], type="l", lwd=2,
     xlab="Latitude", ylab="Precipitation [mm/day]",
     main="90S-90N precipitation along the meridian at 160E: Jan 1979",
     cex.lab=1.5, cex.axis=1.5)
```

One can also write the netCDF `.nc` data as a space–time dataset in the `.csv` format that can be read by Excel. The space–time data can also be used for EOF analysis. The writing process can be done by the following R code.

```
#Write the data as space-time matrix with a header
precst=matrix(0,nrow=10368,ncol=451)
temp=as.vector(precnc[,,1])
head(temp)
for (i in 1:451) {precst[,i]=as.vector(precnc[ , , i])}
#precst is the space-time GPCP data
LAT=rep(Lat, 144)
LON=rep(Lon[1],72)
for (i in 2:144){LON=c(LON, rep(Lon[i],72))}
gpcpst=cbind(LAT, LON, precst)

#Convert the Julian day into calender dates for time
tm=month.day.year(Time, c(month = 1, day = 1, year = 1800))
tm1=paste(tm$year,"-",tm$month)
#tm1=data.frame(tm1)
tm2=c("Lat","Lon",tm1)
colnames(gpcpst) <- tm2
```

```
#Look at a sample section of the space-time data
gpcpst[890:892,1:5]
#          Lat    Lon     1979 - 1   1979 - 2   1979 - 3
#[1,] -26.25 31.25 0.05752099 0.1805309 0.2104454
#[2,] -23.75 31.25 0.10855579 0.2562788 0.2883888
#[3,] -21.25 31.25 0.08718992 0.2738046 0.2569866
write.csv(gpcpst,file="gpcp9716jul.csv")
```

10.5.2 GPCP Climatology and Standard Deviation

Figure 10.15 shows the climatology and standard deviation of the GPCP monthly precipitation data from January 1979 to July 2016. The figure can be generated by the following R code.

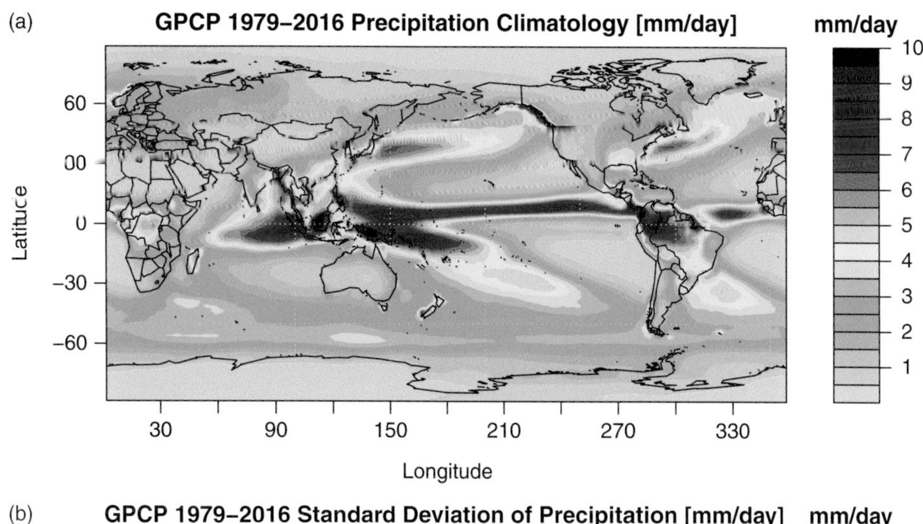

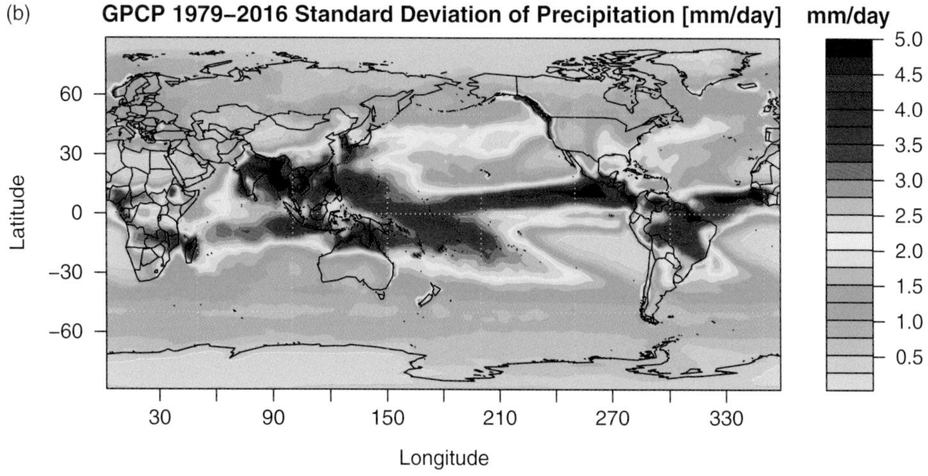

Fig. 10.15 The climatology (a) and standard deviation (b) of the monthly GPCP precipitation from January 1979 to July 2016.

```
#Compute and plot the GPCP climatology from Jan 1979-July 2016
library(maps)
climmat=precnc[,,1]
for(i in 2:451){climmat =climmat + precnc[,,i]}
climmat=climmat/451
mapmat=climmat
mapmat=pmax(pmin(mapmat,10),0)
int=seq(min(mapmat),max(mapmat),length.out=11)
rgb.palette=colorRampPalette(c('bisque2','cyan', 'green', 'yellow',
 'pink','indianred2', 'red','maroon','black'),interpolate='spline')
plot.new()
par(mar=c(4, 4.5, 2, 1))
filled.contour(Lon, Lat, mapmat, color.palette=rgb.palette, levels=int,
               plot.title=title(
               main="GPCP 1979-2016 Precipitation Climatology [mm/day]",
               xlab="Latitude",ylab="Longitude", cex.lab=1.5),
               plot.axes={axis(1,seq(0,360, by=30), cex.axis=1.5);
                 axis(2, seq(-90,90,by=30), cex.axis=1.5);
                 map('world2', add=TRUE);grid()},
               key.title=title(main="mm/day"),
               key.axes={axis(4, seq(0,10, len=11), cex.axis=1.5)})

#Compute and plot standard deviation from Jan 1979-July 2016
sdmat=(precnc[,,1]-climmat)^2
for(i in 2:451){sdmat =sdmat + (precnc[,,i]-climmat)^2}
sdmat=sqrt(sdmat/451)
mapmat=sdmat
mapmat=pmax(pmin(mapmat,5),0)
int=seq(min(mapmat),max(mapmat),length.out=21)
rgb.palette=colorRampPalette(c('bisque2','cyan', 'green', 'yellow',
'pink','indianred2', 'red','maroon','black'),interpolate='spline')
plot.new()
par(mar=c(4.3, 4.5, 2, 1))
filled.contour(Lon, Lat, mapmat, color.palette=rgb.palette, levels=int,
               plot.title=title(
   main="GPCP 1979-2016 Standard Deviation [mm/day]",
   xlab="Longitude",ylab="Latitude", cex.lab=1.5),
               plot.axes={axis(1,seq(0,360, by=30), cex.axis=1.5);
                 axis(2,seq(-90,90,by=30), cex.axis=1.5);
                 map('world2', add=TRUE);grid()},
               key.title=title(main="mm/day"),
               key.axes={axis(4, seq(0,5,len=11), cex.axis=1.5)})
```

The above climatology is the temporal mean of the entire dataset and is equivalent to the annual climatology. The climatology for each month can also be calculated by a simple R code. The one for January is below.

```
#Compute the January climatology
Jmon=seq(3,453,by=12)
Jclim =rowMeans(gpcpst[,Jmon])
```

The above R code for plotting the annual climatology can be used to plot the January climatology with data

```
mapmat=matrix(Jclim,nrow=144)
```

10.6 Chapter Summary

This chapter on R programming focuses on two advanced topics of climate data analysis: EOF analysis and trend patterns.

(i) The EOF analysis viewed from the perspective of spatial pattern, and the PC analysis viewed from the perspective of temporal pattern, have been presented in this book as a result of the linear algebra method of SVD, which is motivated by the viewpoint of the spatiotemporal patterns of physics. Our approach is different from the traditional covariance matrix method, which is considered to be a statistical approach, in which the EOFs are regarded as the eigenvectors of the covariance matrix. In contrast, the Golub–Reinsch SVD algorithm can compute EOFs and PCs without computing the covariance matrix. Hence, one does not need the assumptions of stationarity and ergodicity.[1] Consequently, our SVD approach to the space–time patterns is conceptually simpler than the statistical approach, and it is also more directly related physically to the climatic interpretations of the EOF and PC patterns.

(ii) Section 10.2 uses the SVD analysis to recover two specified space–time patterns and hence demonstrates the basic mathematical structure of SVD.

(iii) Section 10.3 provides an R code to make SVD analysis on the NCEP Reanalysis data. This short but complete R code starts with reading the netCDF reanalysis model data and proceeds to the formation of the space–time data matrix for SVD, and to the digital and graphic output of EOFs and PCs. This R code is a practical and useful tool for analyzing climate model output.

(iv) Section 10.4 describes the R codes for computing the spatial average of the reanalysis data and its temporal trend, and for computing the linear trend for each grid box and displaying the spatial patterns of the trend. The global average and trend patterns are important research tools to explore the mechanisms of climate changes.

(v) Section 10.5 uses gridded precipitation data as an example for presenting some additional methods for analysis and visualization using R.

Therefore, this chapter provides the R codes for practical climate data analysis and visualization. These codes can be efficient tools for comprehensive diagnostic analyses of climate model data. Again, it is impractical to memorize all the R commands included in this chapter. It is probably more productive to treat this chapter as a user's manual or a handbook for your analysis. You can copy and paste our codes and modify them by finding appropriate R commands on the Internet. For more details on the EOF theory, errors, and interpretation, see North et al. (1982), and Monahan et al. (2009). For more details on the SVD method, see Strang (2016).

[1] Both stationarity and ergodicity are rigorously defined concepts of mathematical statistics. Loosely speaking for climate data analysis, stationarity means that mean, variance, skewness, kurtosis, and other moments do not vary in time, and ergodicity means that an expected value may be approximated by a long-time average.

References and Further Readings

[1] Huffman, G. J., R. F. Adler, P. Arkin, A. Chang, R. Ferraro, A. Gruber, J. Janowiak, A. McNab, B. Rudolf, and U. Schneider, 1997: The Global Precipitation Climatology Project (GPCP) combined precipitation dataset. *Bulletin of the American Meteorological Society*, 78, 5–20.

> This is the initial paper on the GPCP products. Many papers have been published on the GPCP data error estimates, data updates, and data applications.

[2] Monahan, A. H., J. C. Fyfe, M. H. Ambaum, D. B. Stephenson, and G. R. North, 2009: Empirical orthogonal functions: The medium is the message. *Journal of Climate*, 22, 6501–6514.

> This review paper is a useful reminder to be careful when interpreting the EOF modes from the perspective of nonlinear climate dynamics.

[3] North, G. R., F. J. Moeng, T. J. Bell, and R. F. Cahalan, 1982: Sampling errors in the estimation of empirical orthogonal functions. *Monthly Weather Review*, 110, 699–706.

> This seminal paper led to the EOF theory becoming a popular method for climate data analysis. Gerald R. North (1938–) is an American theoretical climate scientist who has developed several important theories and methods for climate science, including the space–time EOF method to quantitatively detect and attribute climate changes, multiple-solutions of EBMs and their stabilities, North's rule-of-thumb on EOF accuracy, and the sample design and error estimation methods for the Tropical Rainfall Measuring Mission satellite.

[4] Strang, G., 2016: *Introduction to Linear Algebra*, 5th Edition, Wellesley-Cambridge Press, Wellesley, MA 02482.

> Gilbert Strang (1934–) is an American mathematician and educator. His textbooks and pedagogy have been internationally influential. This text is one of very few basic linear algebra books that include excellent materials on SVD, probability, and statistics.

Exercises

All exercises in this chapter require the use of a computer.

10.1 (a) Use R to compute the December SAT climatology based on the 1961–1990 NCEP/NCAR Reanalysis 1 SAT data on a $2.5^\circ \times 2.5^\circ$ grid.

(b) Use R to plot the climatology map.

10.2 (a) Use R to compute the December SAT standard deviation based on the 1961–1990 NCEP/NCAR Reanalysis 1 SAT data on a $2.5^\circ \times 2.5^\circ$ grid.
(b) Use R to plot the standard deviation map.

10.3 (a) Use R to compute the December SAT anomalies from 1948 to 2015 based on the 1961–1990 climatology from the NCEP/NCAR Reanalysis 1 SAT data on a $2.5^\circ \times 2.5^\circ$ grid.
(b) Use R to plot the anomalies map for December 1997 and December 1998.
(c) Use 100–200 words to describe the El Niño and La Niña phenomena you can observe from the plots in Step (b).

10.4 (a) Use the SVD command of R to calculate the EOFs and PCs for the December SAT anomalies from 1948 to 2015 computed in Exercise 10.3 above.
(b) Use R to plot the squared SVD eigenvalues against mode number for the first 20 modes.
(c) Use R to plot the maps of the first three EOFs.
(d) Use R to plot the time series of the first three PCs.
(e) Use R to verify that all the EOF vectors are orthonormal.
(f) Use R to verify that all the PC vectors are orthonormal.

10.5 (a) Use R to compute the December standardized anomalies from 1948 to 2015 by dividing the anomalies obtained in Exercise 10.3 by the standard deviations obtained in Exercise 10.2.
(b) For the standardized anomalies, repeat the six steps of Exercise 10.4.

10.6 (a) Use R to compute the linear trend from 1948 to 2015 for the December anomalies obtained in Exercise 10.3 on each grid box.
(b) Plot the map of the linear trend.

10.7 (a) Use R to compute the de-trended anomalies by subtracting the linear fit function for each grid box obtained in Exercise 10.6 from the anomalies obtained in Exercise 10.3.
(b) For the de-trended anomalies, repeat the six steps of Exercise 10.4.

10.8 Use the SVD approach to calculate the EOFs and PCs for the Northern Hemisphere's January SAT based on the anomaly data from the NCEP/NCAR Reanalysis. Use the 1981–2010 January mean as the January climatology for each grid box. Plot the squared SVD eigenvalues against mode number for the first 30 modes. The suggested steps are below.

(a) Convert the Reanalysis data into a space–time data matrix.
(b) Extract the Northern Hemisphere's January data using proper row and column indices.
(c) Apply the SVD to the extracted space–time matrix.
(d) Plot the squared eigenvalues.

10.9 Use R to plot the first three EOFs and PCs obtained in Exercise 10.8. Interpret the climatic meaning as well as you can, but limit your answer to no more than 300 words.

10.10 Compute and plot the first three EOFs and PCs for the January SAT anomalies from 1948 to 2018 for a tropic Pacific region (20°S–20°N, 160°E–120°W) with respect to 1971–2000 climatology. Use these EOFs and PCs to describe the spatial and temporal patterns of El Niño, but limit your answer to no more than 300 words.

10.11 (a) Use R to calculate the SVD for the December GPCP precipitation anomaly data from 1981 to 2010, based on the December climatology in the same period from 1981 to 2010.

(b) Use R to plot the first three EOFs.

(c) Use R to plot the first three PCs.

(d) Use 200–500 words to explain the climatic meaning of the EOFs and PCs.

11 R Analysis of Incomplete Climate Data

We are now near the end of our journey. We have discussed many mathematical techniques and tools, and we have demonstrated many capabilities of the programming language R. These include the ability to implement the mathematical tools and to analyze datasets, which might be generated by models or obtained from observations. We have seen that R can carry out many kinds of statistical analyses of data and can produce a wide range of graphical products.

However, between the optimistic beginning and the successful end of almost all research projects, the mischievous gods of science have placed a variety of frustrating obstacles. One of these obstacles is missing data. Data can be missing for many reasons. Instruments can break or malfunction. Communications can fail. Ship tracks or satellite orbits can cause parts of the climate system to be unobserved. Human error can lose or destroy data. And sometimes data that ought to be in a dataset simply cannot be found.

Sometimes, missing data can be created in an effort to develop quality control procedures to flag erroneous data. The history is not perfectly clear, but something like that may have delayed the identification of the Antarctic ozone hole from satellite data. NASA satellite data showing ozone amounts lower than expected were flagged as suspicious. The NASA team later decided that the data were good. Meanwhile, a British team led by Dr. Joseph Farman published its discovery of the ozone hole using ground-based instrumentation. The moral of the story is that one should always devote effort to missing or questionable data.

11.1 The Missing Data Problem

Unlike climate model data which are space–time complete, observed data are often space–time incomplete, i.e., some space–time grid points or boxes do not have data. We call this the missing data problem.

Missing data problems can be of many kinds and can be very complicated. Here we use the NOAAGlobalTemp dataset to illustrate a few methods often used in analyzing datasets with missing data. NOAAGlobalTemp is the merged land and oceanic observed surface temperature anomalies with respect to the 1970–2000 base period climatology, produced by the United States National Centers for Environmental Information in 2015 (Karl et al. 2015).

www.ncdc.noaa.gov/data-access/marineocean-data/
noaa-global-surface-temperature-noaaglobaltemp

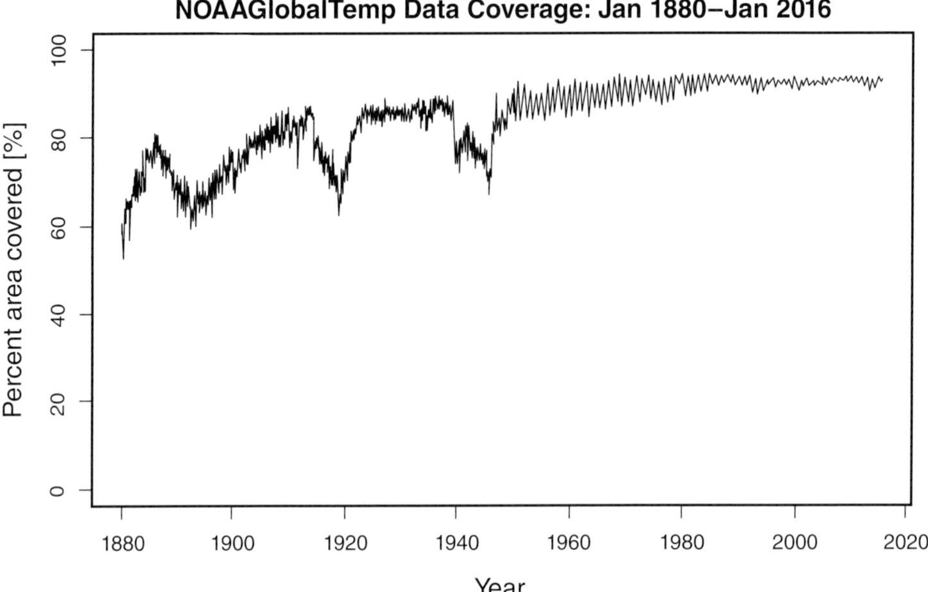

Fig. 11.1 Percentage of the global surface area covered by the NOAAGlobalTemp dataset.

This dataset contains monthly data from January 1880 to the present with $5 \times 5^\circ$ latitude–longitude spatial resolution. The earlier years had many missing data while the recent years are better covered. Figure 11.1 shows the history of the percentage of area covered by the data. One hundred minus this percentage is the percentage of missing data. The minimum coverage is nearly 60%, much of which is due to the good coverage provided by NOAA ERSST (extended reconstructed sea surface temperature) data (Huang et al. 2015, and Smith et al. 2003).

Using software called 4DVD (4-dimensional visual delivery of big climate data) (URL: 4dvd.org) developed at San Diego State University, one can easily see where and when data are missing. Figure 11.2 is a 4DVD screenshot and shows the NOAAGlobalTemp data distribution over the globe for January 1917. The data cover 72% of the global area. The black region includes 28% of the global area and has missing data. The data void regions include the polar areas which could not be accessed at that time, the central tropical Pacific regions which were not on the tracks of commercial ships, central Asia, part of Africa, and the Amazon region. Figure 11.3 shows that the grid box (2.5°S, 22.5°W) in the Amazon region has many missing data. The first datum was for August 1880, next August 1881, the third November 1881, etc., and the record became almost continuous from December 1904.

11.2 Read NOAAGlobalTemp and Form the Space–Time Data Matrix

This section describes how to use R to read the data and convert the data into a standard space–time matrix for various kinds of analyses.

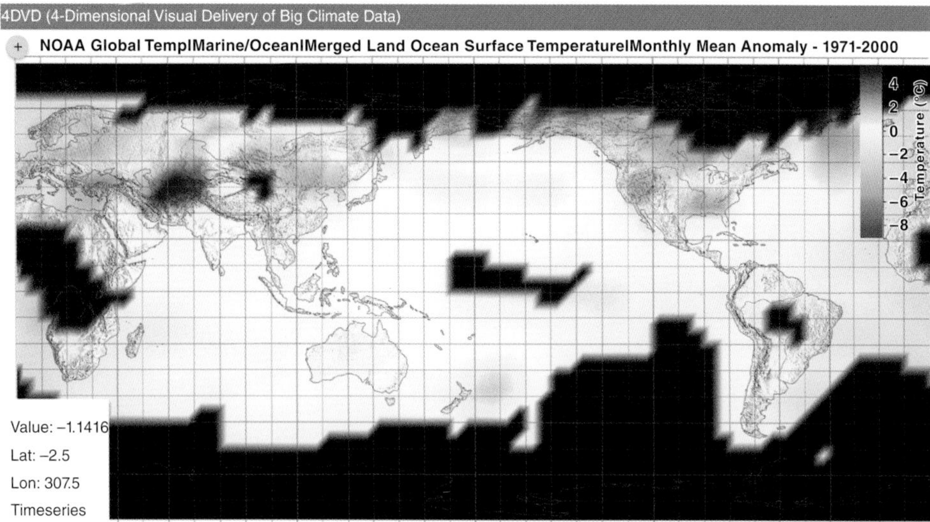

Fig. 11.2 Spatial distribution of the January 1917 of the NOAAGlobalTemp data. The black regions indicate missing data. *Credit: J. Pierret and S. S. P. Shen, 2018*: www.4dvd.org

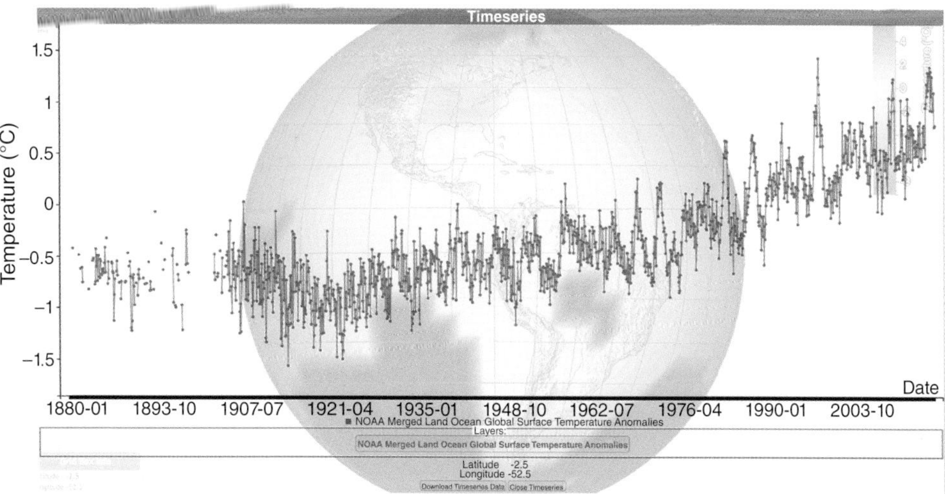

Fig. 11.3 Time series of the monthly temperature anomalies for a grid box over the Amazon region. *Credit: J. Pierret and S. S. P. Shen, 2018*: www.4dvd.org

11.2.1 Read the Downloaded Data

First, we download the NOAAGlobalTemp gridded data from its ftp site
ftp://ftp.ncdc.noaa.gov/pub/data/noaaglobaltemp/operational
The anomalies are with respect to the 1971–2000 climatology.

The ftp site has two data formats: asc and bin. We use the asc format as an example to describe the R analysis. The following R code reads the asc data and makes the conversion.

```
rm(list=ls(all=TRUE))
# Download .asc file
setwd("/Users/sshen/climmath")
da1=scan("data/NOAAGlobalTemp.gridded.v4.0.1.201701.asc")
length(da1)
#[1] 4267130
da1[1:3]
#[1]    1.0 1880.0 -999.9 #means mon, year, temp
#data in 72 rows (2.5, ..., 357.5) and
#data in 36 columns (-87.5, ..., 87.5)
tm1=seq(1,4267129, by=2594)
tm2=seq(2,4267130, by=2594)
length(tm1)
length(tm2)
mm1=da1[tm1] #Extract months
yy1=da1[tm2] #Extract years
head(mm1)
head(yy1)
length(mm1)
length(yy1)
rw1<-paste(yy1, sep="-", mm1) #Combine YYYY with MM
head(tm1)
head(tm2)
tm3=cbind(tm1,tm2)
tm4=as.vector(t(tm3))
head(tm4)
#[1]    1    2 2595 2596 5189 5190
da2<-da1[-tm4] #Remote the months and years data from the scanned data
length(da2)/(36*72)
#[1] 1645 #months, 137 yrs 1 mon: Jan 1880-Jan 2017
da3<-matrix(da2,ncol=1645) #Generate the space-time data
#2592 (=36*72) rows and 1645 months (=137 yrs 1 mon)
dim(da3)
#[1] 2592 1645
```

To facilitate the use of space–time data, we add the latitude and longitude coordinates for each grid box as the first two columns, and the time mark for each month as the first row. This can be done by the following R code.

```
colnames(da3)<-rw1
lat1=seq(-87.5, 87.5, length=36)
lon1=seq(2.5, 357.5,  length=72)
LAT=rep(lat1, each=72)
LON=rep(lon1,36)
gpcpst=cbind(LAT, LON, da3)
head(gpcpst)
dim(gpcpst)
#[1] 2592 1647 #The first two columns are Lat and Lon
#-87.5 to 87.5 and then 2.5 to 375.5
#The first row for time is header, not counted as data.
gpcpst[1:3,1:6] #Part of the data
#        LAT  LON 1880-1 1880-2 1880-3 1880-4
#[1,] -87.5  2.5 -999.9 -999.9 -999.9 -999.9
#[2,] -87.5  7.5 -999.9 -999.9 -999.9 -999.9
#[3,] -87.5 12.5 -999.9 -999.9 -999.9 -999.9

write.csv(gpcpst,file="NOAAGlobalT.csv")
#Output the data as a csv file
```

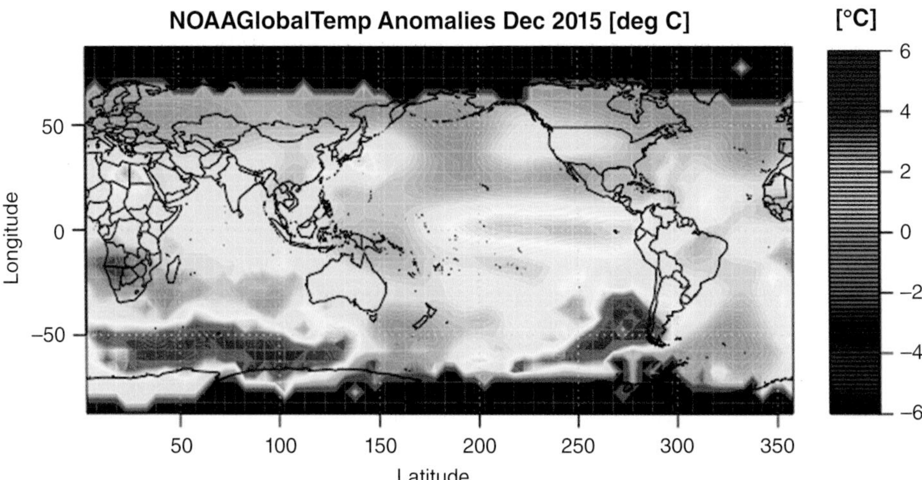

Fig. 11.4 Monthly mean temperature anomalies of December 2015 based on the NOAAGlobalTemp data.

11.2.2 Plot the Temperature Data Map of a Given Month

With this space time data, one can plot a data map for a given month or a data time series for a given location. For example, the following R code plots the temperature data map for December 2015, an El Niño month (see Fig. 11.4).

```
library(maps)#Need to install maps package first if not done before
Lat= seq(-87.5, 87.5, length=36)
Lon=seq(2.5, 357.5, length=72)
mapmat=matrix(gpcpst[,1634],nrow=72)
#column 1634 corresponding to Dec 2015
#Convert the vector into a lon-lat matrix for R map plotting
mapmat=pmax(pmin(mapmat,6),-6) #Put values between -6 and 6
#Matrix flipping is not needed since the data go from 2.5 to 375.5
plot.new()
par(mar=c(4,5,3,0))
int=seq(-6,6,length.out=81)
rgb.palette=colorRampPalette(c('black','blue', 'darkgreen','green',
'yellow','pink','red','maroon'),interpolate='spline')
mapmat= mapmat[,seq(length(mapmat[1,]),1)]
filled.contour(Lon, Lat, mapmat, color.palette=rgb.palette, levels=int,
    plot.title=title(main="NOAAGlobalTemp Anomalies Dec 2015 [deg C]",
                   xlab="Latitude",ylab="Longitude", cex.lab=1.5),
    plot.axes={axis(1, cex.axis=1.5);
               axis(2, cex.axis=1.5);map('world2', add=TRUE);grid()},
    key.title=title(main=expression(paste("[", degree, "C]"))),
    key.axes={axis(4, cex.axis=1.5)})
```

11.2.3 Extract the Data for a Specified Region

If one wishes to study the data over a particular region, say, the tropical Pacific for El Niño characteristics, one can extract the data for the region for a given time interval.

The following code extracts the space–time data for the tropical Pacific region (20°S–20°N, 160°E–100°W) from 1951 to 2000.

```
#Select only the data for the tropical Pacific region
n2<-which(gpcpst[,1]>-20&gpcpst[,1]<20&gpcpst[,2]>160&gpcpst[,2]<260)
dim(gpcpst)
length(n2)
#[1] 160 (=8 latitude bends and 20 longitude bends)
pacificdat=gpcpst[n2,855:1454]
```

Here, we have used a powerful and convenient `which` search command. This very useful command is easier to program and faster than `if` conditions.

Despite the good coverage of ERSST, it still has a few missing data in this tropical Pacific area. Because the missing data are assigned −999.00, they can significantly impact the computing results, such as SVD, when they are used in computing. We assign the missing data to be zero, instead of −999.00. The following code plots the December 1997 temperature data for the tropical Pacific region (20°S–20°N, 160°E–120°W) (see Fig. 11.5).

```
plot.new()
Lat=seq(-17.5,17.5, by=5)
Lon=seq(162.5, 257.5, by=5)
par(mar=c(4,5,3,0))
mapmat=matrix(pacificdat[,564], nrow=20)
int=seq(-5,5,length.out=81)
rgb.palette=colorRampPalette(c('black','blue', 'darkgreen',
                                'green', 'yellow','pink','red','maroon'),
                              interpolate='spline')
#mapmat= mapmat[,seq(length(mapmat[1,]),1)]
filled.contour(Lon, Lat, mapmat, color.palette=rgb.palette, levels=int,
               xlim=c(120,300),ylim=c(-40,40),
plot.title=title(main="Tropic Pacific SAT Anomalies [deg C]: Dec 1997",
                 xlab="Latitude",ylab="Longitude", cex.lab=1.5),
plot.axes={axis(1, cex.axis=1.5); axis(2, cex.axis=1.5);
```

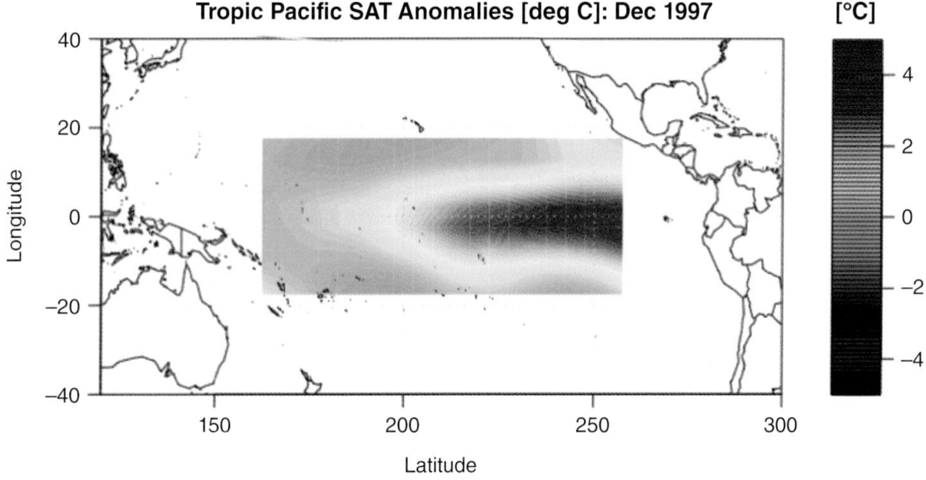

Fig. 11.5 Tropical Pacific SST anomalies of December 1997 based on the NOAAGlobalTemp data.

```
  map('world2', add=TRUE);grid()},
key.title=title(main="[oC]"),
key.axes={axis(4, cex.axis=1.5)})
```

11.2.4 Extract Data from Only One Grid Box

A special case is to extract data for a specified grid box with given latitude and longitude, e.g., the San Diego box (32.5N, 117.5W) or (+32.5, 242.5). This can be easily done by the following R code that includes a simple plotting command.

```
#Extract data for a specified box with given lat and lon
n2 <- which(gpcpst[,1]==32.5&gpcpst[,2]==242.5)
SanDiegoData <- gpcpst[n2,855:1454]
plot(seq(1880,2017, len=length(SanDiegoData)),
     SanDiegoData, type="l",
     xlab="Year", ylab="Temp [oC]",
     main="San Diego temperature anomalies history")
```

11.3 Spatial Averages and Their Trends

Having downloaded and reformatted the data, we now proceed to analyze it. We shall emphasize procedures for dealing with missing data.

11.3.1 Compute and Plot the Global Area-Weighted Average of Monthly Data

The area-weighted average, also called spatial average, of a temperature field $T(\phi,\theta,t)$ on a sphere is mathematically defined as follows

$$\bar{T}(t) = \frac{1}{4\pi}\iint T(\phi,\theta,t)\cos(\phi)d\phi d\theta, \tag{11.1}$$

where ϕ is latitude and θ is longitude, and t is time. The above formula's discrete form for a grid of resolution $\Delta\phi \times \Delta\theta$ is

$$\hat{\bar{T}}(t) = \sum_{i,j} T(i,j,t)\frac{\cos(\phi_{ij})\Delta\phi\Delta\theta}{4\pi}, \tag{11.2}$$

where (i,j) are coordinate indices for the grid box (i,j), and $\Delta\phi$ and $\Delta\theta$ are in radians. If it is a 5° resolution, then $\Delta\phi = \Delta\theta = (5/180)\pi$.

If NOAAGlobalTemp had data in every box, then the global average would be easy to calculate according to the above formula:

$$\hat{\bar{T}}(t) = \sum_{i,j} T(i,j,t)\frac{\cos(\phi_{ij})(5/180)^2}{4\pi^{-1}}. \tag{11.3}$$

However, NOAAGlobalTemp has missing data. We thus should not average the data-void region. One simple and straightforward method is to consider the spatial average problem as a weighted average, which assigns a data box with weight proportional to $\cos\phi_{ij}$ and a data-void box with zero weight. We thus generate a weight matrix `areaw` corresponding to the data matrix `temp` by the following R code.

```
#36-by-72 boxes and Jan1880-Jan2016=1633 months + lat and lon
#Compute the area-weight for each box and each month
#that has data. Thus the area-weight is a matrix.
areaw=matrix(0,nrow=2592,ncol=1647)
dim(areaw)
#[1] 2592 1647
temp=gpcpst
areaw[,1]=temp[,1]
areaw[,2]=temp[,2]
veca=cos(areaw[,1]*pi/180) #convert deg into radian
#create an area-weight matrix equal to cosine for the box with data
#and zero for the box with missing data
for(j in 3:1647) {
  for (i in 1:2592) {if(temp[i,j]> -290.0) {areaw[i,j]=veca[i]} }}
```

Then compute an area-weighted temperature data matrix and its average:

```
#area-weight data matrixs first two columns as lat-lon
tempw=areaw*temp
tempw[,1:2]=temp[,1:2]
#create monthly global average vector for 1645 months
#Jan 1880- Jan 2017
avev=colSums(tempw[,3:1647])/colSums(areaw[,3:1647])
```

Figure 11.6 shows the spatial average of the monthly temperature data from NOAAGlobalTemp from January 1880 to January 2017 and can be generated by the following R code.

```
plot.new()
timemo=seq(1880,2017,length=1645)
par(mar=c(3.5,3.5,3,0.5))
par(mgp=c(2.3,1.0,0.0))
plot(timemo,avev,type="l", cex.lab=1.4,
     xlab="Year", ylab="Temperature anomaly [deg C]",
     main="Area-weighted global average of
     monthly SAT anomalies: Jan 1880-Jan 2017")
abline(lm(avev ~ timemo),col="blue",lwd=2)
text(1930,0.7, "Linear trend: 0.69 [oC] per century",
     cex=1.4, col="blue")
```

11.3.2 Percent Coverage of the NOAAGlobalTemp

As a byproduct of the above weighted average, the matrix *areaw* can be used to calculate the percentage of area covered by the data.

```
rcover=100*colSums(areaw[,3:1647])/sum(veca)
```

The following R code can plot this time series against time, which is the percentage of data-covered area with respect to the entire globe, shown in Fig. 11.1 at the beginning of this chapter.

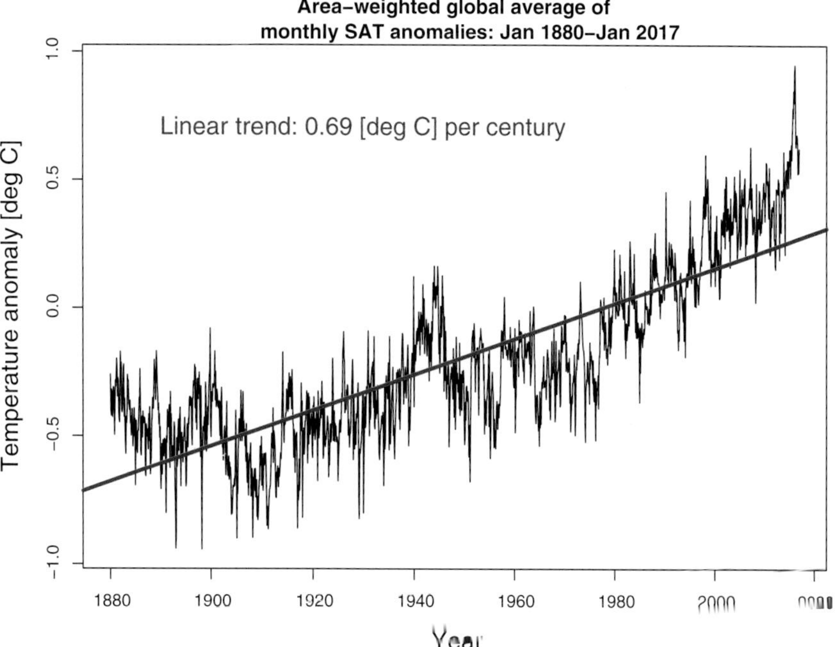

Fig. 11.6 Spatial average of monthly temperature anomalies with respect to 1971–2000 climatology based on the NOAAGlobalTemp data.

```
#Plot this time series
motime=seq(1880, 2017, length=1645)
plot(motime,rcover,type="l",ylim=c(0,100),
     main="NOAAGlobalTemp Data Coverage: Jan 1880-Jan 2017",
     xlab="Year",ylab="Percent area covered [%]")
```

11.3.3 Compare Trends and Variances at Two Different Locations

Figure 11.7 compares the monthly time series, linear trends, and standard deviations of the surface air temperature anomalies at Edmonton, Canada (53.5°N, 113.5°W), and San Diego, USA (32.7°N, 117.2°W), based on the gridded NOAAGlobalTemp data. The figure can be generated by the following R code.

```
#Extract data for a specified box with given lat and lon
n2 <- which(gpcpst[,1]==52.5&gpcpst[,2]==247.5)
dedm <- gpcpst[n2,855:1454]
t=seq(1880,2017, len=length(dedm))
par(mar=c(4,4.5,2.5,1))
plot(t, dedm, type="l", ylim=c(-15,15),col="red",
      cex.lab=1.3, cex.axis=1.3,
      xlab="Year", ylab="Temperature anomalies [deg C]",
      main="Monthly temperature anomalies history of
Edmonton, Canada, and San Diego, USA")
regedm <- lm(dedm ~ t)
abline(regedm,col="red", lwd=3)
```

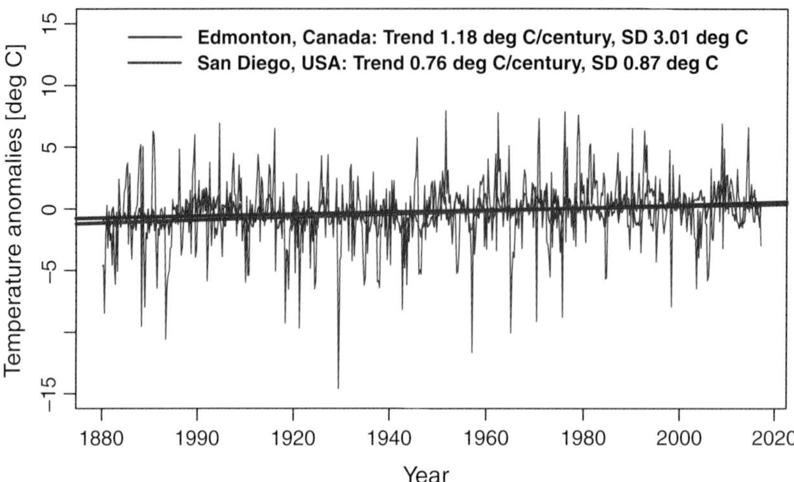

Fig. 11.7 Monthly time series, linear trends, and standard deviations of the surface air temperature anomalies at Edmonton, Canada, and San Diego, USA.

```
#San Diego data
n3 <- which(gpcpst[,1]==32.5&gpcpst[,2]==242.5)
dsan <- gpcpst[n3,855:1454]
lines(t, dsan, type="l", col="blue")
regsan =lm(dsan ~ t)
abline(regsan, col="blue", lwd=3)

legend(1880,16, col=c("red","blue"),lty=1,lwd=2.0,
  legend=c("Edmonton, Canada: Trend 1.18 deg C/century, SD 3.01 deg C",
           "San Diego, USA: Trend 0.76 deg C/century, SD 0.87 deg C"),
       bty="n",text.font=2,cex=1.0)

regedm
#0.01178  #Edmonton trend
regsan
#0.007624  #San Diego trend
sd(dedm)
#[1] 3.006863
sd(dsan)
#[1] 0.8653399
```

The linear trend of San Diego surface air temperature anomalies is 0.76 [°C/century], which is slightly stronger than the linear trend of the global average temperature anomalies 0.69 [°C/century] shown in Fig. 11.6. The standard deviation of San Diego temperature anomalies is 0.87 [°C], which is much larger than that of the global average 0.33 [°C]. The linear trend of Edmonton is 1.18 [°C/century], about 1.5 times of that of San Diego. The standard deviation of Edmonton is 3.01 [°C], about 3.5 times of that of San Diego. These results are consistent with Fig. 9.9, which shows that temperature varies more at a higher latitude than at a lower latitude, and more over the land than over the ocean.

11.3.4 Which Month Has the Strongest Trend?

It is known that climate changes are not uniform across a year. We thus plot the trends of each month from January to December in the period of 1880–2016. Figure 11.8 shows the strongest trend (0.75°/century) in March, and the weakest trend (0.656°/century) in September. This analysis of monthly trends can be especially worthwhile and insightful for hemispheric averages or regional averages, such as the United States. Figure 11.8 can be produced by the following R code.

```
#Display the trends of 12 months using 12 panels on the same figure
#Read directly from the NCEI URL for the NOAA GlobalTemp monthly global
#average data:
#www1.ncdc.noaa.gov/pub/data/noaaglobaltemp/operational/timeseries
#Data file: aravg.mon.land_ocean.90S.90N.v4.0.1.201703.txt
setwd("/Users/sshen/climmath")
aveNCEI=read.table("data/aravg.mon.land_ocean.90S.90N.v4.0.1.201703.txt",
                        header=FALSE)
dim(aveNCEI) #Jan 1880-Feb 2017 #an extra month to be deleted
#[1] 1648   10
par(mar=c(4,5,2,1))
timemo=seq(aveNCEI[1,1],aveNCEI[length(aveNCEI[,1]),1],
           len=length(aveNCEI[,1]) )
#create matrix of 136 yrs of data matrix
# row=year from 1880 to 2016, col=mon 1 to 12
ave=aveNCEI[,3]
myear=length(ave)/12
nyear=floor(myear)
nmon=nyear*12
avem = matrix(ave[1:nmon], ncol=12, byrow=TRUE)

#compute annual average
annv=seq(0,length=nyear)
for(y in 1:nyear){annv[y]=mean(avem[y,])}

#Put the monthly averages and annual ave in a matrix
avemy=cbind(avem,annv)

#Plot 12 panels on the same figure: Trend for each month
dev.off()
quartz(width=10,height=16,pointsize = 16)
#quartz(display = "name", width = 5, height = 5, pointsize = 12,
#        family = "Helvetica", antialias = TRUE, autorefresh = TRUE)
plot.new()
#png(file = 'monthtrend.png') #Automatical saving of a figure
timeyr=seq(aveNCEI[1,1], aveNCEI[1,1]+nyear-1)
par(mfrow = c(4, 3))  # 4 rows and 3 columns
par(mgp=c(2,1,0))
for (i in 1:12) {
  plot(timeyr, avemy[,i],type="l", ylim=c(-1.0,1.0),
       xlab="Year",ylab="Temp [oC]",
       main = paste("Month =", i, split = ""))
  abline(lm(avemy[,i]~timeyr),col="red")
  text(1945,0.7,
       paste("Trend oC/century=",
        round(digits=3, (100*coefficients(lm(avemy[,i]~timeyr))[2]))),
       col="red")
}
dev.off()
```

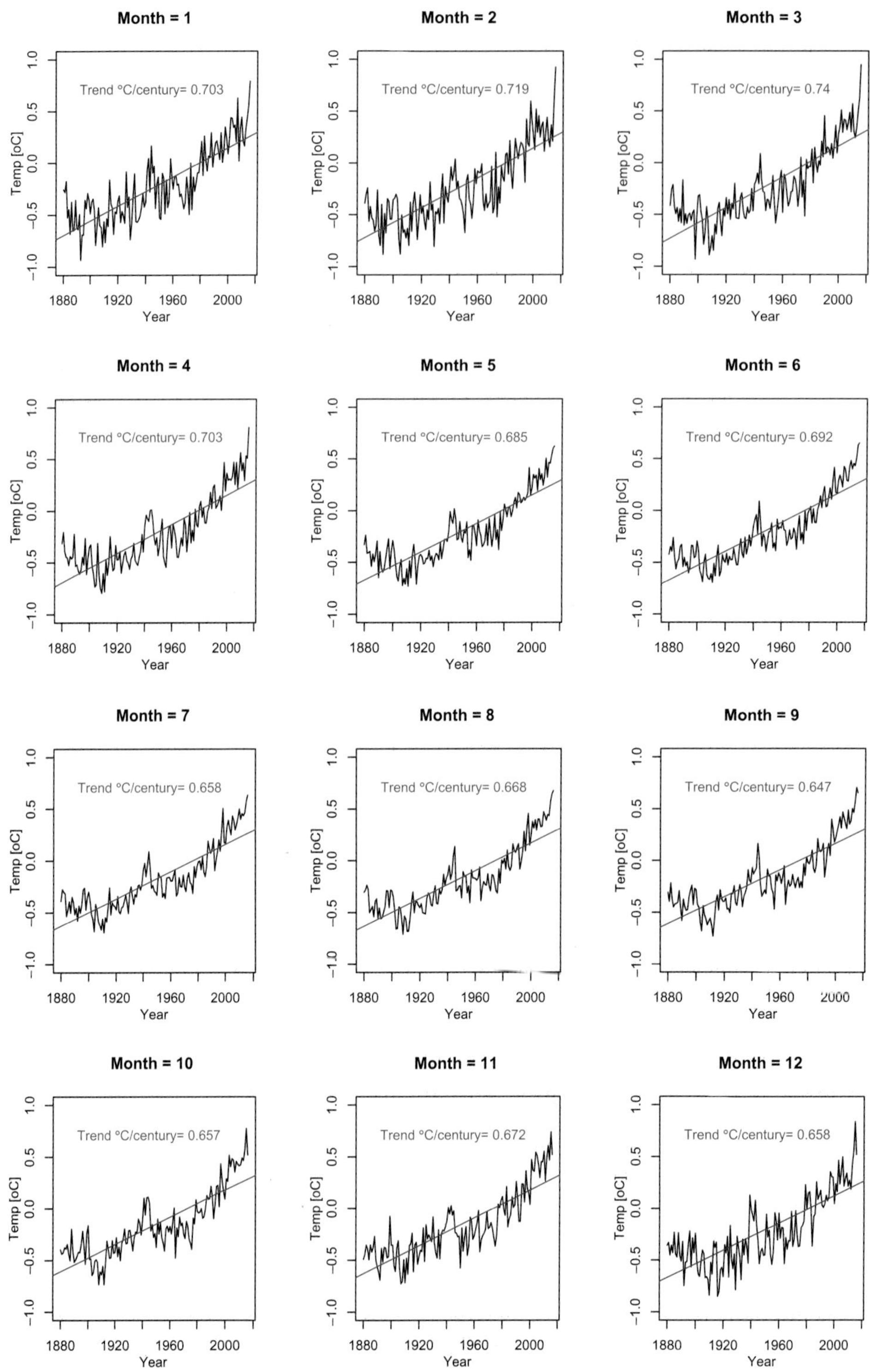

Fig. 11.8 The trend of the spatial average of each month based on the NOAAGlobalTemp data from 1880–2016.

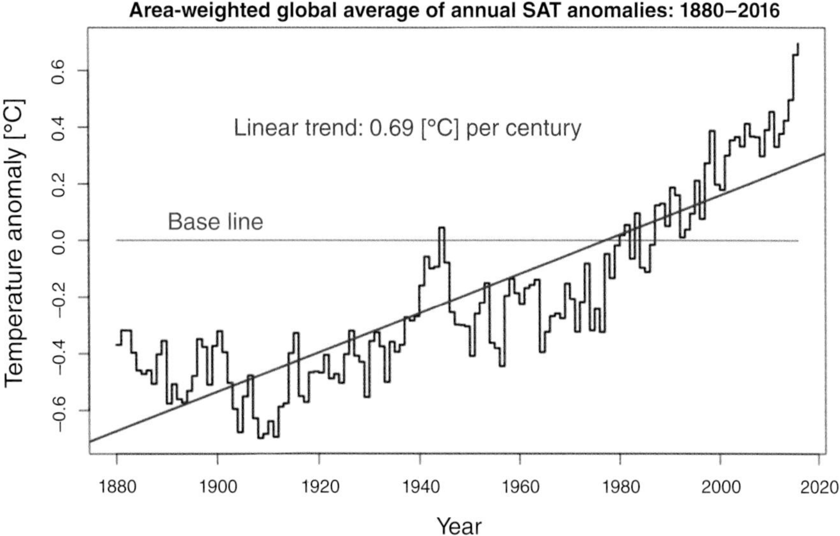

Fig. 11.9 Annual mean of the monthly spatial average anomalies from the NOAAGlobalTemp data.

11.3.5 Spatial Average of Annual Data

The following R code can compute and plot the annual mean. It first converts the vector data of monthly spatial averages to a 12-column matrix. Each column represents a month. The row mean yields the annual mean. The result is shown in Fig. 11.9.

```
par(mar=c(4,4.5,2,1.0))
avem = matrix(avev[1:1644], ncol=12, byrow=TRUE)
#compute annual average
annv=rowMeans(avem)
#Plot the annual mean global average temp
timeyr<-seq(1880, 2016)
plot(timeyr,annv,type="s",
     cex.lab=1.4, lwd=2,
     xlab="Year", ylab="Temperature anomaly [oC]",
main="Area-weighted global average of annual SAT anomalies: 1880-2016")
abline(lm(annv ~ timeyr),col="blue",lwd=2)
text(1940,0.4, "Linear trend: 0.69 [oC] per century",
     cex=1.4, col="blue")
text(1900,0.07, "Base line",cex=1.4, col="red")
lines(timeyr,rep(0,137), type="l",col="red")
```

11.3.6 Nonlinear Trend of the Global Average Annual Mean Data

The global average annual mean temperature obviously does not vary linearly with time. It is thus useful to examine the underlying nonlinear variation of the annual temperature time series. The simplest nonlinear trend exploration is through a polynomial fit. Usually, orthogonal polynomial fits are more efficient and have better fidelity to the

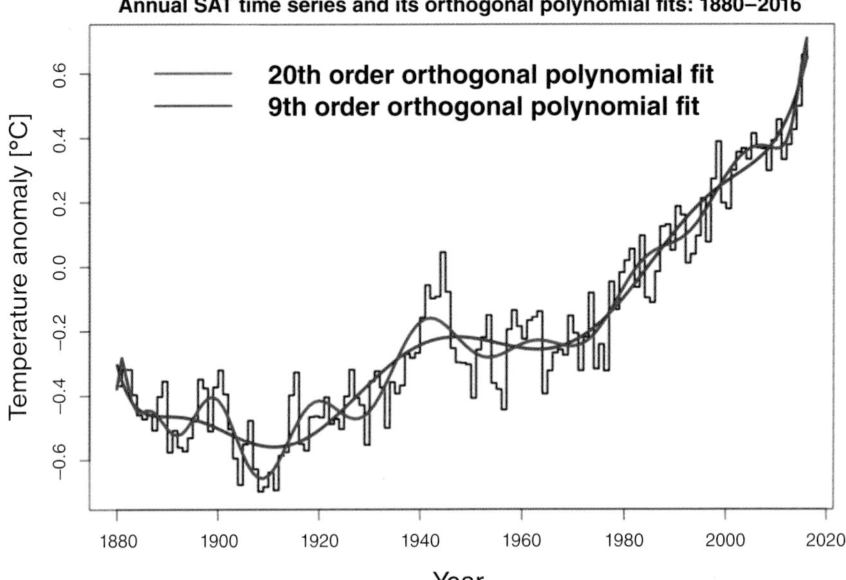

Fig. 11.10 Annual mean time series and its fit by orthogonal polynomials.

data. Figure 11.10 shows two fits by the 9th-order and 20th-order orthogonal polynomials. The choice of 9th-order is because it is the lowest-order polynomial which can reflect the oscillation of temperature from the high in the 1880s to the low in the 1910s, then rising until the 1940s, before decreasing in the 1960s and 1970s. The choice of the 20th-order polynomial fit is because it is the lowest-order orthogonal polynomial that can mimic the detailed climate variations, such as the local highs around 1900 and 1945. We have tried higher-order polynomials which often show an unphysical overfit.

Figure 11.10 can be produced by the following R code.

```
#Polynomial fitting to the global average annual mean
#poly9<-lm(annv ~ poly(timeyr,9, raw= TRUE))
#raw=TRUE means regular polynomial a0+a1x^2+..., non-orthogonal
polyor9<-lm(annv ~ poly(timeyr,9, raw= FALSE))
polyor20<-lm(annv ~ poly(timeyr,20, raw= FALSE))
#raw=FALSE means orthogonal polynomial of 9th order
#Orthogonal polynomial fitting is usually better
plot(timeyr,annv,type="s",
     cex.lab=1.4, lwd=2,
     xlab="Year", ylab="Temperature anomaly [°C]",
     main="Annual SAT data and orthogonal polynomial fits: 1880-2016")
lines(timeyr,predict(polyor9),col="blue", lwd=3)
legend(1880, 0.6,  col=c("blue"),lty=1,lwd=2.0,
       legend=c("9th order orthogonal polynomial fit"),
       bty="n",text.font=2,cex=1.5)
lines(timeyr,predict(polyor20),col="red", lwd=3)
legend(1880, 0.7,  col=c("red"),lty=1,lwd=2.0,
       legend=c("20th order orthogonal polynomial fit"),
       bty="n",text.font=2,cex=1.5)
```

A popular non-parametric fit is the LOWESS (locally weighted scatter plot smoothing), often referred to as the Loess fit. It is basically a weighted piecewise local polynomial fitting. The local fitting property requires many data points to make a reasonable fit. One can use the following one-line R code to generate a nonlinear fit which has a shape similar to the 20th-order polynomial fit.

```
scatter.smooth(annv, timeyr, span=2/18, cex=0.6)
```

11.4 Spatial Characteristics of the Temperature Change Trends

Climate change can be global, but everybody experiences it locally. For this reason and other reasons, we need the ability to analyze the spatial variability of trends in a wide variety of climate parameters. In this section, we provide techniques and R codes for performing this type of analysis.

11.4.1 The Twentieth-Century Temperature Trend

It is widely known that the global average temperature has increased, especially in recent decades since the 1970s. This is known to the general public as "global warming." However, the increase is non-uniform in both space and time, and a few areas have even experienced cooling, such as the 1900–1999 cooling over the North Atlantic off the coast of Greenland. Figure 11.11 shows the uneven spatial distribution of the linear trend of the monthly SAT anomalies from January 1900 to December 1999. Most parts of the world experienced warming, particularly over the land areas. Warming has been especially strong in the Arctic. Canada and Russia thus experienced more warming in the twentieth century, compared to other countries around the world.

Many grid boxes do not have complete data streams from January 1900 to December 1999. Our trend calculation's R code allows some missing data in the middle of the data streams, but it requires data at both the beginning month (January 1900) and the end month (December 1999). When a grid box does not satisfy this requirement, the trend for the box is not calculated. Figure 11.11's large white areas over the polar regions, Pacific, Africa, and Central America do not satisfy the requirement. For the missing data in the middle of a data record for a grid box, our linear regression omits the missing data and carries out the regression with a shorter temperature data record.

We used `lm(temp1[i,243: 1442] ~ timemo1, na.action=na.omit)` to treat the missing data between the beginning month and the end month. The missing data have been replaced by NA. The R command `na.action=na.omit` means that the missing data are omitted in the regression, and the fitted data at the missing data's time locations are omitted too. One can use another command `lm(temp1[i,243: 1442] ~ timemo1, na.action=na.exclude)` to do linear regression with missing data. The slope and intercept results computed by the two commands are the same. The only difference is that the latter outputs NA for the fitted data at the missing data's time locations. An example R code for this difference is as follows:

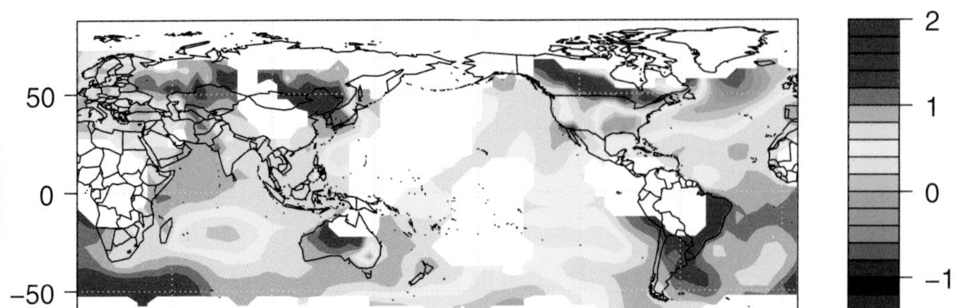

Fig. 11.11 Linear trend of SAT from January 1900 to December 1999. The trend was calculated for each grid box using the NOAAGlobalTemp data, and the procedure required that the box did not have missing data for the first month (January 1900) and the last month (December 1999). The white regions mean that the data did not satisfy our calculation conditions, i.e. these are the regions of insufficient amount of data.

```
x=1:8
y=c(2,4,NA,3,6.8,NA,NA,9)
fitted(lm(y ~ x, na.action=na.exclude))
# 1    2    3    4    5    6    7    8
#2.08 3.04   NA 4.96 5.92   NA   NA 8.80
##
fitted(lm(y ~ x, na.action=na.omit))
# 1    2    4    5    8
#2.08 3.04 4.96 5.92 8.80
```

Figure 11.11 can be produced by the following R code.

```
#Compute the trend for each box for the 20th century
timemo1=seq(1900,2000, len=1200)
temp1=temp
temp1[temp1 < -490.00] <- NA
trendgl=rep(0,2592)
for (i in 1:2592){
  if(is.na(temp1[i,243])==FALSE & is.na(temp1[i,1442])==FALSE)
  {trendgl[i]=lm(temp1[i,243: 1442] ~ timemo1,
                 na.action=na.omit)$coefficients[2]}
  else
    {trendgl[i]=NA}
}
library(maps)
Lat= seq(-87.5, 87.5, length=36)
Lon=seq(2.5, 357.5, length=72)
mapmat=matrix(100*trendgl,nrow=72)
mapmat=pmax(pmin(mapmat,2),-2)
#Matrix flipping is not needed since the data goes from 2.5 to 375.5
plot.new()
par(mar=c(4,5,3,0))
int=seq(-2,2,length.out=21)
```

```
rgb.palette=colorRampPalette(c('black','blue', 'darkgreen','green',
'yellow','pink','red','maroon'),interpolate='spline')
#mapmat= mapmat[,seq(length(mapmat[1,]),1)]
filled.contour(Lon, Lat, mapmat, color.palette=rgb.palette, levels=int,
plot.title=title(main="Jan 1900-Dec 1999 temp trends: [oC/century]",
                 xlab="Latitude",ylab="Longitude", cex.lab=1.5),
plot.axes={axis(1, cex.axis=1.5); axis(2, cex.axis=1.5);
                              map('world2', add=TRUE);grid()},
               key.title=title(main=" "),
               key.axes={axis(4, cex.axis=1.5)})
```

The above trend for a specific grid box is calculated in the following way.

```
i=600
timemo1 = seq(1900, 1999, length=1200)
lm(temp1[i,243:1442] ~ timemo1, na.action=na.omit)
#0.009423   #trend
lm(temp1[i,243:1442] ~ timemo1, na.action=na.exclude)
#0.009423   #trend
temp1[i,243:1442]
# 1900-1  1900-2  1900-3  1900-4  1900-5
#-0.7457      NA -1.4406 -1.0936 -0.8193
```

11.4.2 Twentieth-Century Temperature Trend Computed under a Relaxed Condition

If we relax our trend calculation condition and allow a trend to be computed for a grid box when the box has less than one third of its data missing, then the trends can be computed for more grid boxes. Figure 11.12 shows the trend map computed under this relaxed condition.

Figure 11.12 uses °C per decade as the unit, while Fig. 11.11 uses °C per century. The patterns of the two figures are consistent, which implies that the relaxed condition for trend calculation has not led to spatially inconsistent trends. Thus, Fig. 11.12 can be regarded as an accurate spatial extension of Fig. 11.11.

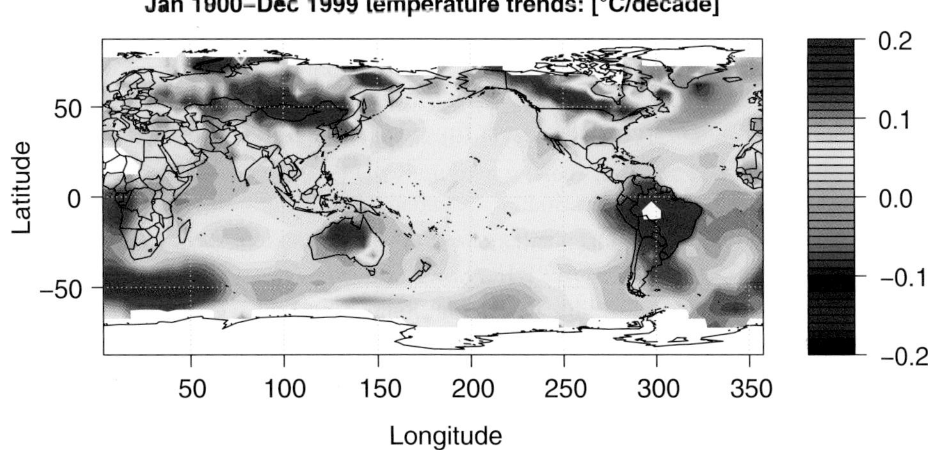

Fig. 11.12 Linear trend of SAT from January 1900–December 1999. The trend was calculated for each grid box using the NOAAGlobalTemp data when the box has less than one third of data missing.

Figure 11.12 can be generated by the following R code.

```
#Trend for each box for the 20th century: Version 2: Allow 2/3 data
#Compute the trend
timemo1=seq(1900,2000, len=1200)
temp1=temp[,243:1442]
temp1[temp1 < -490.00] <- NA
temptf=is.na(temp1)
bt=et=rep(0,2592)
for (i in 1:2592) {
  if (length(which(temptf[i,]==FALSE)) !=0)
  {
    bt[i]=min(which(temptf[i,]==FALSE))
    et[i]=max(which(temptf[i,]==FALSE))
  }
}
##
trend20c=rep(0,2592)
for (i in 1:2592){
  if(et[i]-bt[i] > 800)
  {trend20c[i]=lm(temp1[i,bt[i]:et[i]] ~ seq(bt[i],et[i]),
                na.action=na.omit)$coefficients[2]}
  else
  {trend20c[i]=NA}
}
#plot the 20C V2 trend map
plot.new()
#par(mar=c(4,5,3,0))
mapmat=matrix(120*trend20c,nrow=72)
mapmat=pmax(pmin(mapmat,0.2),-0.2)
int=seq(-0.2,0.2,length.out=41)
rgb.palette=colorRampPalette(c('black','blue', 'darkgreen','green',
'yellow','pink','red','maroon'),interpolate='spline')
filled.contour(Lon, Lat, mapmat, color.palette=rgb.palette, levels=int,
    plot.title=title(main="Jan 1900-Dec 1999 temp trends: [oC/decade]",
                     xlab="Latitude",ylab="Longitude", cex.lab=1.5),
    plot.axes={axis(1, cex.axis=1.5); axis(2, cex.axis=1.5);
               map('world2', add=TRUE);grid()},
    key.title=title(main="[oC]"),
    key.axes={axis(4, cex.axis=1.5)})
```

11.4.3 Trend Pattern for the Four Decades of Consecutive Warming: 1976–2016

The long recent period of extended global warming (four decades from 1976–2016) exhibits a warming that is greater than the previous period of extended global warming from the 1910s to the early 1950s, which also lasted about four decades. Figure 11.13 shows the strong global warming trend from January 1976 to December 2016. It shows that during this period, the world became warmer on every continent except Antarctica.

The trend data for Fig. 11.13 can be calculated using the following R code.

```
timemo2=seq(1976,2017, len=492)
temp1=temp
temp1[temp1 < -490.00] <- NA
trend7616=rep(0,2592)
for (i in 1:2592){
  if(is.na(temp1[i,1155])==FALSE & is.na(temp1[i,1646])==FALSE)
```

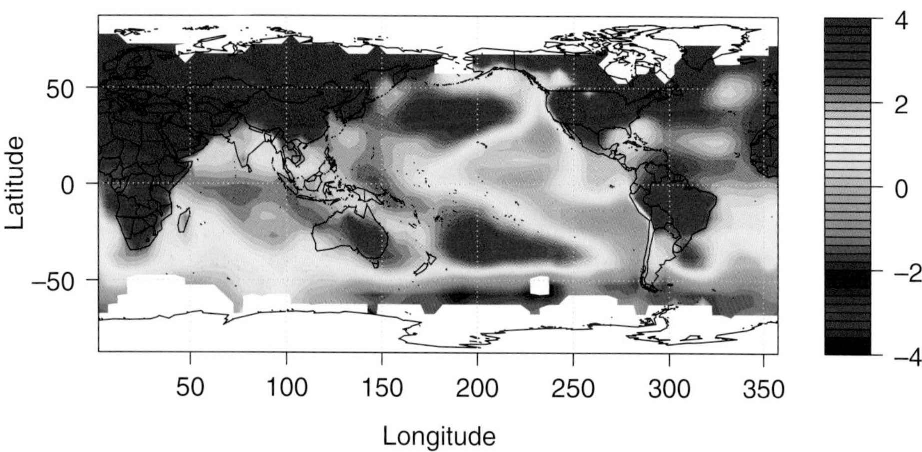

Fig. 11.13 Linear trend of SAT from January 1976–December 2016. The white regions are those with an insufficient amount of data.

```
  {trend7616[i]=lm(temp1[i,1155:1646] ~ timemo2,
                   na.action=na.omit)$coefficients[2]}
  else
  {trend7616[i]=NA}
}
```

The R code for plotting Fig. 11.13 is almost identical to that for the twentieth century trend of Fig. 11.12 and is omitted here.

11.5 Chapter Summary

This chapter has described the R codes for coping with two missing data problems: the spatial average, and the spatial pattern of the temporal trends. These are two problems frequently encountered in the analysis of observational climate data. These simple codes have practical value for research and applications, and they may be used to substitute for the traditionally long and tedious codes for solving the same problems. Because of their simplicity, our codes can be easily used in classrooms and other settings in such a way that comprehensive climate data can be viewed and analyzed by many more people, including users other than climate science professionals.

(i) Our code was developed for the NOAAGlobalTemp dataset of the monthly gridded surface temperature anomalies since January 1880. Section 11.1 shows the statistics of the missing data.
(ii) Section 11.2 shows (a) the use of the 4DVD (4-Dimensional Visual Delivery of big climate data) software package, (b) the reading of `.asc` data file by `scan("datafilename.asc")`, (c) the formation of the space–time data matrix, (d) plotting the NOAAGlobalTemp data for a given month, (e) the extraction of the

NOAAGlobalTemp data for a specified region by a `which()` command, and (f) plotting the extracted data on the specified region.

(iii) Section 11.3 (a) shows the definition of the global average and its approximation, (b) describes several short R codes to make temperature trend analyses and to visualize the global average temperature data from different perspectives, and (c) fits the global average temperature by both linear and nonlinear models using least squares regression.

(iv) Section 11.4 computes the temporal trend for each grid box when missing data are considered.

The methods and R codes we have presented are taken from current work in climate research. R codes have made these tasks more simple and easy than was previously possible and have enabled more people to analyze and visualize the observed data for their own applications.

This chapter ends our book, which we hope has provided you with practical and easy-to-use mathematical theories and computing tools that will be helpful in your own work. The value of these theories and applications in practical use has been our main criterion for selecting topics and materials for this book. The usefulness of the materials presented here, either as a user manual or textbook or reference, or simply as a toolbox to support your research and other applications, makes our book different from many of the more traditional science and mathematics books. From this perspective, we earnestly hope that our efforts will have helped to modernize and improve the curriculum of mathematical education for climate science.

References and Further Readings

[1] Huang, B., V. F. Banzon, E. Freeman, J. Lawrimore, W. Liu, T. C. Peterson, T. M. Smith, P. W. Thorne, S. D. Woodruff, and H. M. Zhang (2015): Extended reconstructed sea surface temperature version 4 (ERSST v4). Part I: upgrades and intercomparisons. *Journal of Climate*, 28, 911–930.

The Extended Reconstructed Sea Surface Temperature (ERSST) dataset has been popular in the climate science community. It is updated every few years.

[2] Karl, T. R., A. Arguez, B. Huang, J. H. Lawrimore, J. R. McMahon, M. J. Menne, T. C. Peterson, R. S. Vose, and H. M. Zhang (2015): Possible artifacts of data biases in the recent global surface warming hiatus. *Science*, 348, 1469–1472.

This paper released the NOAAGlobalTemp dataset that includes the monthly data of both the sea surface temperature and the land surface air temperature on a $5^\circ \times 5^\circ$ latitude–longitude grid from January 1880. Thomas R. Karl (1951–) is an American meteorologist, who led the US National Climatic Data Center for more than two decades beginning in the 1990s.

[3] Smith, T. M. and R. W. Reynolds (2003): Extended reconstruction of global sea surface temperatures based on COADS data (18541997). *Journal of Climate*, 16, 1495–1510.

This paper released Version 1 of the ERSST dataset. Richard W. Reynolds and Thomas M. Smith have been pioneers in SST reconstruction since the 1980s.

Exercises

All exercises in this chapter require the use of a computer.

11.1 (a) Following the R code for generating Fig. 11.6 for the monthly global average SAT anomalies, write an R code to generate a similar figure but for the Northern Hemisphere (NH)'s SAT anomalies from January 1880 to December 2016, based on the gridded 5-deg NOAAGlobalTemp.
(b) Plot the time series of the NH average and the corresponding linear trend on the same figure similar to Fig. 11.6.

11.2 Use R to generate the plots similar to those in Fig. 11.8 but for NH.

11.3 Perform the same tasks as Exercise 11.1, but for the Southern Hemisphere (SH).

11.4 Use R to generate the plots similar to those in Fig. 11.8 but for SH.

11.5 Compute and plot the spatial average of the annual mean SAT anomalies for NH from 1880 to 2016.

11.6 Do the same as the previous problem, but for SH.

11.7 (a) Use R to compute the zonal spatial average of the January NOAAGlobalTemp anomaly data for each 10-degree latitude zone from 1880 to 2016.
(b) Compute the linear trend for each zonal average of the January anomalies obtained in (a) from 1880 to 2016.
(c) Make a bar chart for the trends obtained in (b).

11.8 Perform the same tasks as Exercise 11.7, but for the July data.

11.9 Perform the same tasks as Exercise 11.7, but for the annual mean data.

11.10 Use R to generate a linear trend map similar to Fig. 11.12, but only for the January data from 1900 to 1999.

11.11 Use R to generate a linear trend map similar to Fig. 11.13, but only for the January data from 1976 to 2016.

11.12 Use R to generate a linear trend map similar to Fig. 11.12, but only for the July data from 1900 to 1999.

11.13 Use R to generate a linear trend map similar to Fig. 11.13, but only for the July data from 1976 to 2016.

11.14 Plot and compare the maps of the January SAT anomalies' linear trends from 1948 to 2016 based on the gridded January SAT anomalies relative to the 1971–2000 climatology period for two datasets: the NCEP/NCAR Reanalysis data and the NOAAGlobalTemp data. Use 200–500 words to describe your results.

11.15 (a) Plot the time series of the spatially averaged annual mean SAT anomalies for the contiguous United States using the NOAAGlobalTemp data from 1880 to 2016.

(b) Add a linear trend line to the time series plot in (a). Mark the trend value on the figure with the unit [°C per century].

Appendix A Dot Product of Two Vectors

A.1 Two Definitions for the Dot Product

Figure A.1 shows that the work done by a force **F** on a body which has made a displacement **d** is

$$W = Fd\cos\theta, \tag{A.1}$$

where $F = |\mathbf{F}|$ is the magnitude of the force, $d = |\mathbf{d}|$ is the magnitude of the displacement, and θ is the angle between the force and the displacement.

Thus, the work is determined both by vector magnitudes and by their directions. A convenient and general notation of equation (A.1) is

$$W = \mathbf{F} \cdot \mathbf{d}, \tag{A.2}$$

which is called the dot product of two vectors.

In two-dimensional space, if $\mathbf{F} = (F_1, F_2)$ and $\mathbf{d} = (d_1, d_2)$, then

$$W = \mathbf{F} \cdot \mathbf{d} = F_1 d_1 + F_2 d_2. \tag{A.3}$$

This formula can be proven to be the same as formula (A.1). The proof can be developed in many ways. A simple proof is to orient the x-axis along the direction of **d**, which yields $\mathbf{F} = (F\cos\theta, F\sin\theta)$ and $\mathbf{d} = (d, 0)$. Formula (A.3) then becomes

$$W = F_1 d_1 + F_2 d_2 = F\cos\theta \times d + F\sin\theta \times 0 = Fd\cos\theta. \tag{A.4}$$

This is formula (A.1).

Formula (A.3) for a dot product has an easy extension to n-dimensional space while formula (A.1) has a good geometric interpretation. Climate science uses both formulas, with the choice between them depending on specific applications and the corresponding context and data.

The n-dimensional extension of the dot product of two vectors $\mathbf{u} = (u_1, u_2, \ldots, u_n)$ and $\mathbf{v} = (v_1, v_2, \ldots, v_n)$ is

$$\mathbf{u} \cdot \mathbf{v} = u_1 v_1 + u_2 v_2 + \cdots + u_n v_n = \sum_{k=1}^{n} u_k v_k. \tag{A.5}$$

Although the angle between the two vectors in an n-dimensional space cannot be plotted for $n > 3$, one can still imagine the angle for any two n-dimensional vectors.

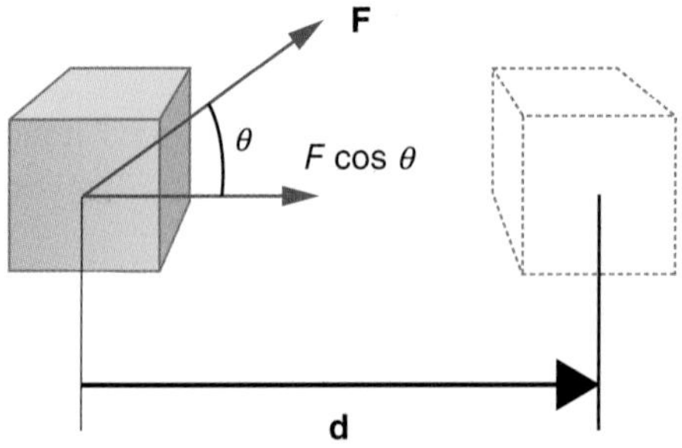

Fig. A.1 Work done by force **F** and displacement **d**. *Credit*: www.quora.com/At-what-angle-is-the-work-done-by-a-body-at-its-minimum-W-F%C3%97S-Cos-thetha

The correlation between two datasets can be considered as a dot product, which is shown below. The datasets $\{x_i\}_{i=1}^n$ and $\{y_i\}_{i=1}^n$ can be written as two n-dimensional vectors

$$\mathbf{x} = (x_1, x_2, \ldots, x_n), \tag{A.6}$$

$$\mathbf{y} = (y_1, y_2, \ldots, y_n). \tag{A.7}$$

Then, the statistical correlation between these two datasets is defined as

$$r_{xy} = corr(\mathbf{x}, \mathbf{y}) = \frac{\sum_{i=1}^n (x_i - \bar{x})(y_i - \bar{y})}{\left[\sum_{i=1}^n (x_i - \bar{x})^2 \sum_{i=1}^n (y_i - \bar{y})^2\right]^{1/2}}. \tag{A.8}$$

This definition can be written as a dot product

$$r_{xy} = \mathbf{x}_{Na} \cdot \mathbf{y}_{Na}, \tag{A.9}$$

where

$$\mathbf{x}_{Na} = \frac{\mathbf{x}_a}{|\mathbf{x}_a|} \tag{A.10}$$

is called the normalized anomalies of the data $\mathbf{x}$, and

$$\mathbf{x}_a = (x_1 - \bar{x}, x_2 - \bar{x}, \ldots, x_n - \bar{x}) \tag{A.11}$$

is called the anomalies of the data $\mathbf{x}$, and

$$\bar{x} = \frac{x_1 + x_2 + \cdots + x_n}{n} \tag{A.12}$$

is the data mean. The data mean, anomalies, and normalized anomalies can be defined in the same manner for the dataset $\mathbf{y}$. Therefore, the correlation between two datasets is the dot product of the two corresponding normalized anomaly vectors as shown in Eq. (A.9).

With the dot product notation, the linear regression coefficients between two datasets $\{x_i\}_{i=1}^n$ and $\{y_i\}_{i=1}^n$ can also be written in the form of dot products:

$$y = \left[\bar{y} - \frac{\mathbf{x}_a \cdot \mathbf{y}_a}{|\mathbf{x}_a|^2}\bar{x}\right] + \left[\frac{\mathbf{x}_a \cdot \mathbf{y}_a}{|\mathbf{x}_a|^2}\right] x. \tag{A.13}$$

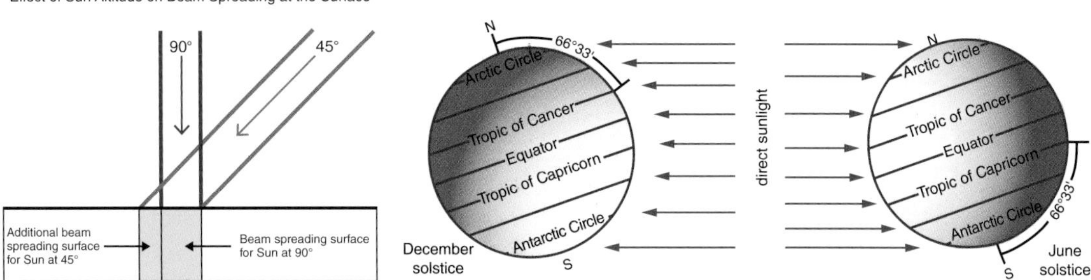

Fig. A.2 Schematic illustration of solar radiation flux to the Earth's surface and seasonality. *Credit (left): Climate Literacy Lab 1, Georgia State University. Credit (right): Encyclopedia Britannica.*

A.2 Solar Power Flux to the Earth's Surface and Seasonality

Figure A.2 shows that the solar power flux to the Earth's surface for a given location is a negative dot product of the solar radiation vector **S** with the normal vector **N** of the Earth's surface at this location:

$$R = -\mathbf{S} \cdot \mathbf{N} = S \cos \gamma \tag{A.14}$$

where the negative sign means the incoming power flux to the Earth, the radiation vector **S** points to the Earth from the Sun, the unit normal vector **N** points out of the Earth and is perpendicular to the Earth's tangent surface at the given location, $S = |\mathbf{S}|$ is the magnitude of the radiation vector **S** and is called the solar constant approximately equal to $1,368$ [W m^{-2}], and γ is the incident acute angle between the incoming solar radiation ray and the negative normal vector of the Earth's surface at the location. The reason the above formula holds is that the same total solar power coming through an area A with an incident angle γ is spread over a larger area $A/\cos\gamma$. Thus, the solar power going through a unit area, i.e., the power flux, is smaller and given by $S\cos\gamma$. As shown in Fig. A.2, the incident angle over the equator is almost zero, and hence the equator has the largest annual mean solar power flux. Because of the large tilt angle 23.5° between the Earth's rotation axis and a line perpendicular to the plane of the Earth's orbit around the Sun, the incident angle varies with season. The Northern Hemisphere faces the Sun more directly in June and has a small incident angle, and hence has a larger solar power flux in boreal (which means northern) summer than in boreal winter.

A.3 Divergence Theorem for the Mass Continuity Equation in Climate Models

Mass must be conserved in a climate model. For a given spatial domain V, any mass increase inside the domain must be equal to the amount of external mass flowing into the domain. The rate of mass increase inside the domain is described by a triple integral

$$M_{inc} = \iiint_V [-\nabla \cdot (\rho \mathbf{u})] dV \tag{A.15}$$

where ρ is density and $\mathbf{u} = (u, v, w)$ is the three-dimensional velocity field of a flow, and

$$\nabla \cdot (\rho \mathbf{u}) = \frac{\partial}{\partial x}(\rho u) + \frac{\partial}{\partial y}(\rho v) + \frac{\partial}{\partial z}(\rho w) \tag{A.16}$$

is called the divergence of the vector $\rho \mathbf{u}$.

If we regard the differentiation operator

$$\nabla = \left(\frac{\partial}{\partial x}, \frac{\partial}{\partial y}, \frac{\partial}{\partial z} \right) \tag{A.17}$$

as a vector, then the divergence of a vector is symbolically equal to the dot product of the operator ∇ and the vector.

This mass increase rate must be equal to the rate of mass flowing into the volume

$$M_{flow} = -\oint_{\partial V} \rho \mathbf{u} \cdot \mathbf{n} dS, \tag{A.18}$$

where $\mathbf{n}$ is the unit normal vector for a point on the surface of the volume and points out of the volume, and dS stands for a small integration area of the surface of the domain V.

Setting $M_{inc} = M_{flow}$ leads to the divergence theorem expressed below

$$\iiint_V \nabla \cdot (\rho \mathbf{u}) dV = \oint_{\partial V} \rho \mathbf{u} \cdot \mathbf{n} dS. \tag{A.19}$$

The divergence theorem means that the volume integral of a vector's divergence in a three-dimensional spatial domain is equal to the surface integral of the vector's flux through the volume's boundary. The proof of this theorem can be found from many sources, including the Internet. An application of this theorem to the mass flow field $\rho \mathbf{u}$ yields the equation of continuity applicable to the atmosphere or the ocean:

$$\frac{\partial \rho}{\partial t} + \nabla \cdot (\rho \mathbf{u}) = 0. \tag{A.20}$$

Appendix B Cross Product of Two Vectors

In addition to the dot product of two vectors described in Appendix A, there is another type of vector product: the cross product of two vectors, which is a third vector that is perpendicular to both of the two vectors.

Climate applications of the cross product of two vectors include (i) the Coriolis force whose expression includes a cross product of the angular velocity vector $\mathbf{\Omega}$ of the rotating reference frame and the flow velocity vector $\mathbf{u}$ relative to the rotating reference frame, and (ii) the vorticity vector of a flow which is defined symbolically as a cross product of the operator ∇ and the flow velocity $\mathbf{u}$.

B.1 Definition of the Cross Product of Two Vectors

Given two vectors

$$\mathbf{a} = (a_1, a_2, a_3), \quad \text{and} \tag{B.1}$$

$$\mathbf{b} = (b_1, b_2, b_3), \tag{B.2}$$

their cross product

$$\mathbf{c} = \mathbf{a} \times \mathbf{b} \tag{B.3}$$

is defined to be a vector that is perpendicular to both $\mathbf{a}$ and $\mathbf{b}$ with its direction determined by the right-hand rule as shown in Figure B.1, and whose magnitude is

$$|\mathbf{a} \times \mathbf{b}| = |\mathbf{a}||\mathbf{b}| \sin\theta, \tag{B.4}$$

where $|\mathbf{a}|$ is the magnitude of $\mathbf{a}$, $|\mathbf{b}|$ is the magnitude of $\mathbf{b}$, and θ is the acute angle between $\mathbf{a}$ and $\mathbf{b}$.

Thus, if the magnitudes of $\mathbf{a}$ and $\mathbf{b}$ and the angle θ are known, formula (B.4) and the right-hand rule uniquely determine the cross product of the two vectors.

However, sometimes, the two vectors are expressed in terms of their components as in formulas (B.1) and (B.2), and the cross product is also required to be expressed in terms of its components. In this situation, the cross product can be computed symbolically by a 3×3 determinant, denoted by det:

$$\mathbf{a} \times \mathbf{b} = \det \begin{bmatrix} \mathbf{i} & \mathbf{j} & \mathbf{k} \\ a_1 & a_2 & a_3 \\ b_1 & b_2 & b_3 \end{bmatrix}. \tag{B.5}$$

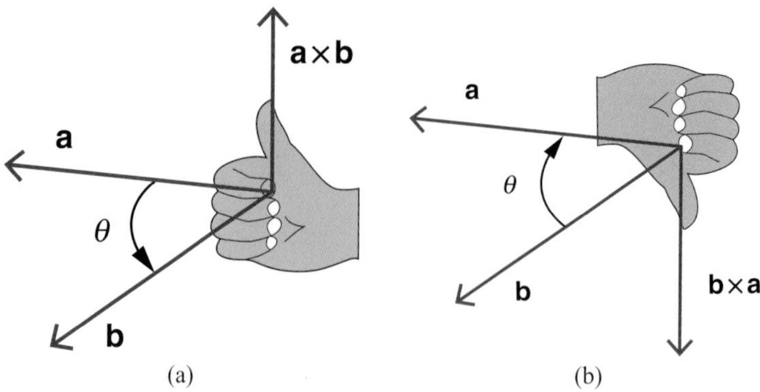

Fig. B.1 (a) The right-hand rule for the product of two vectors. For $\mathbf{a} \times \mathbf{b}$, the right-hand rule is that the right hand grabs the two vectors from **a** to **b** with four fingers and sticks up the thumb that points to the desired direction of the cross product. (b) $\mathbf{b} \times \mathbf{a}$ points to the opposite direction of $\mathbf{a} \times \mathbf{b}$.

The determinant, like eigenvalues, is an index of a square matrix, defined in a certain way. For example, the determinant of a 2×2 matrix is defined as follows:

$$\det \begin{bmatrix} a & b \\ c & d \end{bmatrix} = ad - bc. \tag{B.6}$$

An analytic calculation of the determinant of a 3×3 matrix or a higher-order matrix is often done by Laplace's formula of expansion, which reduces the 3×3 determinant to three 2×2 determinants.[1]

A numerical computation of the determinant of matrix A can be done by a simple R command: `det(A)`.

Expanding this determinant in Eq. (B.5) according to the first row using Laplace's formula, we obtain that

$$\mathbf{a} \times \mathbf{b} = \mathbf{i} \det \begin{bmatrix} a_2 & a_3 \\ b_2 & b_3 \end{bmatrix} - \mathbf{j} \det \begin{bmatrix} a_1 & a_3 \\ b_1 & b_3 \end{bmatrix} + \mathbf{k} \det \begin{bmatrix} a_1 & a_2 \\ b_1 & b_2 \end{bmatrix}, \tag{B.7}$$

or

$$\mathbf{a} \times \mathbf{b} = \mathbf{i}(a_2 b_3 - b_2 a_3) - \mathbf{j}(a_1 b_3 - b_1 a_3) + \mathbf{k}(a_1 b_2 - b_1 a_2). \tag{B.8}$$

Another situation is that with the given three components as in formulas (B.1) and (B.2), we wish to express the cross product in terms of magnitude and the right-hand rule. In this case, we need to compute the magnitudes of **a** and **b** and the acute angle θ between the two vectors:

$$|\mathbf{a}| = \sqrt{a_1^2 + a_2^2 + a_3^2}, \tag{B.9}$$

[1] Since determinant is only used as a notation in our book, we do not cover the details of this topic. Interested readers are referred to a conventional elementary linear algebra book, which often has a chapter on determinants.

$$|\mathbf{b}| = \sqrt{b_1^2 + b_2^2 + b_3^2}, \tag{B.10}$$

$$\cos\theta = \frac{\mathbf{a} \cdot \mathbf{b}}{|\mathbf{a}||\mathbf{b}|}, \tag{B.11}$$

where the dot product $\mathbf{a} \cdot \mathbf{b}$ is

$$\mathbf{a} \cdot \mathbf{b} = a_1 b_1 + a_2 b_2 + a_3 b_3. \tag{B.12}$$

The angle θ can be found using the inverse cosine function.

B.2 Coriolis Force

An application of the cross product in climate science is the calculation of Coriolis force, which is defined as

$$\mathbf{F}_c = -2m\mathbf{\Omega} \times \mathbf{u}, \tag{B.13}$$

where m is the mass of the fluid in a control volume, $\mathbf{u}$ is the velocity of the control volume, and $\mathbf{\Omega}$ is the angular velocity of the Eulerian frame at the location of interest and is latitude-dependent. See Chapter 8 for a more detailed description of the Coriolis force.

B.3 Vorticity

The vorticity of a flow vector field $\mathbf{u}(x, y, z, t)$ is defined as

$$\zeta = \nabla \times \mathbf{u}. \tag{B.14}$$

This mathematical operation for the differential operator ∇ is called the curl of $\mathbf{u}$. Here $\mathbf{u} = (u, v, w)$ is the velocity vector depending on spatial coordinates x, y, z and time t. From the definition of the cross product, the vorticity computation can be symbolically written as

$$\nabla \times \mathbf{u} = \det \begin{bmatrix} \mathbf{i} & \mathbf{j} & \mathbf{k} \\ \partial/\partial x & \partial/\partial y & \partial/\partial z \\ u & v & w \end{bmatrix}. \tag{B.15}$$

An expansion of the determinant according to the first row yields the vorticity around each axis x, y, or z. For example, the vorticity around the vertical z-axis is

$$\zeta_z = \begin{bmatrix} \partial/\partial x & \partial/\partial y \\ u & v \end{bmatrix} \mathbf{k} \tag{B.16}$$

or

$$\zeta_z = \left(\frac{\partial v}{\partial x} - \frac{\partial u}{\partial y} \right) \mathbf{k}. \tag{B.17}$$

If we regard a simplified hurricane or tornado as an idealized phenomenon involving rotary motion around a vertical axis, then the vorticity around such a vertical z-axis, is determined by these partial derivatives of the horizontal velocity components.

B.4 Stokes' Theorem

The divergence theorem in Appendix A reveals a relationship between a volume integral over a 3D solid domain and a surface integral on the domain's boundary, which is a closed surface. Stokes' theorem explores a relationship between a surface integral over a non-closed surface and a line integral along the surface's boundary, which is a closed contour.

B.4.1 Stokes' Theorem

Stokes' theorem[2] describes a relationship between a surface integration of the curl $\nabla \times \mathbf{u}$ of a flow field $\mathbf{u}(x,y,z,t)$ over an open 3D surface S and a line integral of the flow $\mathbf{u}(x,y,z,t)$ along the 3D closed boundary contour C of the surface

$$\iint_S (\nabla \times \mathbf{u}) \cdot \mathbf{n} dS = \oint_C \mathbf{u} \cdot d\mathbf{r}. \tag{B.18}$$

Here, the vector $\mathbf{n}(x,y,z)$ is the unit normal vector of the surface S and varies according to the location on S. The direction of $\mathbf{n}(x,y,z)$ is determined by the right-hand rule according to the rotating direction of C. The area differential of the surface S is dS. The differential of the position vector $\mathbf{r} = (x,y,z)$ along the closed contour C is $d\mathbf{r} = (dx,dy,dz)$, and is in the tangential direction of C. See Fig. B.2 for the notations and their corresponding geometry.

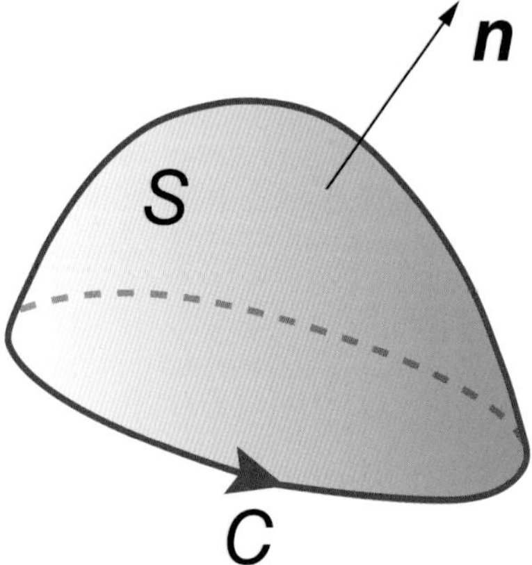

Fig. B.2 The flow field around a closed contour is related to the flux of the vorticity through the surface bounded by the contour by Stokes' theorem.

[2] Stokes' theorem was named after George Gabriel Stokes (1819–1903), who was an Anglo-Irish mathematician and physicist and made fundamental contributions to theoretical fluid dynamics, including the Navier–Stokes equations, which express Newton's second law of motion for fluids.

B.4.2 Green's Theorem

A special case of Stokes' theorem is that the 3D surface S becomes a flat region on a 2D plane. We can set the xy-coordinates on this plane. Then, $\nabla = (\partial/\partial x, \partial/\partial y, 0)$, $\nabla \times \mathbf{u} = (0, 0, \partial v/\partial x - \partial u/\partial y)$, $\mathbf{n} = (0, 0, 1)$,

$$(\nabla \times \mathbf{u}) \cdot \mathbf{n} = (0, 0, \partial v/\partial x - \partial u/\partial y) \cdot (0, 0, 1) = \partial v/\partial x - \partial u/\partial y,$$

$dS = dxdy$, $\mathbf{u} = (u, v, 0)$, and $d\mathbf{r} = (dx, dy, 0)$, $\mathbf{u} \cdot d\mathbf{r} = udx + vdy$. Consequently, the 3D statement of Stokes' theorem becomes a 2D formula:

$$\iint_S \left(\frac{\partial v}{\partial x} - \frac{\partial u}{\partial y} \right) dxdy = \oint_C udx + vdy. \tag{B.19}$$

This is called Green's theorem after the British mathematical physicist George Green (1793–1841).

Green published this theorem in 1828, before the formal appearance of Stokes' theorem in 1854. William Thomson (1824–1907), George Stokes (1819–1903), and Hermann Hankel (1839–1873) were involved in the original development, presentation, and proof of Stokes' theorem. Stokes' theorem may be considered as a 3D extension of the 2D Green's theorem, and can be proved using Green's theorem

B.4.3 GPS-Planimeter as a Smartphone App

When $u = -y/2, v = x/2$, Green's theorem yields

$$\begin{aligned} \oint_C (-y/2)dx + (x/2)dy &= \iint_S \left(\frac{\partial (x/2)}{\partial x} - \frac{\partial (-y/2)}{\partial y} \right) dxdy \\ &= \iint_S dxdy \\ &= A \end{aligned} \tag{B.20}$$

where A is the area of the region S inside the closed contour C. Thus, the boundary coordinates (x, y) data on the boundary contour C can determine the line integral, hence the area of the region. This agrees with our intuition. A smartphone can record the GPS coordinate data of a closed contour C, and hence can use Green's theorem to calculate the area inside the closed contour. The device that calculates the area inside a closed contour is called a planimeter, or platometer. The mathematical principle of a planimeter is Green's theorem. Many GPS-planimeter smartphone apps are now freely available.

If the xy-coordinate data $(x_i, y_i)_{i=1}^{n+1}$ of a closed contour are known with $x_1 = x_{n+1}$ and $y_1 = y_{n+1}$, then the following simple R code can calculate the enclosed area:

```
s=rep(0,n)
for (i in 1:n){s[i]=-y[i]*x[i+1] + x[i]*y[i+1]}
A=0.5*sum(s)
#Example of computing the area of an ellipse by a planimeter
n=1000
a=4
b=2
t=seq(0,2*pi,length=n+1)
```

```
x=a*cos(t)
y=b*sin(t)
s=rep(0,n)
for (i in 1:n){s[i]=-y[i]*x[i+1] + x[i]*y[i+1]}
A=0.5*sum(s)
A
#[1] 25.13258
#The aera of an ellipse according to formula: pi ab
pi*a*b
#[1] 25.13274
```

B.4.4 Boundary Data and Integration inside the Boundary

A common feature among the divergence theorem, Green's theorem, Stokes' theorem, and Definition D.2 for an integral as the height increment of a curve (also known as Part II of the Fundamental Theorem of Calculus in conventional calculus textbooks) is that they all reveal a relationship between the integral of the derivative of a function over a domain and the function values on the domain's boundary. This feature can help derive many important results in mathematical physics, such as the vorticity and circulation formulas in oceanography, and Maxwell's equations in electromagnetism.

(i) If the domain is a 3D solid, then the boundary is a closed surface, and the relationship is about a volume integral over the 3D solid domain and a surface integral on the boundary surface. The divergence theorem formulates this kind of relationship. Because the dimension of the 3D solid domain is L^3 and that of the boundary surface is L^2, the integrand of the volume integral must have a dimension XL^{-1} if the integrand of the surface integral has a dimension X. The first-order derivative of a function of dimension X has a dimension XL^{-1}. Divergence is an example of a first-order derivative. Thus, dimensional analysis helps one explore this kind of relationship, and the divergence theorem is only an example of the relationship.

(ii) If the domain is a 3D surface, then the boundary is a closed 3D contour, and the relationship is about a surface integral on the 3D surface and a line integral along the closed 3D contour. Stokes' theorem formulates this kind of relationship. Following dimensional analysis similar to the above, the integrand of the surface integral may be a kind of derivative of the integrand along the 3D closed contour. Curl is an example of a first-order derivative. Stokes' theorem is an example of the relationship.

(iii) If the domain is a 2D region, then the boundary is a closed 2D contour, and the relationship is about a double integral over the region and a line integral along the closed contour. Green's theorem formulates this kind of relationship.

(iv) If the domain is a 1D interval, then the boundary is the two end points of the interval. Because an integral cannot be defined on points, only the function values at the two boundary points can be considered. The relationship is about an integral over the interval and the values of a function at the two boundary points. Again, because of dimensional analysis, the integrand of the integral over the interval should have a dimension XL^{-1} if the function values at the end points have a dimension X. Integration of a derivative over the domain may be an appropriate mathematical formulation to balance the dimensions on both sides of an equation. The exact formulation is Definition D.2, which defines the height increment of a curve $f(x)$ over an interval as the integration of the slope $f'(x)$ of the curve on the same interval, i.e.,

$$f(b) - f(a) = \int_a^b f'(x)dx. \tag{B.21}$$

Appendix C Spherical Coordinates

The Earth is approximately a sphere in the geometric height coordinate, and for some purposes it may be regarded as a perfect sphere in the geopotential height coordinate. Spherical coordinates are naturally used in climate model equations and in many formulas of climate science. This appendix describes the relationship between the spherical coordinates (r, ϕ, θ) and Cartesian coordinates (x, y, z), and it explains the representation of differentials in spherical coordinates. These formulas are frequently encountered not only in climate model equations, but also in climate data analysis, such as in the area-factor used in computing covariance matrices and empirical orthogonal functions (EOFs).

C.1 Transform between the Spherical Coordinates and Cartesian Coordinates

Figure C.1(a) shows the spherical coordinates (r, ϕ, θ) for a point on a sphere and its correspondence on a map of latitude–longitude (ϕ–θ) grid, where r is the radius of the sphere being considered, ϕ is the latitude defined as an angle from the equator to the point and is positive in the Northern Hemisphere and negative in the Southern Hemisphere, and θ is the longitude defined as an angle starting from the Greenwich meridian and is positive to the east of the Greenwich meridian, and negative to the west of that meridian.

The spherical coordinates and their corresponding ranges are as follows:

$$r: \ [0, \infty) \tag{C.1}$$

$$\phi: \ [-\pi/2, \pi/2] \ \text{ or } \ [-90°, 90°] \tag{C.2}$$

$$\theta: \ [-\pi, \pi] \ \text{ or } \ [-180°, 180°]. \tag{C.3}$$

Figure C.1(b) shows right triangles that can be used to compute Cartesian coordinates (x, y, z) from the spherical coordinates (r, ϕ, θ). The z-value is the projection of the position vector $\mathbf{r}$ on the vertical z-axis via the right triangle ΔOMZ: $\overline{OZ} = r \sin \phi$. The x-value is the result of two projections. First, project the position vector $\mathbf{r}$ via the vertical right triangle ΔOPM onto the horizontal plane to obtain $\overline{OP} = r \cos \phi$, which in turn is projected onto the x-axis via the right triangle ΔOXP to obtain $x = \overline{OP} \cos \theta$. Similarly, $y = \overline{OP} \sin \theta$. These

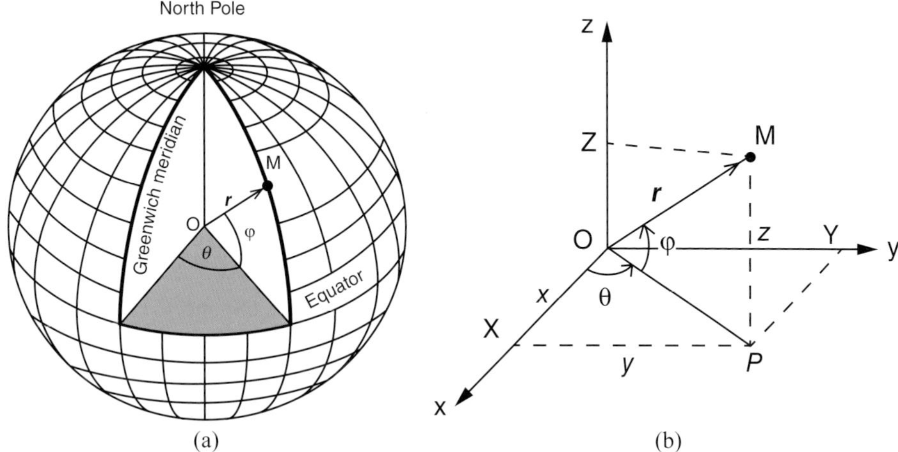

Fig. C.1 (a) A correspondence between the spherical coordinates (r, ϕ, θ) on a sphere and the latitude–longitude $(\phi\text{–}\theta)$ grid on a map. (b) The right triangles show the correspondence between the spherical coordinates (r, ϕ, θ) and Cartesian coordinates (x, y, z).

projections lead to the conversion of the spherical coordinates (r, ϕ, θ) to the Cartesian coordinates (x, y, z) as follows:

$$x = r\cos\phi\cos\theta, \tag{C.4}$$

$$y = r\cos\phi\sin\theta, \tag{C.5}$$

$$z = r\sin\phi. \tag{C.6}$$

C.2 Area and Volume Differentials in Spherical Coordinates

Differentials are a fundamental concept of calculus (described in Appendix D), including linear differential as a small segment of a line, area differential, volume differential, and high-dimensional differential.

In three-dimensional Cartesian coordinates, an area differential $dxdy$ on a horizontal plane with a fixed z is the area of a small rectangle with edges dx and dy. The volume differential $dxdydz$ is the volume of a small cuboid with edges dx, dy, and dz.

Similarly, in spherical coordinates, the area differential on a sphere with a fixed r has edges $rd\phi$ in the meridional direction and $r\cos\phi d\theta$ in the zonal direction, as shown in Fig. C.2. Thus, the area differential is

$$dS = r^2\cos\phi\, d\phi d\theta. \tag{C.7}$$

Climate data analysis often uses the concept of area-factor $\sqrt{\cos\phi}$, which originates from this $\cos\phi$ in the spherical coordinates. The area-factor is used to take account of the fact that uniform latitude–longitude grid boxes have smaller areas at higher latitudes.

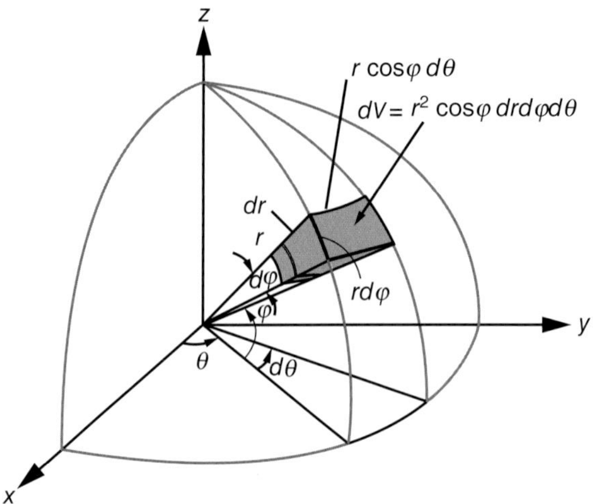

Fig. C.2 The differential area $dS = r^2 \cos\phi \, d\phi d\theta$ on a sphere of radius r due to the differentials $d\phi$ and $d\theta$, and the differential volume $dV = r^2 \cos\phi \, dr d\phi d\theta$ due to the differentials $d\phi, d\theta$, and dr.

An integration of the area differential is the area of a given surface. The area of the entire sphere of radius R can thus be calculated from the following integral:

$$\begin{aligned}
A &= \iint dS \\
&= \int_{\theta=-\pi}^{\pi} \int_{\phi=-\pi/2}^{\pi/2} R^2 \cos\phi \;\; d\phi d\theta \\
&= \int_{\theta=-\pi}^{\pi} \left(\int_{\phi=-\pi/2}^{\pi/2} R^2 \cos\phi \; d\phi \right) d\theta \\
&= R^2 \int_{\theta=-\pi}^{\pi} [\sin\phi]_{-\pi/2}^{\pi/2} d\theta \\
&= R^2 \int_{\theta=-\pi}^{\pi} 2 d\theta \\
&= 4\pi R^2.
\end{aligned} \tag{C.8}$$

In climate data analysis, we sometimes treat the polar region separately due to a lack of data or to bad data quality. For example, a global precipitation dataset might consider only the region between latitudes 75°S and 75°N, omitting the polar regions from latitude 75°N and S to the North and South Poles. The area of the omitted regions is actually very small, only about 3.4% of each hemisphere. This can be demonstrated by the integration of the area-differential over the polar region:

$$\begin{aligned}
A_P &= \iint dS \\
&= \int_{\theta=-\pi}^{\pi} \int_{\phi=0.4167\pi}^{\pi/2} R^2 \cos\phi \;\; d\phi d\theta
\end{aligned}$$

$$
\begin{aligned}
&= R^2 \int_{\theta=-\pi}^{\pi} [\sin\phi]_{0.4167\pi}^{\pi/2} d\theta \\
&= R^2 \int_{\theta=-\pi}^{\pi} (1 - 0.9659)\, d\theta \\
&= 0.0341 \times 2\pi R^2.
\end{aligned} \tag{C.9}
$$

Here, $2\pi R^2$ is the area of the northern hemisphere when R is regarded as the radius of Earth.

A small increment of the radius of the sphere from this area differential leads to a small volume with base dS and height dr. Thus, the volume differential in the spherical coordinates is

$$
dV = r^2 \cos\phi \; d\phi d\theta dr. \tag{C.10}
$$

The volume of a ball of radius R can be found from the integration of this volume differential as follows:

$$
\begin{aligned}
V &= \iiint dV \\
&= \int_{r=0}^{R} \int_{\theta=-\pi}^{\pi} \int_{\phi=-\pi/2}^{\pi/2} R^2 \cos\phi \; d\phi d\theta dr \\
&= \int_{r=0}^{R} 4\pi r^2 \, dr \\
&= \frac{4}{3}\pi R^3.
\end{aligned} \tag{C.11}
$$

Appendix D Calculus Concepts and Methods for Climate Science

This appendix describes calculus using the simple Descartes' direct approach without limits, which is a different approach from that of a conventional calculus textbook. A derivative may be regarded intuitively as resembling the slope (i.e., the steepness) of a mountain road. Similarly, an integral resembles the hiker's total elevation increase, which is a result of the accumulation (called integration) of the small elevation increases of each step. The slope and elevation are a pair of key features experienced by the mountain road hiker, and they form a derivative–antiderivative pair, which is a basic concept of calculus. Each step of the mountain hiker represents the local slope of the mountain road, measured by a small distance forward, and also by a small distance upward. The calculus method is to use the small steps to find various kinds of local and integrated results using these three quantities: the local slope denoted by $f'(x)$, the small distance forward denoted by dx, and the small distance upward denoted by dy.

This appendix emphasizes the calculus concepts and their mathematical implementation from various perspectives, such as that of statistics and that of climate science applications. In this appendix, we use R and WolframAlpha to do the actual and sometimes tedious calculations of derivatives and integrals. The appendix contains both single-variable calculus and multivariate calculus with climate science examples. It also includes vector calculus through the description of the line integral, surface integral, and volume integral. Two theorems involving these integrals, both of which are very useful in climate science, are found in Appendix A for the divergence theorem and in Appendix B for Stokes' theorem.

D.1 Descartes' Direct Calculus for Functions of a Single Variable

"Calculus" is not a commonly used word in daily life. The *Oxford English Dictionary* indicates that the word comes from the mid-seventeenth-century Latin and literally means small pebble (such as those used on an abacus) for counting. The dictionary gives three meanings of "calculus": a branch of mathematics that deals with derivatives and integrals, a particular method of calculation, and a hard mass formed by minerals. Obviously here we are interested in the first meaning. We introduce derivatives directly using René Descartes' (1596–1650) method of tangents. Then derivatives and antiderivatives can be introduced simultaneously, using derivative–antiderivative (DA) pairs. Our introduction to integrals is directly from the DA pairs and the height increment of an antiderivative curve. This is different from the conventional introduction of calculus. Thus, this section introduces

Descartes' direct (DD) calculus and describes the basic calculus concepts of derivative and integral in a direct and non-traditional way, without the need to introduce limits: The derivative of a curve is defined and computed from the slope of a tangent line to a curve, and the integral of the curve's slope is defined as the height of a curve. This direct approach to calculus has three distinct features: (i) it defines derivative and (definite) integral without using limits, (ii) it defines derivative and antiderivative simultaneously via a DA pair, and (iii) it posits the fundamental theorem of calculus as a natural consequence of the definitions of derivative and integral. The first D in DD calculus attributes this approach to Descartes for his method of tangents and the second D to DA-pair. The Descartes' DA-pair calculus, or Descartes' direct calculus, or simply DD calculus makes many traditional procedures unnecessary, when introducing calculus to climate science students. The DD calculus has few intermediate procedures, which can help dispel the mystery of calculus as perceived by the general public.

D.1.1 Slope and DA Pairs

When driving on a steep highway, we often see a grade warning sign like the one in Figure D.1. The 9% grade means a slope equal to 0.09 for a highway. The elevation will decrease 90 feet when the horizontal distance increases 1,000 ft. The grade or slope is calculated by

$$m = \tan\theta = \frac{H}{L} = \frac{90}{1000} \tag{D.1}$$

i.e., the ratio of the opposite side to the adjacent side of the right triangle in Figure D.1. This subsection will discuss the slope of a curve, also called the derivative of a function, the antiderivative, and the DA pair.

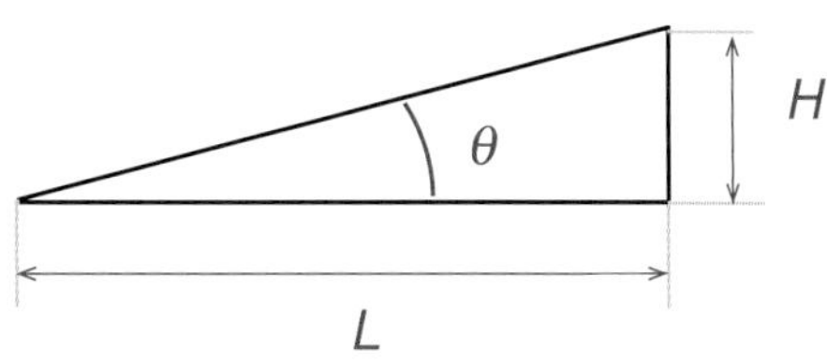

Fig. D.1 Highway grade sign and a right triangle to show slope.

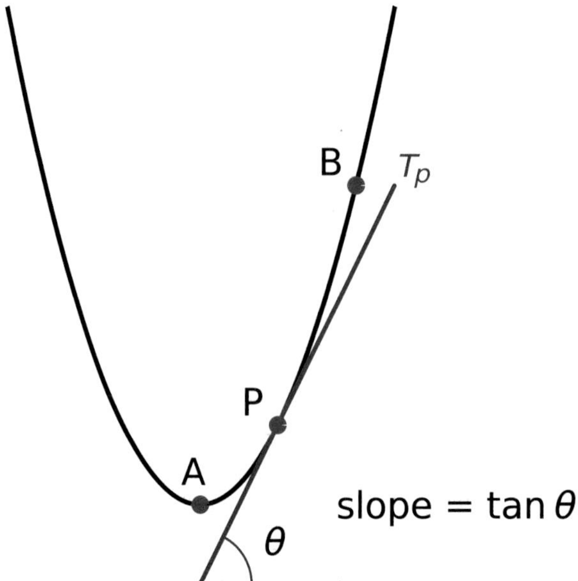

Fig. D.2 A curve; three points A, P, and B; and a tangent line T_P without coordinates.

D.1.1.1 The Slope of a Curve Varies from Point to Point

The slope of a curve at a given point is defined as the slope of a tangent line at this point. Figure D.2 shows three points: P, A, and B. T_P represents the tangent line at point P whose slope is defined as the slope of the curve at P. The tangent line's slope is used to measure the curve's steepness. The steepness of a mountain road changes from point to point, and so does the slope of a curve.

Calculus studies (i) the slope $\tan\theta$ of a curve at various points, and (ii) the height increment H from one point to another, say, A to B, as shown in Fig. D.3.

Our geometric intuition tells us that height H and slope $\tan\theta$ are related because H increases rapidly if the slope is large for an upward trend. A core concept of calculus is to describe the relationship between these two quantities: height and slope. In plain language, height is an integral, and slope is a derivative; or more precisely, height is an integral of a derivative, and slope is a derivative of an integral.

D.1.1.2 Calculate the Slope of a Parabola using the Tangent Line Concept

Let us introduce the coordinates x and y and use a function $y = f(x)$ to describe the curve. We start with an example $f(x) = x^2$.

The tangent line of the curve at point $P(x_0, y_0)$ can be described by a point–slope equation

$$y - y_0 = m(x - x_0), \tag{D.2}$$

where $y_0 = x_0^2$, and m is the slope to be determined by the condition that the tangent line touches the curve at only one point. In fact, "tangent" is a word derived from Latin and means "touch."

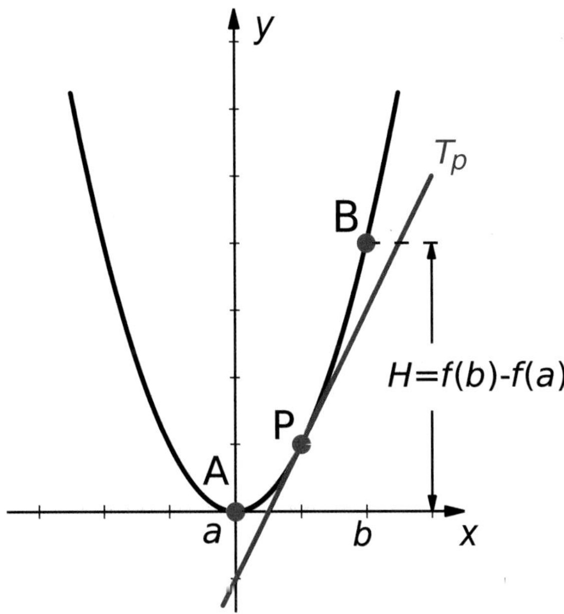

Fig. D.3 A curve; three points A, P, and B; and a tangent line T_P on the xy-coordinate plane.

The tangent line (D.2) and the curve

$$y = x^2 \tag{D.3}$$

have a common point $P(x_0, y_0)$ (See Figure D.3).

Substituting (D.3) into (D.2) to eliminate y, we have

$$x^2 - x_0^2 = m(x - x_0), \tag{D.4}$$

then

$$(x - x_0)(x + x_0) - m(x - x_0) = 0, \tag{D.5}$$

or

$$(x - x_0)(x + x_0 - m) = 0. \tag{D.6}$$

The two solutions of this quadratic equation are

$$x_1 = x_0 \tag{D.7}$$

and

$$x_2 = -x_0 + m. \tag{D.8}$$

Because $P(x_0, y_0)$ is a tangent point, the two solutions must be a repeated root, also called a double root, meaning that

$$x_1 = x_2. \tag{D.9}$$

This yields

$$x_0 = -x_0 + m \tag{D.10}$$

i.e.,

$$m = 2x_0. \tag{D.11}$$

This implies that the slope m of the curve $y = x^2$ at x_0 is $2x_0$, at a is $2a$, at 3 is $2 \times 3 = 6$, and in general, at x is $2x$.

Note that we have rigorously derived the slope formula for the parabola $y = x^2$ without introduction of the concept of limits, while most calculus textbooks would present limits first and then use the limit concept to calculate slope.

According to the point–slope equation $y - y_0 = m(x - x_0)$ and the dimensional analysis in Chapter 1, the dimension of the slope is

$$[m] = \frac{[y - y_0]}{[x - x_0]} = \frac{[y]}{[x]}. \tag{D.12}$$

In the case of the mountain hiking trail, both the vertical height y and the horizontal distance x have the length dimension L. Then, the slope of a mountain trail is dimensionless, and is equal to $\tan\theta$. Here, θ is the angle from the horizontal line to the tangent line measured counterclockwise.

If y is the distance traveled by a car with dimension $[y] = L$ and x is the time used during the trip with dimension $[x] = T$, then the slope has the speed dimension because

$$[m] = [y]/[x] = L/T = LT^{-1} \tag{D.13}$$

is the dimension of speed. Usually, if x is time, the slope is called the rate of change. The speed is the rate of change of the spatial position. Acceleration is the rate of change of speed.

D.1.1.3 Definition of Derivative and DA Pair

The slope measures the steepness of the curve $y = f(x)$, i.e., the rate of the curve's height increase or decrease. The slope, or rate, varies from point to point. The slope is thus a derived quantity from the original function $y = f(x)$ and is called "derivative." We have the following definition.

Definition D.1 (Derivative as slope, and DA pair) The slope $2x$ is called the derivative of x^2. Furthermore, x^2 is called an antiderivative of $2x$. And $(2x, x^2)$ is called a derivative–antiderivative (DA) pair.

For a general function $y = f(x)$, the slope of the curve $y = f(x)$ is called the *derivative* and is denoted by $f'(x)$ or y', and $f(x)$ is called the antiderivative of $f'(x)$. Thus, $(f'(x), f(x))$ is a DA pair. Other notations for the same derivative include

$$\frac{dy}{dx}, \ \frac{df}{dx}, \ \frac{d}{dx}f(x), \ \text{and} \ \dot{y}. \tag{D.14}$$

One may start with $f(x)$ and form a DA pair $(f(x), F(x))$, then $F(x)$ is an antiderivative of $f(x)$, denoted by $'f(x)$ or $'y$, i.e., the prime is ahead of the function notation. Therefore, generically, $(f(x), 'f(x))$ is a DA pair, similar to $(f'(x), f(x))$ being a DA pair. Other notations for the same antiderivative may include the following:

$$a\lfloor f,\ a\lfloor y,\ ydx,\ fdx,\ \underset{\cdot}{f},\ \underset{\cdot}{y},\ \text{and}\ \int f(x)dx.^{1} \tag{D.15}$$

You may wish to invent your own convenient notations and terminology, which are probably not simpler than these. For example, one may use slope function and antislope function, or slope–antislope pair (SA pair). Or slope–height pair since anti-slope is the height.

The word "derivative" is late Middle English (1400–1450), from Latin "derivativus" and "derivare," meaning "a word derived from another." In calculus, it means that the derivative function $f'(x)$ is derived from its original function $f(x)$.

The dimension of the derivative $\frac{dy}{dx}$ may be symbolically regarded as

$$\left[\frac{dy}{dx}\right] = \frac{[dy]}{[dx]} = \frac{[y]}{[x]} \tag{D.16}$$

which is the derivative's dimension $[m]$ discussed earlier.

The antiderivative's dimension is the derivative's dimension $[m]$ times the dimension of x, i.e., $[m][x]$. Thus, the dimension of a DA pair $(f'(x), f(x))$ is $([f'(x)], [f'(x)][x]) = ([m], [m][x])$.

For a DA pair in the form $(g(x), G(x))$ such that $G'(x) = g(x)$, then $[(g(x), G(x))]$ is still $([m], [m][x])$ where $[m]$ is now $[g(x)]$. Thus, it is generic that the dimension of a DA pair is $([m], [m][x])$.

If $[m] = 1$ is the dimensionless rate, then $[m][x] = L$ if $[x] = L$ is the distance dimension. The corresponding DA pair is the slope–height pair.

If $[m] = L$ is the length dimension, then $[m][x] = L^2$ is the dimension of area if $[x] = L$ is also the distance dimension. This DA pair is the length–area pair.

If $[m] = L/T$ is the speed dimension, then $[m][x] = L$ is the dimension of length if $[x] = T$ is the time dimension. This DA pair is the speed–distance pair.

If $f(x) = C$ is a constant, then $y = C$ represents a horizontal line whose slope is 0 for any x. Hence $(C)' = 0$, and $(0, C)$ is a DA pair.

If $f(x)$ is a linear function, then $y = \alpha + \beta x$ represents a straight line whose slope is β for any x, hence $(\alpha + \beta x)' = \beta$, i.e., $(\beta, \alpha + \beta x)$ is a DA pair.

Thus, the derivative's geometric meaning is the slope of the curve $y = f(x)$: $f'(x)$ is large at places where the curve $y = f(x)$ is steep. At a flat point, such as the maximum or minimum point of $f(x)$, the slope is zero since the tangent lines at these points are horizontal.

In addition to the geometric meaning, a derivative may have a physical meaning, such as speed, or a biological meaning, such as growth rate, as well as the meaning of the rate

[1] The antiderivative notation $\int f(x)dx$ is commonly used in conventional calculus textbooks, and is often given a name of indefinite integral. However, the name "indefinite integral" is unnecessary, and a student might confuse it with the concept of *integral*, which is one of the two key parts of calculus.

of change in almost any scientific field and anyone's daily life. As an example, for a car driven at $v = 60$ mph (miles per hour) for two hours, the total distance traveled is $s = v \times t = 60 \times 2 = 120$ miles. Here, (v, vt) or (v, s) is a DA pair for a general time t.

Free fall is another example. An object's free fall has its distance of falling equal to $s = (1/2)gt^2$ and its falling speed is $v = gt$, where $g = 9.8[\text{m s}^{-2}]$ is the Earth's gravitational acceleration. Galileo Galilei (1564–1642) discovered this time-square relationship for the distance. Since derivative t^2 with respect to t is $2t$, we have $ds/dt = \frac{1}{2}g(2t) = gt$. Thus, $(gt, (1/2)gt^2)$ or (v, s) is a DA pair.

Similarly, the momentum mv and kinetic energy $mv^2/2$ of a small water parcel m in the ocean moving at speed v form a DA pair: $(mv, mv^2/2)$, because $(d/dv)(mv^2/2) = mv$. Of course, this holds for the motion of almost anything, including atmospheric air and oceanic water.

Another example of a DA pair is acceleration and velocity (a, v), which are both functions of time t. Acceleration is the time derivative of velocity $a = dv/dt$, and velocity is an antiderivative of acceleration.

In general, the meaning of derivative is the rate of change of the function $f(x)$ with respect to the independent variable x, which can typically be either time or spatial location.

Apparently, derivative and antiderivative are inverse operations to each other. For example, $('f)' = f$, and $'(f')$ and f may differ by a constant because the derivative of a constant is zero.

D.1.1.4 Calculate the Derivative of $y = x^3$ as the Slope of a Tangent Line

Let us now return to the derivative calculation. The above tangent line approach for finding the slope for x^2 can be applied to the function $y = x^3$. In that case, we need to solve the following simultaneous equations

$$y - y_0 = m(x - x_0), \tag{D.17}$$

$$y = x^3. \tag{D.18}$$

Eliminating y, we have

$$x^3 - x_0^3 = m(x - x_0). \tag{D.19}$$

The factorization of this equation yields

$$(x - x_0)(x^2 + xx_0 + x_0^2 - m) = 0. \tag{D.20}$$

This factorization implies $x_1 - x_0 = 0$ and $x_2^2 + x_2x_0 + x_0^2 - m = 0$. The cubic equation has three solutions. Because $P(x_0, y_0)$ is the tangent point, x_0 is a repeated root of these two equations: $x_1 = x_2 = x_0$, which leads to

$$m = 3x_0^2. \tag{D.21}$$

Thus, we have established that the derivative of x^3 is $3x^2$, and x^3 is an antiderivative of $3x^2$. In other words, $(3x^2, x^3)$ forms a DA pair.

D.1.1.5 Commonly Used DA Pairs and Their Computer Calculations

Following the tangent line approach, the above examples have demonstrated four DA pairs

$$(0,C),(1,x),(2x,x^2),(3x^2,x^3). \tag{D.22}$$

The tangent line approach can be applied to any power function x^n, where n is a positive integer. The DA pair for x^n is

$$(nx^{n-1},x^n). \tag{D.23}$$

This formula actually holds for any real number n with the exception of $n = 0$. For example, $(x^{1/2})' = (1/2)x^{-1/2}$. Proof of this statement is not elementary and is beyond the scope of this book.

Fortunately, the calculation of a great many derivatives and antiderivatives can easily be done using computers or smartphones. Many open-source computer software packages and smartphone apps are available to perform this kind of calculation. WolframAlpha and R are two examples. At the website www.wolframalpha.com, you can enter a derivative command and a function, such as

```
derivative  x^(3/2)+4x^3
```

The software will give you its derivative, plus other information, such as the graph of the derivative function (see Fig. D.4). You can also use WolframAlpha via a smartphone application.

To find an antiderivative, you can use a similar command

```
antiderivative  x^(3/2)+4x^3
```

You can use R to find derivatives

```
D(expression(x^2),"x")
2 * x
```

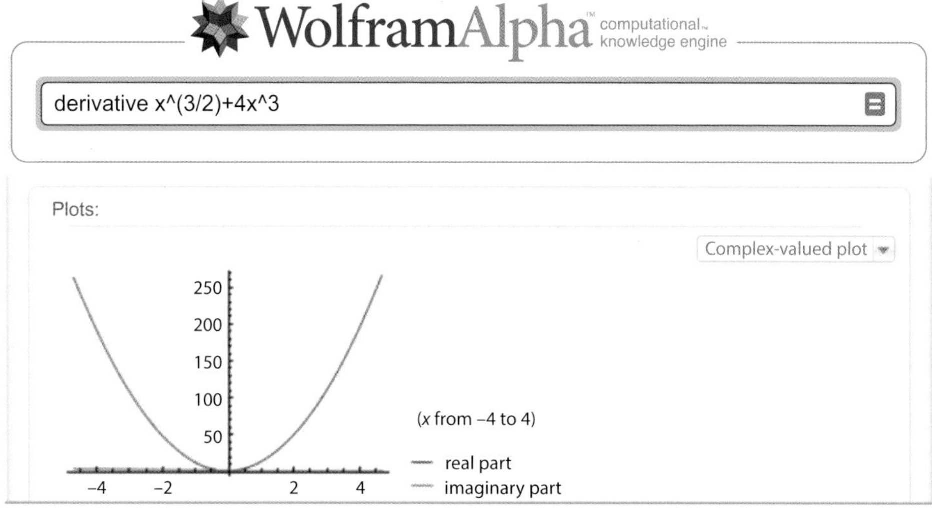

Fig. D.4 WolframAlpha derivative example.

Chapter 2 of this book is on R basics, and Chapters 9–11 are on using R to make complex analysis and visualization of climate data.

Using R, one can easily find DA pairs for commonly used functions. Here is a partial list:

1. Exponential function (e^x, e^x).
2. Natural logarithmic function $(\frac{1}{x}, \ln x)$.
3. Sine function: $(\cos x, \sin x)$.
4. Cosine function: $(-\sin x, \cos x)$.
5. Tangent function: $(\sec^2 x, \tan x)$.

D.1.2 Height Increment and Integrals

When we trace a curve, we consider not only the slope, but also the ups and downs of the curve, i.e., the increment or decrement of the curve from one point to another. When we drive over a mountain road, we care about both steepness (i.e., slope) and elevation. Apparently, the slope and height increment are related. The slope has already been defined as a derivative in the preceding section. In this section, the height increment is defined as an integral, because the height increment or elevation increment is an integration process, or an accumulation process, measured by both speed and time. When we walk from the bottom of a mountain to its peak, we have gained the total height of the mountain by adding together all of our steps. The total height of the mountain is an integration of steps. A step gains more height at a very steep location, and gains less height at a less steep location. Therefore, the height of a mountain is naturally linked to its slope.

The atmosphere's convective available potential energy (CAPE) is an indicator of atmospheric instability and can be a useful measure of the potential for severe weather. CAPE is the amount of energy a parcel of air would have if lifted from a lower to a higher level under certain precise conditions and subject to specific assumptions.

For a function $y = f(x)$, its increment from $A(a, f(a))$ to $B(b, f(b))$ is $f(b) - f(a)$ as shown in Fig. D.3. Another notation for the increment is $f(b) - f(a) = f(x)|_a^b$. This height increment is employed in formulating the integral definition below.

Definition D.2 (Integral as height increment of a curve) The function's increment $f(b) - f(a)$ from $A(a, f(a))$ to $B(b, f(b))$ is defined as the integral of the derivative function $f'(x)$ in the interval $[a,b]$ and is denoted by $I[f'(x), a, b] = f(b) - f(a)$. Here, $f'(x)$ is called the integrand, and $[a,b]$ is called the integration interval.

The conventional notation for integral is $\int$, introduced by Gottfried W. Leibniz (1646–1716):

$$\int_a^b f'(x)dx = I[f'(x), a, b] = f(b) - f(a). \tag{D.24}$$

In the 1600s, the handwritten form of S was often very long like $\int$. Leibniz thought of integral as a summation. Thus, the $\int$ symbol means summation. He treated summation as "integral" or "integration." Our book uses both notations: $\int_a^b f'(x)dx$ and $I[f'(x), a, b]$, or $\int_c^d f(x)dx$, or $I[f, c, d]$.

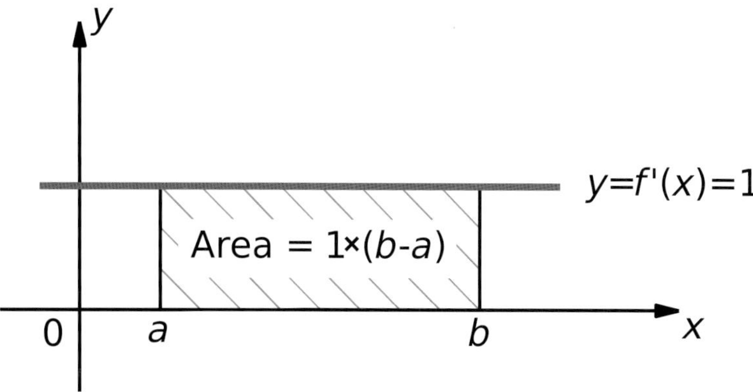

Fig. D.5 The area of a rectangle under a horizontal line.

Example D.1 Given $f(x) = x$, $f'(x) = 1$, and $[a,b] = [0,2]$, we have

$$\int_a^b f'(x)\,dx = \int_0^2 1\,dx = x|_0^2 = 2 - 0 = 2. \tag{D.25}$$

The area between $y = 1$ and $y = 0$ in the interval $[0,2]$ is also 2 (see Fig. D.5 for $f'(x) = 1, a = 0$, and $b = 2$).

Example D.2 If we integrate speed $v(t)$, we will then get the distance $I[v(t),a,b]$ traveled from time $t = a$ to $t = b$. If $v(t)$ is a constant, say, 60 mi/hour, and if $a = 14:00$ and $b = 16:00$, then the integral $I[60,14,16] = 60t|_{14}^{16} = 60 \times (16 - 14) = 120$ mi is the total distance traveled from 2:00 p.m. to 4:00 p.m. Here $60t$ is an antiderivative of 60. If we plot v as a function of t, then 120 is equal to the area of the rectangle bounded by $v = 60$, $v = 0, t = 14$, and $t = 16$ (see Fig. D.5 for $f'(x) = 60$, $a = 14$, and $b = 16$)

Example D.3 Given $f(x) = (1/2)x^2$, $f'(x) = x$, and $[a,b] = [0,1]$, we have

$$I[f'(x),a,b] = I[x,0,1] = (1/2)x^2|_0^1 = (1/2)(1^2 - 0^2) = 1/2. \tag{D.26}$$

The area under the integrand $y = x$ but above the x-axis in $[0,1]$ is 1/2 (see Fig. D.6 for $a = 0$ and $b = 1$).

Example D.4 In the free fall problem, the speed is a linear function of time, $v = gt$, and the integration $I[gt,0,x] = (1/2)gx^2$ is the distance traveled from time zero to time x. The region bounded by $v = gt$, $v = 0$, $t = 0$, and $t = x$ is a right triangle with base equal to x, height gx, and area $(1/2) \times x \times gx = (1/2)gx^2$. In this example, we have chosen to use x as an arbitrary right bound for the region. This x can be any number, such as 1, 2, or 2.5.

In the above four examples, the area under a curve is equal to an integral. As a matter of fact, this interpretation of an integral as being equal to an area is generally true. For an irregular region, we can simply use the integral $I[f'(x),a,b]$ as the definition of the area of the region bounded by $y = f'(x)$, x-axis, $x = a$, and $x = b$. The next section will justify this definition. As for the calculation of an integral, if one knows the relevant

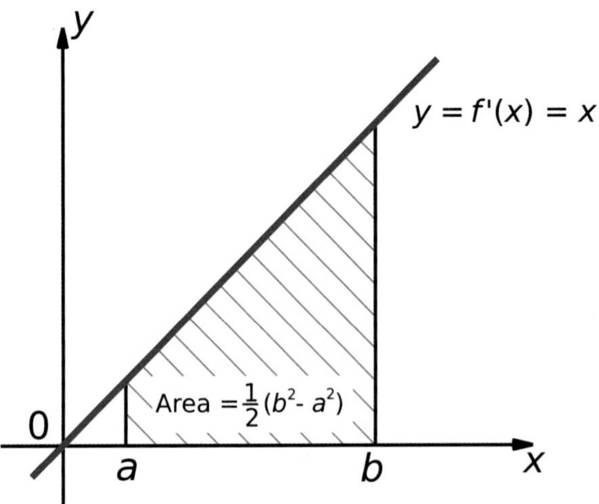

Fig. D.6 The area of a triangle under a straight line.

DA pair, the integral is a simple substitution $f(b) - f(a)$. Otherwise, one can use a computer to find the antiderivative, or to directly evaluate $I[f'(x), a, b]$. Just as is the case for derivative software, there are many free online computer programs and smartphone apps for calculating integrals. Figure D.7 shows the WolframAlpha calculation of the integral $I[x^4 + 2x, 0, 1]$ using the command

```
integrate x^4+2x from 0 to 1
```

The result is 6/5. i.e., $I[x^4 + 2x, 0, 1] = 6/5$.

The definition of an integral states that the integral of a function in an interval is the increment of its antiderivative in the same interval. One can write

$$I[g(t), a, b] = G(b) - G(a) \tag{D.27}$$

where $G(t)$ is an antiderivative of $g(t)$. Another way to express the above is

$$I[G'(u), a, b] = G(b) - G(a). \tag{D.28}$$

In the above two expressions, the independent variables t and u in functions $g(t)$ and $G(u)$ are the integration variables, also called dummy variables. The integral values do not depend on the choice of dummy variables. One can use any symbol to represent this variable. In practical applications, if the independent variable is time (e.g., speed is a function of time), t is often used as the dummy variable.

Also according to the integral definition, the integral of $f'(t)$ in the interval $[a, x]$ is

$$I[f'(t), a, x] = f(x) - f(a). \tag{D.29}$$

Taking the derivative of both sides of this equation with respect to x, we have

$$(I[f'(t), a, x])'_x = f'(x) \tag{D.30}$$

integrate x^4+2x from x=0 to 1

Definite integral:

$$\int_0^1 (x^4 + 2x)\, dx = \frac{6}{5}$$

Visual representation of the integral:

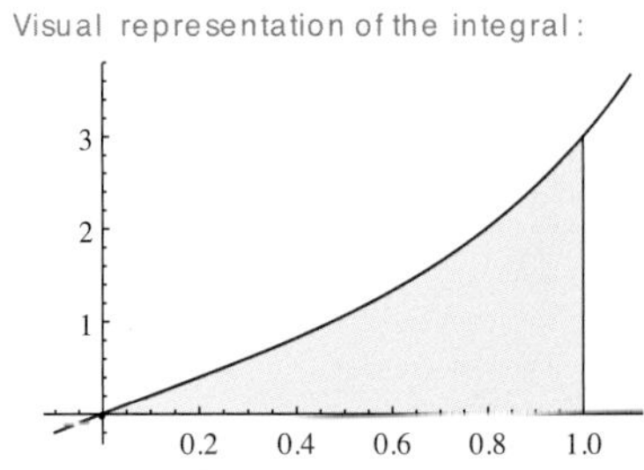

Riemann sums:

left sum	$\frac{(n-1)(36n^3-9n^2+n+1)}{30n^4} = \frac{6}{5} - \frac{3}{2n} + O\left(\left(\frac{1}{n}\right)^2\right)$

(assuming n subintervals of equal length)

Fig. D.7 WolframAlpha integral example.

since $(f(a))'_x = 0$ due to $f(a)$ being a constant with respect to x and having a slope of zero. Here we have used a subscript x to indicate that the independent variable is x and the derivative is with respect to x.

Equation (D.30) is often called Part II of the Fundamental Theorem of Calculus (FTC), while the definition of an integral is actually often referred to as Part I of the FTC. Part II of the FTC states that an antiderivative can be explicitly expressed by an integral. Thus, the FTC makes a computable and close connection between slope and height increment, and confirms our geometric intuition that the height increment of a curve in an interval is related to the slope of the curve. How are they related? Answer: by an integration of the slope in the interval.

Dimensional analysis can also help explain the relationship between the slope and height involved in an integral. The dimension of $\int_a^b f'(x)dx$ is the dimension $[f'(x)dx] = [f'(x)][dx] = [f'(x)][x] = [f(x)]$. Namely, an integral's dimension is the integrand's dimension times the integration variable's dimension. If the integrand is the dimensionless slope and the integration variable is length, then the integral is also length. If the integrand is speed and the integration variable is time, then the integral is also distance equal to the speed times time. If the integrand is length and the integration variable is also length, then the integral has the area dimension L^2. That may be why some people regard an integral as a height, and others regard an integral as an area.

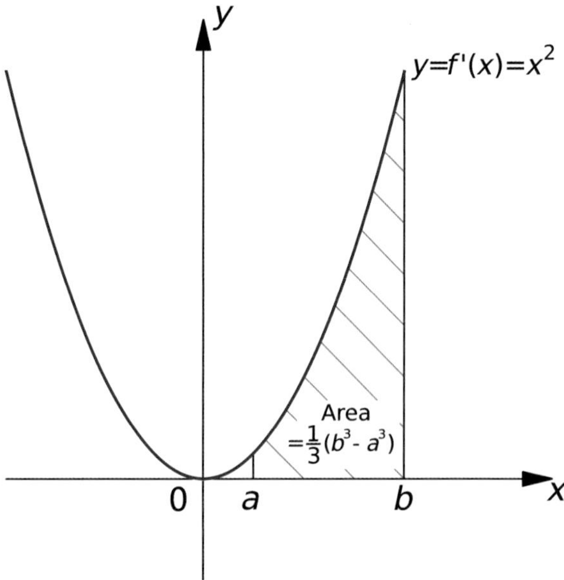

Fig. D.8 The area of a "triangle" with a curved "hypotenuse" when $a = 0$.

Example D.5 $I[x,0,1] = (x^2/2)|_0^1 = 1^2/2 - 0^2/2 = 1/2$, because $(x,x^2/2)$ is a DA pair. The area between $y = x$ and $y = 0$ over $[0,1]$ is 1/2 ($=I[x,0,1]$) (see Fig. D.6).

Example D.6 $I[x^2,0,1] = (x^3/3)|_0^1 = 1^3/3 - 0^3/3 = 1/3$, because $(x^2,x^3/3)$ is a DA pair, or simply since x^2's antiderivative is $x^3/3$. The area of the right triangle with a curved "hypotenuse" bound by $y = x^2$, $y = 0$, and $x = 1$ is 1/3 ($=I[x^2,0,1]$) (see Fig. D.8). The WolframAlpha command for this calculation is `integrate x^2 from 0 to 1`.

D.1.3 Discussion and Mathematical Rigor of Direct Calculus

Two points are discussed here. First, is our definition of area by an integral reasonable and mathematically rigorous? Second, in addition to using computer programs to calculate the DA pairs of complicated functions, can one provide a systematic procedure of hand calculation?

First, how do we know that our definition of area using an integral is reasonable? According to the *Oxford English Dictionary*, "area" is defined as "the extent or measurement of a surface or piece of land." The word "area" comes from the mid-sixteenth-century Latin, literally meaning a "vacant piece of level ground." The units of an area are "square feet," "square meters," etc., meaning that the area of a region is equal to the number of equivalent squares, each side equal to a foot, a meter, or another length unit, that fit into the region. For the area between $y = f'(x)$ and $y = 0$ over $[a,b]$, the simplest measure is to use an equivalent rectangle of length $L = b - a$ and width W (see Fig. D.9). That is, the excess area above

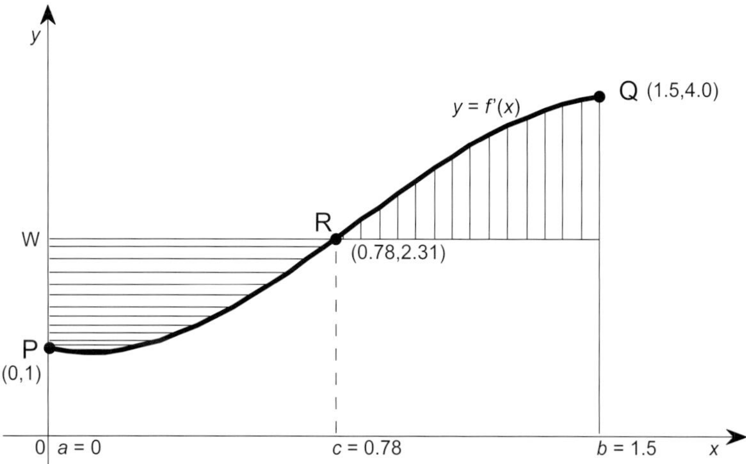

Fig. D.9 An area interpretation of an integral.

$y = W$ (the vertically striped region) is moved to fill the deficit area (the horizontally striped region). The corresponding description by a mathematical formula is below:

$$I[f'(x), a, b] = L \times W = (b-a) \times W. \tag{D.31}$$

This can be true as long as we have

$$W = \frac{I[f'(x), a, b]}{b-a} = \frac{f(b) - f(a)}{b-a}. \tag{D.32}$$

We call this W the mean value of $f'(x)$ over the interval $[a,b]$. Geometrically, $W = (f(b) - f(a))/(b-a)$ is the slope of the secant line that connects points A and B of Fig. D.10. If $y = f(x)$ is not a straight line, then there must be a point m in $[a,b]$ whose slope is less than $(f(b)-f(a))/(b-a)$, and another point M whose slope is larger than $(f(b)-f(a))/(b-a)$, i.e.,

$$f'(m) \leq \frac{f(b - f(a)}{b-a} \leq f'(M). \tag{D.33}$$

Between $f'(m)$ and $f'(M)$, $(f(b)-f(a))/(b-a)$ must meet the mid-ground slope $f'(c)$ at a point c in $[a,b]$, i.e.,

$$W = f'(c) = \frac{f(b) - f(a)}{b-a}. \tag{D.34}$$

This is the mean value theorem (MVT) of calculus. It states that

Theorem D.3 (MVT) *There exists c in (a,b) such that $f'(c) = \frac{f(b)-f(a)}{b-a}$ if $f'(x)$ has a value for every x in (a,b).*

Geometrically, MVT means that there is at least one point c whose tangent line is parallel to the secant line AB. Of course, this requirement is met if $y = f(x)$ is a straight line, in which case c can be any point in $[a,b]$.

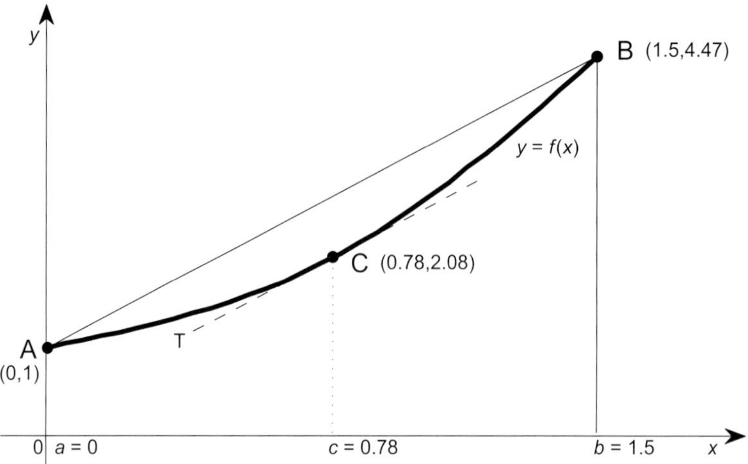

Fig. D.10 Illustration of the mean value theorem: There is a tangent line parallel to the secant line connecting points *A* and *B*.

In climate science, the mean value $W = (T_a(t_2) - T_a(t_1))/(t_2 - t_1)$ may be regarded as an approximate average rate of temperature anomaly change at an Earth location when considering the temperature variation as a function of time $T(t)$. A temperature anomaly at a given point and time $T_a(x,t)$ is defined as the temperature $T(x,t)$ minus this point's normal or mean temperature $T_c(x)$, i.e., $T_a(x,t) = T(x,t) - T_c(x)$. However, the temperature change is not monotonic and often has a large fluctuation. The mean rate of change thus should consider not only the end points, but also the points in between. For example, the mean rate of temperature change, such as the rate of global average annual surface air temperature during the twentieth century, is usually calculated by a linear regression method, which is described in Chapters 3 and 11. In this sense, linear regression may be regarded as an extension of the mean value theorem from only the two end points to multiple points.

A strict mathematical derivation for MVT would require one to prove the above statement "$(f(b) - f(a))/(b - a)$ must meet the mid-ground slope $f'(c)$ at one point c in (a,b)," namely, it proves the existence of the point c. To provide such a proof of this intermediate value theorem can be done but is beyond the scope of this introductory text.

Therefore, the integral $I[f'(t),a,x] = f(x) - f(a)$ is the increment of the antiderivative from a to x, and is also the area for the region between the integrand derivative function and $y = 0$ in the interval $[a,x]$, i.e., the region bounded by $y = f'(t), y = 0, t = a$ and $t = x$. The conventional definition of an integral is from the aspect of an area that is defined as a sum of many rectangles of increasingly narrow widths, under the condition of each width approaching zero. For many non-mathematicians, the condition of each width approaching zero, which involves the concept of a limit, introduces unnecessary complexity and confusion. In contrast, the geometric meaning of our direct integral is the height increment of the antiderivative, not the area underneath an integrand. The area is only regarded as an additional geometric interpretation according to the intermediate value theorem. Under this interpretation of area, we have the following example.

Example D.7 $I[\sqrt{1-x^2},0,1]$ is the area of a quarter unit round disc and is thus equal to $\pi/4$, since $y=\sqrt{1-x^2}$ represents a quarter circle in the first quadrant.

Calculating the slope using the factorization method works for polynomial functions, but the procedure is tedious. The procedure may not even work for a function which is not a polynomial or a fraction of polynomials, such as $y=\sin x$, $y=\exp(x)$, or $y=\ln x$. The MVT provides another way to calculate the slope by using the slope of a secant line. In the above MVT, if B moves very close to A, then the mean value $(f(b)-f(a))/(b-a)=f'(c)$ in Theorem 1 is approaching the slope at A, since c is approaching a, forced by b approaching a. The formal expression is

$$\lim_{b\to a}\frac{f(b)-f(a)}{b-a}=f'(a). \tag{D.35}$$

This can also be considered a definition of a derivative and is called defining a derivative by a limit. This procedure is an efficient way to calculate derivatives by hand and to derive many traditional derivative formulas in calculus. The procedure is particularly effective in deriving the derivative formulas of non-polynomial functions, such as analytically deriving $\frac{d}{dx}\sin x=\cos x$ (Suzuki 2005).

D.1.4 Descartes' Method of Tangents and Brief Historical Note

What is the origin of the direct calculus ideas described above? A great many papers and books have discussed the historical development of calculus. Here we recount the development by a few major historical figures to sort out the origin of the main ideas of the calculus method outlined above. We cite only a few modern references and two works by Newton. Our focus is on (i) Descartes' method of tangents that is the earliest systematic way of finding the slope of a curve without using a limit, and (ii) Wallis's formulas of area, which were the earliest form of FTC. We do not attempt to present a complete list of important works on the history of calculus.

In 1638, René Descartes (1596–1650) derived his method of tangents and included the method in his 1649 book, *Geometry* (see Cajori 1985, p. 176). Descartes' method of tangents is purely geometric, constructing a tangent circle at a given point of a curve with the center on the x-axis (Cajori 1985, pp. 176–177). Graphically, it is easier to draw a tangent circle than a tangent line using a compass and rulers. The tangent circle can be constructed using an adjustable radius line segment and moving the center of the circle on the x-axis so that the circle touches the curve at only one point. Then a tangent line can be drawn as the line perpendicular to the radial line of the circle at the tangent point. The radial line connects the tangent point and the circle's center.

Descartes' method of tangents also has an analytic description. The tangent circle is determined by the given point $P(x_0,y_0)$ on the curve and the moving center on the x-axis $(a,0)$. The circle's equation is

$$(x-a)^2+(y-0)^2=(x_0-a)^2+(y_0-0)^2. \tag{D.36}$$

The tangent condition requires that this equation and the curve's equation $y = f(x)$ have a double root at $P(x_0, y_0)$. This can determine a and hence the tangent circle. The radial line is determined by $P(x_0, y_0)$ and $(a, 0)$ and has its slope $m_R = y_0/(x_0 - a)$. The tangent line of the circle at point $P(x_0, y_0)$ is the tangent line of the curve at the same point. The slope of the tangent line is calculated as $m_T = -1/m_R = (a - x_0)/y_0$. In the above procedure, the concept of limit is not used.

Example D.8 We can use Descartes' method of tangents to find the slope of $y = \sqrt{x}$ at $(1, 1)$: Substituting $y = \sqrt{x}$ into Eq. (D.36), we obtain

$$(x-a)^2 + x = (1-a)^2 + 1. \tag{D.37}$$

This can be simplified to

$$x^2 + (1-2a)x + 2(a-1) = 0. \tag{D.38}$$

Knowing that $x = 1$ is a solution of this equation helps factorize the left-hand side

$$(x-1)(x+2-2a) = 0. \tag{D.39}$$

The double root condition for a tangent line requires $x_1 = x_2 = 1$. This leads to

$$x_2 + 2 - 2a = 1 + 2 - 2a = 0. \tag{D.40}$$

Hence, $a = 3/2$. The slope of the radial line is $m_R = 1/(1 - 3/2) = -2$. The slope of the tangent line is thus $m_T = 1/2$.

Although Descartes' method of tangents may seem complicated in calculation, its concept is simple, clear, and unambiguous, and its geometric procedure is sound. It does not involve small increments of an independent variable (as developed by Fermat in the 1630s), and hence it does not involve limits or infinitesimals. According to the point–slope equation of a tangent line presented earlier, the complexity of Descartes' method of tangents is unnecessary to calculate a slope. However, the point–slope form of a line was not known during Descartes' lifetime. According to Range (2011), the point–slope form of a line was first introduced explicitly by Gaspard Monge (1746–1818) in a paper published in 1784. Thus, Monge's point–slope method of tangent appeared more than 100 years after Descartes' method of tangents.

Pierre de Fermat (1601–1665) developed a method of tangents that is similar to the modern method of differential quotient

$$\frac{f(x+E) - f(x)}{E} \tag{D.41}$$

and uses a small increment E (i.e., infinitesimal), which is ultimately set to be zero when the infinitesimal is forced to disappear from the denominator (Ginsburg et al. 1998). His small increment E is equivalent to the modern notation Δx or h. Fermat's method of tangents is more efficient for calculation from the point of view of limit, while Descartes' method of tangents is geometrically more direct and easier to plot by hand, and Monge's method of tangents is geometrically more direct. Fermat's method is also very useful in deriving mathematical models for physics, such as equations for climate dynamics. Some of the most important equations in climate models are often derived from considering the balance of all forces and energy over a small volume of mass: $dV = dxdydz$. Integrating these equations yields climate model solutions. The so-called calculus method often refers to this process of cutting into small pieces and then integrating them together. Fermat's approach is very good for calculus calculations and applications, while Descartes' approach is simple for understanding the concept of calculus. Their results are the same.

Fermat also used a sequence of rectangular strips to calculate the area under a parabola. His strips have variable width, which enabled him to use the sum of geometric series. This method of calculating an area can be traced back to Archimedes (287–212 BC). Archimedes used what is called the "method of exhaustion" to find the area under a parabola. He used infinitely many triangles inscribed inside the parabola, and he also utilized the sum of a geometric series.

Bonaventura Cavalieri (1598–1647) used rectangular strips of equal width to calculate the area under a straight line (i.e., a triangle) and under a parabola.

Around 1655–1656, John Wallis (1616–1703) derived algebraic formulas that represent the areas under the curve of simple functions, such as $y = kt$ and $y = kt^2$, from 0 to x (Ginsburg et al. 1998). Considering the existing work on tangents (i.e., slopes or derivatives) at that time, and considering the DA pair concept here, we thus may conclude that Wallis had already explicitly demonstrated, before Newton, the relationship between slope and area using examples, i.e., FTC.

Isaac Newton (1642–1727) was admitted to Trinity College, Cambridge in 1661 and quickly made himself a master of Descartes' geometry. He learned much mathematics from his teacher and friend Isaac Barrow (1630–1677), who knew the method of tangents by both Descartes and Fermat and who also knew how to calculate areas under some simple functions. Barrow understood that differentiation and integration were inverse operations, i.e., FTC (Cajori 1985). Newton summarized the past work on tangents and area calculation, introduced many applications of the two operations, and made the tangent and area methods a systematic mathematical theory. Newton's method of tangents followed that of Fermat and had a small increment that eventually approaches zero. That is to say, he used a sequence of secant lines to approach a tangent line as is done in the modern calculus definition of derivative. Although the "method of limits" is frequently attributed to Newton (see his book (Newton 1729) entitled *The Mathematical Principles of Natural Philosophy* (p. 45)), he was not as adamant as Leibniz about letting an infinitesimal be zero at the end of a calculation. Newton was dissatisfied with the omitted small errors. He wrote that "in mathematics the minutest errors are not to be neglected" (see Cajori 1985, p. 198). As a matter of fact, Newton's vision was correct. When including the second-order $(\Delta x)^2$ in stochastic calculus, one can derive a new type of calculus, such as Itô calculus which was developed in the last century and applies to fast-oscillation functions that have no tangent line at any point defined by Descartes' approach.

Newton's "method of fluxions" was intended to solve two fundamental mechanics problems that are equivalent to the two geometrical problems of slope and height increment of a curve pointed out earlier in this appendix:

"(i). The length of the space described being continually (i.e., at all times) given; to find the velocity of the motion at any time proposed.
(ii). The velocity of motion being continuously given; to find the length of the space described at any time proposed" (see Cajori 1985, p. 193 and Newton 1736, p. 19).

The solution to these two problems also led to the FTC, as geometrically interpreted in previous sections. The above statement of the two problems is directly cited from Newton's book *The Method of Fluxions* (Newton 1736, p. 19), which was translated into English from Latin and published by John Colson. Newton's work transformed the calculus due to Descartes, Fermat, and others into a systematic area of mathematics, and a very powerful mathematical methodology for science and engineering.

Gottfried Wilhelm Leibniz (1646–1716) produced a profound work similar in many respects to Newton's that summarized the method of tangents and the method of area using a systematic approach. His approach and his beautiful descriptions have been passed on and are widely used today, including his notations for derivative and integration.

Our description of various calculus methods and approaches has demonstrated that if we avoid calculating the area underneath a curve and instead define an integral by the height increment, we can readily extend Descartes' method of tangents to establish the theory of differentiation and integration by considering the slope (i.e., grade), DA pair, and height increment. The no-limit approach to calculus outlined in this appendix is directly traceable to Descartes' original ideas, and is different from those of Fermat, Newton, and Leibniz. The FTC itself is attributable to Wallis's original ideas. Ginsburg et al. (1998) concluded that the question of whether Leibniz plagiarized Newton's work on calculus is not really a valid question since the essential calculus ideas had already been developed by others before the calculus accomplishments of either Newton or Leibniz. Both of them summarized the work of earlier mathematicians and developed differentiation and integration into a systematic branch of mathematics by using the methods of infinitesimals and limits. After their work, calculus became a very useful tool in engineering, natural sciences, and numerous other fields.

In addition to the mathematicians already mentioned, there are many others who contributed to the development of calculus, including Gregory of St. Vincent (1584–1667), Gilles Personne de Roberval (1602–1675), Blaise Pascal (1623–1662), Christiaan Huygens (1629–1695), and Leonhard Euler (1707–1783). Augustin-Louis Cauchy (1789–1857) has been credited with the rigorous development of calculus from the definition of limits. Karl Weierstrass (1815–1897) corrected some mistakes made by Cauchy and introduced the delta-epsilon language we use today in mathematical analysis.

D.1.5 Summary and Discussion of DD Calculus

We have used some simple ideas and basic mathematics to introduce DD calculus. Geometrically, derivatives were introduced directly from the slope of a tangent line. Algebraically, derivatives and antiderivatives were introduced simultaneously as a DA pair. Then the integral was introduced as the height increment of the antiderivative function. This increment was geometrically interpreted as the area of the region bounded by the integrand function, the horizontal axis, and the integration interval. A justification of this interpretation was given to demonstrate that this definition of area was reasonable and mathematically rigorous up to the proof of the intermediate value theorem (IVT) which, like the Euclidean axioms, may be regarded as intuitively true for most people other than professional mathematicians. At the end, we pointed out that Fermat's small increment was an efficient approach to calculate derivatives by hand and could help derive derivative formulas for complicated functions other than polynomials. The increment approach was later systematically shaped into the limit approach in modern calculus. The limit approach to calculus is an excellent computing method for finding derivatives by hand. In the pre-computer era, this limit approach was obviously critical in calculating derivatives of a variety of functions. In the current computer era, the limit approach to calculus is less essential and may be unnecessary for non-mathematics majors.

Although the ideas of direct calculus described in this appendix come from practical applications, we have ensured logical consistency and mathematical rigor. Advanced mathematical analysis regarding the structure of the real line and the sequence approach to a compact set are not topics for this introductory text. These analytical approaches mainly due to Cauchy and Weierstrass have certainly enriched calculus as begun by Archimedes, Descartes, Fermat, Wallis, Newton, Leibniz, and others. However, we have shown that it is possible to introduce the basic concepts and calculation methods of calculus directly, without using the concept of limits at all.

One can also regard calculus as an extension of the trigonometry of a regular right triangle to the trigonometry of a "curved-hypotenuse" right "triangle." For a regular right triangle, the slope (i.e., derivative = tangent of the angle) of the hypotenuse is a derivative, and the vertical increment (i.e., an integral) is equal to the opposite side, which is the integral of the derivative (see Fig. D.11). It is obvious that the vertical increment and the slope are related and have the following relationships:

$$\tan\theta = \frac{BC}{AC} \qquad \text{(derivative)} \tag{D.42}$$

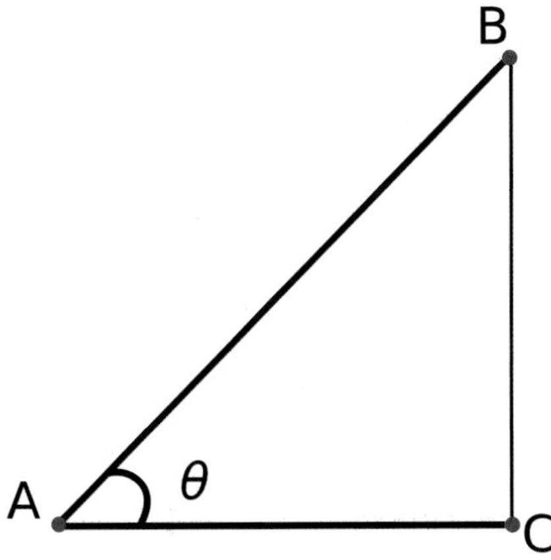

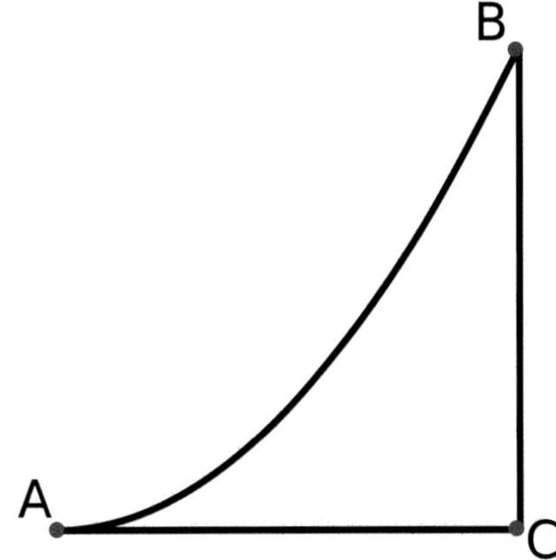

Fig. D.11 Calculus explained by triangles with straight and curved hypotenuses and

$$BC = \tan\theta \times AC \qquad \text{(integral)}. \tag{D.43}$$

These two formulas are the FTC for the regular right triangle. The extension is from this straight line hypotenuse to the curved hypotenuse, such as a parabola or an exponential function. For the curved hypotenuse, the slope varies at different points and the total height increment is an integral. That is,

$$m = f'(x) \qquad \text{(derivative)} \tag{D.44}$$

and

$$BC = \int_a^b f'(x)\,dx \qquad \text{(integral)}. \tag{D.45}$$

D.2 Calculus from a Statistics Perspective

This section provides another approach to establishing the essential basis of calculus theory from the concept of a mean, such as an average of some climate data. This approach is from a statistics perspective and can help calculus learners understand calculus ideas and analyze a function defined by data or by sampling values from a given function, rather than from an explicit mathematical formula. The basics of this approach are two averages: the arithmetic mean and the graphic mean. The arithmetic mean is used to define integral. Area is used to interpret the meaning of an integral. Antiderivative is introduced from integral, and derivative-antiderivative pair is introduced as a mathematical operation entity. The graphic mean can be thought of as the average speed in a time interval and is used to interpret the meaning of a derivative. We will define the concept of graphic mean more precisely in Section D.2.4.

A mathematical function $y = f(x)$ can be represented in four ways: (i) a description in words or text, (ii) a table of at least two columns of data, (iii) a graph in the xy-coordinates plane, and (iv) an explicit mathematical formula. Conventional calculus textbooks almost exclusively deal with Case (iv): functions defined by explicit formulas. The graphical representations (i.e., Case (iii)) are conventionally regarded as only a supporting tool for the function formula. Case (ii), data-represented functions, is usually considered as belonging to the category of statistics, which analyzes data and makes inferences about the implications of the data. Hence, most calculus textbooks do not deal with this case even today, in the modern era of computers. However, the rapid development of data-based information in our daily lives motivates us to broaden the boundaries of traditional statistical methods, including interconnections between statistics and calculus. Today, advances in computer speed and the widespread availability of personal computers and smartphones make it possible to take advantage of new and innovative approaches to calculus for functions represented in any of the four ways. This section will develop the ideas of calculus from a statistics perspective. We will build the calculus concept for the function of Case (ii) and make the concept applicable to the functions of all the four cases.

We use two types of averages: arithmetic mean and graphic mean. The arithmetic mean and the law of large numbers (LLN) are used to define an integral, and the graphic mean is used to define a derivative.

We do not intend to use this approach as a replacement for the conventional way of teaching calculus. Instead, this approach may be used as supplemental material in teaching calculus using the conventional approaches and methods. It thus provides a new perspective for explaining the essential calculus concepts to students. Furthermore, we do not claim that this approach is superior to other approaches to calculus. Every approach to calculus has its advantages and disadvantages. Ours is not an exception. Our calculus from a statistics perspective simply provides a set of supplemental materials to explain the concepts of calculus from a non-traditional perspective.

To simplify the description of our approach to the integral and derivative concept, we limit our functions $y = f(x)$ to those with positive x and y values and having a defined slope on each point of curves of the functions, when we discuss the Case (iv) function $y = f(x)$. These limits do not affect the rigor of the mathematics developed here and can be easily removed when a more sophisticated development of the mathematics of calculus is introduced.

D.2.1 Arithmetic Mean, Sampling, and Average of a Function

For any given location on Earth, the variation of temperature with respect to time forms a functional relationship. Consider the arithmetic mean for the annual surface air temperature (SAT) (in units [°C]) as given by observational meteorological data for the years 1951–2010 at Fredericksburg (38.32°N, 77.45°W), which is 80 kilometers southwest of Washington DC, the capital of the United States (see Table D.1). The data pairs $(x_i, f_i)(i = 1,2,\ldots,n)$ represent the tabular form of a function (the Case (ii) function) with x_i for time and f_i for temperature, and n is 60 for this case. The arithmetic mean is defined as

Table D.1 Annual mean SAT at Fredericksburg from 1951–2010.

Annual mean surface air temperature at Fredericksburg (38.32°N, 77.45°W) Virginia, United States. The temperature units are [°C]. The data are from the US Historical Climatology Network. https://www.ncdc.noaa.gov/ushcn

1951	13.1	1961	12.9	1971	13.2	1981	12.5	1991	14.3	2001	13.7
1952	13.4	1962	12.3	1972	12.8	1982	12.7	1992	12.5	2002	14.3
1953	14.4	1963	12.4	1973	13.7	1983	12.8	1993	13.2	2003	13.0
1954	13.7	1964	13.2	1974	13.3	1984	13.0	1994	13.2	2004	13.6
1955	12.6	1965	12.7	1975	13.0	1985	13.6	1995	13.2	2005	13.7
1956	13.1	1966	12.4	1976	12.6	1986	13.4	1996	12.6	2006	14.2
1957	13.3	1967	12.3	1977	13.1	1987	13.4	1997	13.0	2007	14.1
1958	12.1	1968	12.8	1978	12.3	1988	12.6	1998	14.7	2008	13.7
1959	13.6	1969	12.5	1979	12.5	1989	12.9	1999	13.9	2009	13.2
1960	12.6	1970	13.1	1980	12.7	1990	14.3	2000	13.0	2010	14.2

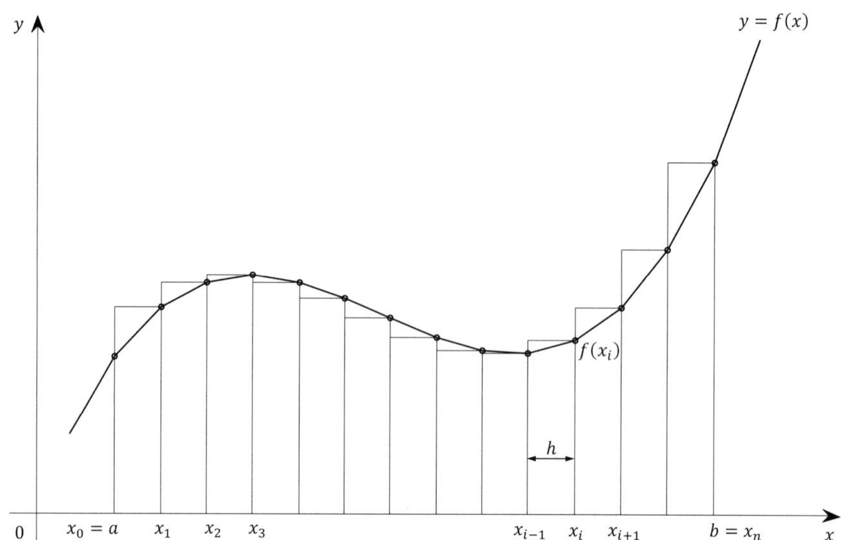

Fig. D.12 Uniform sampling for a function $y = f(x)$ in $[a, b]$. The area under the curve is approximated by the sum of the narrow rectangular strips

$$\bar{f} = \frac{\sum_{i=1}^{n} f_i}{n}. \tag{D.46}$$

For the SAT data in Table D.1, the mean is 13.2[°C].

Next we explore the data that are samples from the Case (iv) functional values. The samples of independent and dependent variables are $(x_i, y_i), i = 1, 2, \ldots, n$, where the Case (iv) function is denoted by $y = f(x)$. The sampling of x is usually done using one or more of three conventional procedures.

1. Uniform sampling: The distance between each pair of neighborhood samples is the same, i.e., $h = x_i - x_{i-1}$ is the same for each i. The sample points can be determined by $x_i = x_{i-1} + h, i = 1, 2, \ldots, n$. An example is the SAT sampling in Table D.1 where h is one year. Another example is the uniform sampling of a function shown in Fig. D.12.
2. Random sampling: Random sampling is, as the name suggests, a probabilistic process and can be determined in a predefined way using statistics software. Many software tools for generating random numbers are available in the public domain for free, such as WolframAlpha: www.wolframalpha.com. The command `RandomReal[{0,2},10]` yields 10 random real numbers in the interval $[0, 2]$. A trial of this command is $\{1.25685,\ 1.02584,\ 1.26752,\ 0.813785,\ 0.224482,\ 0.905491,\ 1.73265,\ 1.44104,\ 1.95108,\ 0.926807\}$
   Of course, each trial yields a different random result.
3. Convenience sampling: This sampling may not be either as evenly spaced as the uniform sampling or completely random, but the samples are taken from convenient and practical locations. For the Himalaya mountains, the weather station locations are often at lower elevations which can be easily accessed by people, because it is impractical to place a station at the highest peaks of the Himalayas. Thus, convenience sampling is often used

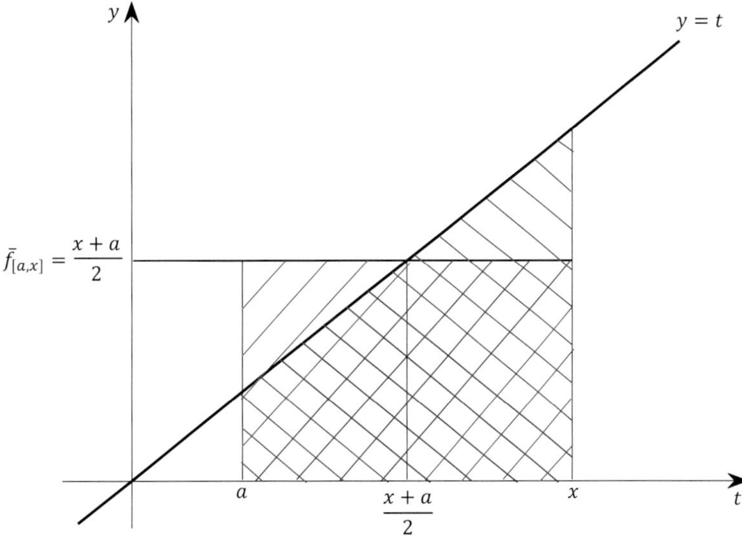

Fig. D.13 Arithmetic mean of a linear function.

in practice, but suffers possible drawbacks of bias, such as heat island bias effects for temperature due to urban expansion. When given a Case (iv) function, this sampling is unnecessary, but for a function of the other three cases in practical applications, this sampling may be essential to describe a function for applications.

Let us explore two examples of calculating the arithmetic means of sample data from a function: $y = x$ and $y = x^2$.

From our intuition or assisted by a figure (see Fig. D.13 with $a = 0$), we can infer that the mean of the function $y = x$ in $[0,1]$ is 1/2. This result can be simulated by different sampling methods.

1. Uniform sampling: We use a sample of size 100 with the first sample at $x = 0.01$ and last sample at $x = 1$. These 100 points divide the interval $[0,1]$ into 100 equal sub-intervals of length 1/100:
$$x_1 = 0.01, x_2 = 0.02, \ldots, x_i = \frac{i}{100}, \ldots, x_{100} = 1.$$
$$y_i = x_i, \quad i = 1, 2, \ldots, 100.$$
The arithmetic mean of data $y_i, i = 0, 1, \ldots, 100$ is equal to
$$\hat{\bar{y}} = \frac{\sum_{i=1}^{100} y_i}{100} = \frac{\sum_{i=1}^{100} (i/100)}{100} = \frac{\sum_{i=1}^{100} i}{100 \times 100} = 0.505, \tag{D.47}$$
which is very close to the exact average value 0.5.
2. Random sampling: Random sampling usually does not sample the end points. Our 100 samples are at the internal points of $[0,1]$, i.e., over $(0,1)$. WolframAlpha command `RandomReal[{0,1},100]` generates 100 random numbers whose mean is 0.5312, which is close to the exact result 0.5. Again, the result 0.5312 is different each time the

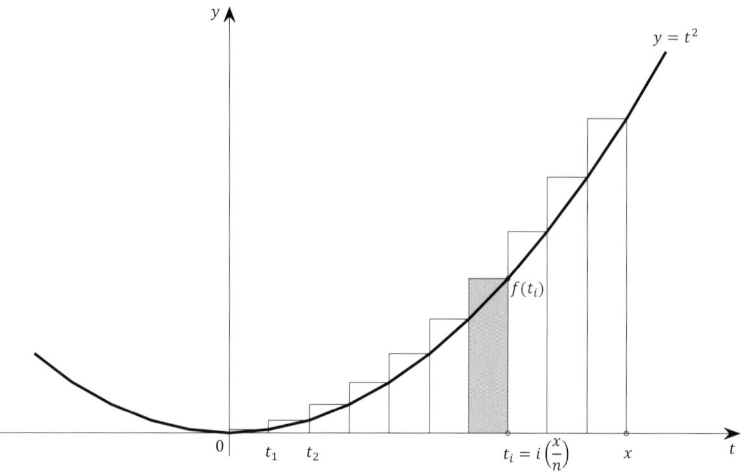

Fig. D.14 Uniform sampling of a parabolic function.

command is implemented due to the random nature of the RandomReal generator. It is intuitive that larger samples should be likely to lead to more accurate approximation. When using 1,000 samples, a result is 0.4981, which is very close to 0.5.

Next we consider $y = x^2$ in $[0,1]$ (see Fig. D.14). The uniform sampling of using 100 points as above yields the following:

$$\hat{y} = \frac{\sum_{i=1}^{100} y_i^2}{100} = \frac{\sum_{i=1}^{100} (i/100)^2}{100} = \frac{\sum_{i=0}^{100} i^2}{100^2 \times 100}$$
$$= \frac{100 \times (100+1)(2 \times 100 + 1)/6}{100^2 \times 100}$$
$$= \frac{101 \times 201}{6 \times 100^2} = 0.33835$$

The random sampling with WolframAlpha uses a command `RandomReal[{0,1},10]^2` to generate the data, and the text result can be copied and used to calculate the mean by using WolframAlpha. A result is 0.36726. The accurate mean value should be 0.33333, which can be computed from n samples when n is very large.

WolframAlpha has limited statistical computing power. The open source statistics software R and the commonly used MS Excel have more statistical computing power and can be used for practical applications and research. An Excel calculation of the sample mean for $y = x^2$ in $[0,1]$ for different numbers of samples is given in Table D.2, which indicates that, in general, the accuracy of the mean improves as the sample size n increases. As a matter of fact, the improvement of the accuracy can be quantified by the error $\sigma/\sqrt{n}$ where σ is the standard deviation of the population being sampled. This is a result of the Central Limit Theorem (CLT) in statistics (see a basic statistics text, e.g., Johnson and Bhattacharyya (1996) or Wackerly et al. (2002)). The CLT states that the mean of samples with sufficiently large sample size is normally distributed. The CLT further asserts that (i) the expected value of the sample mean is the same as the population mean, and (ii) the

Table D.2 Convergence of the sample mean as the sample size increases

Sample Size n	10	100	1,000	10,000	100,000	1,000,000
Uniform Sampling	0.3850	0.3384	0.3338	0.3334	0.3333	0.3333
Random Sampling						
Trial 1	0.4350	0.3505	0.3463	0.3344	0.3315	0.3332
Trial 2	0.3560	0.3058	0.3284	0.3325	0.3331	0.3332
Trial 3	0.4640	0.3518	0.3355	0.3317	0.3357	0.3332
Average	0.4183	0.3360	0.3367	0.3329	0.3335	0.3332

variance of the sample mean is $(1/n)$th of the population variance. These two assertions are widely assumed in practical statistical applications and computer calculations, such as Monte Carlo simulations. This normal distribution result can be intuitive to students and may be accepted as an axiom. However, there is no need to consider the CLT in this first introduction of calculus from a statistics perspective. This is somewhat similar to the statement that the first introduction of conventional calculus does not need to include a proof of the existence of a limit from a rigorous argument.

Let $\hat{\bar{f}}[n]$ denote the mean from n samples. Then, it is "almost always true" that $\hat{\bar{f}}[n]$ approaches the true mean of the function as n increases. Here, "almost always true" means that the statement being true has a probability equal to one. The probability for this statement to be false is zero. This intuitive conclusion is the strong law of large numbers (LLN) (see Wackerly et al. 2002), which asserts that the event of the following limit being true has a probability equal to one:

$$\lim_{n\to\infty} \hat{\bar{f}}[n] = \bar{f}. \tag{D.48}$$

Here $\bar{f}$ is the limit of the sequence $\{\hat{\bar{f}}[1], \hat{\bar{f}}[2], \hat{\bar{f}}[3], \ldots\}$, or simply $\{\hat{\bar{f}}[n]\}$.

For example, when x^2 is uniformly sampled by n points over $[0,1]$, the mean is

$$\hat{\bar{f}}[n] = \frac{\sum_{i=1}^{n}(i/n)^2}{n} = \frac{\sum_{i=0}^{n} i^2}{n \times n^2} = \frac{n(n+1)(2n+1)/6}{n^3} = \frac{n+1}{n} \times \frac{2n+1}{6n}.$$

As $n \to \infty$, the first factor goes to 1 and the second to $2/6 = 1/3$. Thus, the above mean approaches $1/3 \approx 0.33333333$ as $n \to \infty$.

The above leads to the definition of average of a function $y = f(x)$ in an interval $[a,b]$

$$\bar{f} = \lim_{n\to\infty} \frac{\sum_{i=1}^{n} f(x_i)}{n}, \tag{D.49}$$

where $\{x_i\}_{i=1}^{n}$ are sampled from $[a,b]$ by uniform sampling, random sampling, and possibly convenience sampling. We also call $\bar{f}$ the arithmetic mean, or simply the mean, of the function $y = f(x)$.

D.2.2 Definition of Integral, Antiderivative, and DA Pair

Definition D.4 (Definition of integral) If $\bar{f}$ is the average of the function $f(x)$ over the interval $[a,b]$, then $(b-a)\times\bar{f}$ is defined as the integral of $f(x)$ over $[a,b]$, denoted by

$$I[f,a,b]=(b-a)\bar{f}. \tag{D.50}$$

This serves as the definition of the definite integral in conventional calculus. Since we will not introduce the concept of indefinite integral, we thus treat "integral" here as the "definite integral." This definition is equivalent to the integral definition described in DD calculus.

From the examples of mean in the preceding Section D.2.1, we can calculate the following integrals:

$$I[x,0,1]=(1-0)\times(1/2)=1/2,$$

and

$$I[x^2,0,1]=(1-0)\times(1/3)=1/3.$$

Then, what is the graphic meaning of integral from the above definition of integral? We use uniform sampling to make an interpretation. Since $\hat{\bar{f}}[n]\approx\bar{f}$, we have

$$I[f,a,b]\approx(b-a)\hat{\bar{f}}[n]=(b-a)\frac{\sum_{i=1}^{n}f_i}{n}=\sum_{i=1}^{n}\frac{b-a}{n}f_i.$$

Here $f_i=f(x_i)$ is the sample value of the function at the sampling location x_i (see Fig. D.12). Since this is uniform sampling, $x_{i+1}-x_i=h=(b-a)/n$. Thus, each term in the above sum

$$\frac{b-a}{n}f_i$$

is the area of a rectangular strip with base $h=(b-a)/n$ and height f_i, as shown in Fig. D.12. Thus, $(b-a)\hat{\bar{f}}[n]$ is the area under the up-and-down stairs, formed by n rectangular strips whose heights are determined by the function values from a uniform sampling $f(x_i)=f(ih), i=1,2,\ldots,n$. This sum approaches the true area, denoted by S, of the region under the curve $y=f(x)$ in $[a,b]$. The ever-improving approximation as $n\to\infty$ is a process of taking a limit and is denoted by

$$\lim_{n\to\infty}(b-a)\hat{\bar{f}}[n]=S=I[f,a,b]. \tag{D.51}$$

Dividing both sides by $b-a$ leads to the limit of $\hat{\bar{f}}[n]$ as $n\to\infty$

$$\lim_{n\to\infty}\hat{\bar{f}}[n]=\bar{f}. \tag{D.52}$$

As mentioned earlier, the existence of this limit is guaranteed by LLN, and the probability for this limit to fail is zero.

Therefore, the geometric meaning of the integral $I[f,a,b]$ is the area of the region bounded by $y=f(x)$, $y=0$, $x=a$, and $x=b$.

Following the idea of derivative–antiderivative (DA) pair introduced earlier, we define the integral $I[f,a,x]$ as the *antiderivative* of $f(x)$ and is denoted by $F(x)$, where a is an arbitrary constant and x is regarded as a variable. We also call $f(x)$ the derivative of $F(x)$, and denote it as $F'(x) = f(x)$. Because x is regarded as a variable, the antiderivative $F(x)$ is a function. The function pair $(f(x), F(x))$ is called a *DA pair*.

From the geometric meaning of integral, $F(x) = I[f,0,x]$ is the area of the region bounded by $y = f(t)$, $y = 0$, $t = 0$, and $t = x$ over the $t-y$ plane (see Fig. D.15).

Example D.9 Evaluate the antiderivative of $f(x) = 1$: The area of the rectangle bounded by $y = 1$, $y = 0$, $t = 0$, and $t = x$ over the t–y plane is x. Thus, $F(x) = x$, and $(1,x)$ is a DA pair.

Example D.10 Evaluate the antiderivative of $f(x) = x$: The area of the triangle bounded by $y = t$, $y = 0$, $t = 0$, and $t = x$ over the t–y plane is $x^2/2$ (see Fig. D.13). Thus, $F(x) = x^2/2$, and $(x, x^2/2)$ is a DA pair, i.e., $(x^2/2)' = x$ and $I[t,0,x] = x^2/2$.

Example D.11 Evaluate the antiderivative of $f(x) = x^2$: This is a problem of calculating the area of a curved triangle (see Fig. D.14). We can use uniform sampling as we have done above for $y = t^2$ in the interval $[0,1]$, but now for the interval $[0,x]$.

$$\hat{\bar{f}}[n] = \frac{\sum_{i=1}^{n}(xi/n)^2}{n} = x^2\frac{\sum_{i=0}^{n} i^2}{n \times n^2} = x^2\frac{n(n+1)(2n+1)/6}{n^3} = x^2\frac{n+1}{n} \times \frac{2n+1}{6n}.$$

This mean approaches $x^2/3$ as $n \to \infty$. Hence,

$$F(x) = (x-0) \times x^2/3 = x^3/3.$$

We thus have a DA pair $(x^2, x^3/3)$, i.e., $(x^3/3)' = x^2$ and $I[t^2,0,x] = x^3/3$.

Again, one can also use www.wolframalpha.com and other software packages to find derivatives. For example, in the pop-up box of WolframAlpha, type `derivative x^5/5` and press enter. The software returns x^4. This verifies that $(x^5/5)' = x^4$.

D.2.3 Calculation of an Integral $\int_a^b f(x)\,dx$

We can use an antiderivative to calculate an integral. From the area meaning of an integral, we have the following formula:

$$\int_c^d f(x)\,dx = F(d) - F(c). \tag{D.53}$$

This is also denoted by $\int_c^d f(x)\,dx = F(x)|_c^d$. Namely, the area of the region bounded by $y = f(x)$, $y = 0$, $x = c$, and $x = d$ is equal to the difference of the area bounded by $y = f(x)$, $y = 0$, $x = a$, and $x = d$ minus the area bounded by $y = f(x)$, $y = 0$, $x = a$, and $x = c$. This is shown in Fig. D.15: $\int_c^d f(x)\,dx$ is the area CDD′C′, equal to area ODD′O′minus area OCC′O′.

In traditional calculus, this way of calculating an integral is called Part II of the Fundamental Theorem of Calculus (FTC), and the DA pair is called Part I of FTC.

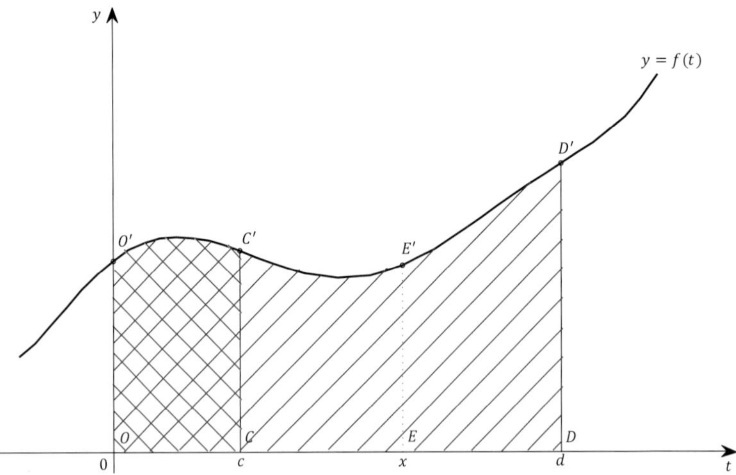

Fig. D.15 Areas and their differences under a curve: an illustration for FTC.

Example D.12 Evaluate $\int_0^1 x^2\, dx$.

Because the antiderivative of x^2 is $F(x) = x^3/3$, we have the following

$$\int_0^1 x^2\, dx = x^3/3|_0^1 = 1^3/3 - 0^3/3 = 1/3. \tag{D.54}$$

Example D.13 Evaluate $\int_0^\pi \sin^2(x)\ dx$: In www.wolframalpha.com, the command `integrate sin^2 x` yields the antiderivative

$$F(x) = (1/2)(x - \sin x \cos x). \tag{D.55}$$

Thus,

$$\begin{aligned}\int_0^\pi \sin^2 x\, dx &= (1/2)(x - \sin x \cos x)|_0^\pi \\ &= (1/2)(\pi - \sin\pi\cos\pi) - (1/2)(0 - \sin 0 \cos 0) \\ &= \pi/2.\end{aligned} \tag{D.56}$$

One can also use www.wolframalpha.com to calculate the integral directly by using the command `integrate [sin^2 x, 0, pi]` or `integral [sin^2 x, 0, pi]`. This command directly returns the value $\pi/2$.

D.2.4 Using Average Speed and Graphic Mean to Interpret the Meaning of a Derivative

In the preceding subsections, we have introduced the concepts and calculation techniques of integral and derivative from a statistics perspective. Further, the meaning of the integral of a function in $[a, b]$ is well interpreted as the value increment of its antiderivative function from $x = a$ to $x = b$. This subsection addresses the remaining task of interpreting the meanings

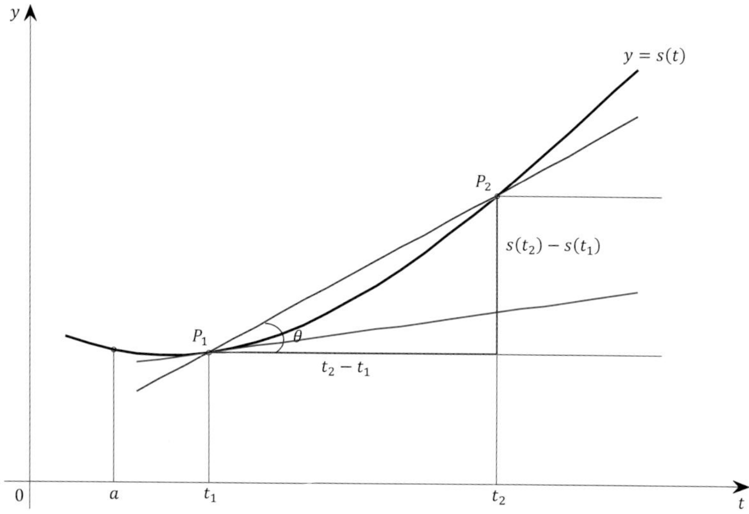

Fig. D.16 Illustration of a graphic mean.

of derivative of a function and makes connections with the conventional way of defining a derivative.

If one drives along a highway from Mile 6 at 2:00 p.m. and arrives at Mile 98 at 4:00 p.m., one's average speed is $(98-6)/(4-2)=46$ mph. If we use $s(t)$ to represent the location of the car at time t, then the distance traveled by the car from time t_1 to time t_2 is $s(t_2)-s(t_1)$. The average speed is

$$\bar{v}=\frac{s(t_2)-s(t_1)}{t_2-t_1}. \tag{D.57}$$

This kind of average is called a *graphic mean*, in contrast to the arithmetic mean discussed earlier. Figure D.16 gives a schematic illustration of the function $y=s(t)$. The graphic mean is thus the slope of the secant line, i.e., the purple line that connects P_1 and P_2 in Fig. D.16. Namely, $\bar{v}=\tan\theta$.

The instantaneous speed at a time, say t_1, is approximately the average speed in an ever smaller interval $[t_1,t_2]$, which means as t_2 goes to t_1. The limit expression is

$$\lim_{t_2\to t_1}\frac{s(t_2)-s(t_1)}{t_2-t_1}=v(t_1). \tag{D.58}$$

This is regarded as the *slope* of $y=s(t)$ at t_1, i.e., the slope of the tangent line (i.e., the red line in Fig. D.16) of $y=s(t)$ at t_1. We would like to show that $s(t)$ is an antiderivative of $v(t)$, i.e., $s(x)=I[v(t),0,x]$. We divide $[0,x]$ uniformly into n sub-intervals of width $h=x/n$. Within interval $[x_i,x_{i+1})$, the distance traveled is approximately $v(x_i)h$. The total distance traveled in $[0,x]$ is $\sum_{i=1}^{n}v(x_i)h$, which goes to $s(x)-s(0)=I[v,0,x]$ as $n\to\infty$. Thus, $(v(t),s(t))$ is a DA pair, i.e., the derivative of distance with respect to time is speed, and the integral of speed is the total distance traveled. Equation (D.58) provides another way of defining derivative in addition to the derivative introduced via the DA pair. This definition of derivative by limit in (D.58) is the most popular way of introducing derivatives in conventional classrooms today. It is also the main idea of Isaac Newton's approach to fluxions (see Newton (1736) in his book entitled *Method of Fluxions*).

Therefore, a derivative of $F(x)$ is a limit of its graphic mean. One can use the formulas of DA pairs given in the last section to find the derivative of a given function. For example, to find the derivative of x^2, we use the DA pair $(x^n, x^{n+1}/n+1)$. With $n = 1$, the pair becomes $(x, x^2/2)$. So $(x^2/2)' = x$, hence $(x^2)' = 2x$.

D.2.5 Summary of Calculus from a Statistics Perspective

We have used the arithmetic mean to define the integral. This definition leads to the DA pair concept, the area meaning of the integral, and FTC. We have used the graphic mean to interpret a derivative as the slope of a curve or the speed of a moving object. The graphic mean also provides another way of calculating the derivative as a limit, an application of Fermat's method of tangents. Computer calculation for DA pairs is used here in place of the traditional derivation of derivative formulas using the concepts of limit and graphic mean. Before today's popularity of computers and the Internet, the method of introducing the derivative by a limit was very useful in the calculation of calculus problems, since the method can easily be used to derive various kinds of derivative formulas for non-polynomial functions, compared to the DA pair approach $I[f, 0, x]$. However, with today's easy access to notebook computers, smartphones, and publicly available software, finding derivatives and integrals for a given function can readily be done via the Internet.

Thus, present-day calculus teaching may well shift its emphasis from limit-based hand calculations to web-based calculations, concepts, interpretations, and applications. For many people, the concept of limit is a difficult, subtle, and puzzling notion. As an intermediate procedure, it may be an obstacle to learning calculus. Although in the statistics approach to calculus described here, we introduced the limit notation to state the existence of the arithmetic mean following LLN, we hardly used the concept of limit in the formulation of derivations and actual calculations. Statistical calculus has a rigorous background that needs LLN, which requires a proof, but its conclusion has a certain degree of plausibility by intuition and can be treated as an axiom like those in Euclidean geometry. Thus, the calculus based on LLN implies that, with today's useful computer technology, the calculus basics and the methods of calculus can be developed directly with an intuitively plausible minimum of theory and without the need of unnecessary, intermediate, and complex procedures. In this way, the basic concepts and methods of calculus for a single variable can readily be taught in high school classes or in a university-level workshop of only a few hours.

D.3 Differentiation Methods and Higher Derivatives

It is tedious to calculate the derivatives by definition. Before computers and smartphones became popular, people used a set of differentiation rules to hand-calculate derivatives. These rules can all be rigorously proved following the definition of the derivative.

D.3.1 Rules of Differentiation

Several commonly used rules are listed below:

$$\frac{d}{dx}(f(x)+g(x)) = f'(x)+g'(x) \quad \text{(Addition rule)}, \tag{D.59}$$

$$\frac{d}{dx}(f(x)-g(x)) = f'(x)-g'(x) \quad \text{(Subtraction rule)}, \tag{D.60}$$

$$\frac{d}{dx}f(x)g(x) = f'(x)g(x)+f(x)g'(x) \quad \text{(Product rule)}, \tag{D.61}$$

$$\frac{d}{dx}cg(x) = cg'(x) \quad \text{(Constant product rule)}, \tag{D.62}$$

$$\frac{d}{dx}f(g(x)) = \frac{df}{du}\frac{du}{dx}, \quad u=g(x) \quad \text{(Chain rule)}, \tag{D.63}$$

$$\frac{d}{dx}(f(x)/g(x)) = [f'(x)g(x)-f(x)g'(x)]/g^2(x) \quad \text{(Division rule)} \tag{D.64}$$

The constant product rule above is a special case of the product rule when $f(x)=c$, and $f'(x)=0$.

The chain rule is used for a composite function, which is a function of a function. For example, $y=\sqrt{1-x^2}$ is a composite function: $x \to u(x) = 1-x^2 \to y = \sqrt{u}$. The derivative chain goes from outside to inside:

$$\frac{dy}{dx} = \frac{dy}{du}\frac{du}{dx} = \frac{d}{du}u^{1/2}\frac{d}{dx}(1-x^2) = 1/2u^{-1/2}(-2x) = \frac{-x}{\sqrt{1-x^2}}. \tag{D.65}$$

The division rule above is a combination of the product rule and chain rule:

$$\begin{aligned}\frac{d}{dx}(f(x)/g(x)) &= \frac{d}{dx}\left(f(x)\times(g(x))^{-1}\right) = \frac{df(x)}{dx}\times(g(x))^{-1}+f(x)\frac{d}{dx}(g(x))^{-1}\\ &= f'(x)/g(x)+(-1)(g(x))^{-1-1}g'(x)\\ &= [f'(x)g(x)-f(x)g'(x)]/g^2(x).\end{aligned} \tag{D.66}$$

Example D.14 Derivative of a trigonometric function for a sinusoidal periodic motion:

$$\frac{d}{dt}2\sin(3t-8) = 2\frac{d}{dt}\sin(3t-8) = 2\times\cos(3t-8)\frac{d}{dt}(3t-8) = 6\cos(3t-8). \tag{D.67}$$

Example D.15 Derivative of an exponential decay for heat energy dissipation:

$$\begin{aligned}&\frac{d}{dt}5\exp(-t/4-3) = 5\frac{d}{dt}\exp(-t/4-3)\\ &= 2\times\exp(-t/4-3)\frac{d}{dt}(-t/4-3) = -(1/2)\exp(-t/4-3).\end{aligned} \tag{D.68}$$

Example D.16 Derivative of an oscillational exponential decay for modeling energy dissipation of a mechanical vibration.

$$\begin{aligned}
&\frac{d}{dt}3\exp(-0.3t)\cos(2t-1)\\
&\quad=3\left[\cos(2t-1)\left(\frac{d}{dt}\exp(-0.3t)\right)+\exp(-0.3t)\frac{d}{dt}\cos(2t-1)\right]\\
&\quad=3\left[\cos(2t-1)(-0.3\exp(-0.3t))+\exp(-0.3t)(-2\sin(2t-1))\right]\\
&\quad=-3\exp(-0.3t)(0.3\cos(2t-1)+2\sin(2t-1)).
\end{aligned} \tag{D.69}$$

D.3.2 Higher Derivatives

A derivative is a function. Differentiating a derivative function yields the derivative's derivative, called the second derivative, denoted by

$$\frac{d}{dx}f'(x)=f''(x)=\frac{d^2f}{dx^2}=\frac{d^2}{dx^2}f(x). \tag{D.70}$$

Example D.17 The second derivative of a free-fall distance is the gravitational acceleration. The derivative expression is explained below. The free-fall distance is $y=(1/2)gt^2$, where g is the gravitational acceleration constant, and t is the falling time. The first derivative is

$$\frac{dy}{dt}=\frac{d}{dt}\left((1/2)gt^2\right)=gt. \tag{D.71}$$

The second derivative is

$$\frac{d^2y}{dt^2}=\frac{d}{dt}\frac{dy}{dt}=\frac{d}{dt}gt=g. \tag{D.72}$$

Similarly, one can define the third and fourth derivatives, and so on.

The dimension of the second-order derivative is

$$\left[\frac{d^2y}{dx^2}\right]=\frac{[y]}{[x]^2}. \tag{D.73}$$

If y is length and x is time, then $\left[d^2y/dx^2\right]=L/T^2$ is the dimension of acceleration. The above free-fall problem has already verified this result. If both x and y are length, then $\left[d^2y/dx^2\right]=1/L$ is a measure of the curvature of the curve $y=f(x)$ at a specific point. Curvature measures the rate or speed of a smooth turn. A circle of radius L has its curvature equal to $1/L$ for every point on the circle. A point moving along a small circle turns very quickly, and the circle has a large curvature. A large circle has slow turns and hence a small curvature.

The positive second-order derivative of a function at a point means that the function's curve is concave-up (CUP). The curve around this point is like a cup standing upright and being able to hold water. The negative second-order derivative of a function at a point means that the function's curve is concave-down (COW). The curve around this point looks like an upside-down cup and cannot hold water.

Normally, a curve is concave-up at its minimum point, which is at the bottom of an upright cup; while a curve is concave-down at its maximum point, which is at the top of an upside-down cup.

Similar to the above analysis, one may conclude that the nth-order derivative's dimension is $[y]/[x]^n$.

D.4 Calculus for Functions of Two and More Variables

Although they were introduced for functions of a single variable, the calculus concepts and rules are similar for multivariate functions, which have two or more variables. For example, the normal or average or climatological global atmospheric surface air temperature is a function of latitude and longitude: $T = f(\phi, \theta)$ where ϕ is latitude and θ is longitude. This is a function of two variables. The visualization of this function is often accomplished by color maps with today's computer technology, as shown in Fig. D.17, which is the surface air temperature climatology equal to the 1961–1990 mean.

D.4.1 Introduction of Functions of Multiple Variables

For a given value of latitude and longitude, one can identify the climatological average temperature at that place. The equator has a higher temperature than the polar regions. Thus, there is a poleward gradient (or slope), the meridional gradient, denoted by $\partial T/\partial \phi$. This is a partial derivative notation for functions of multiple variables. A partial derivative

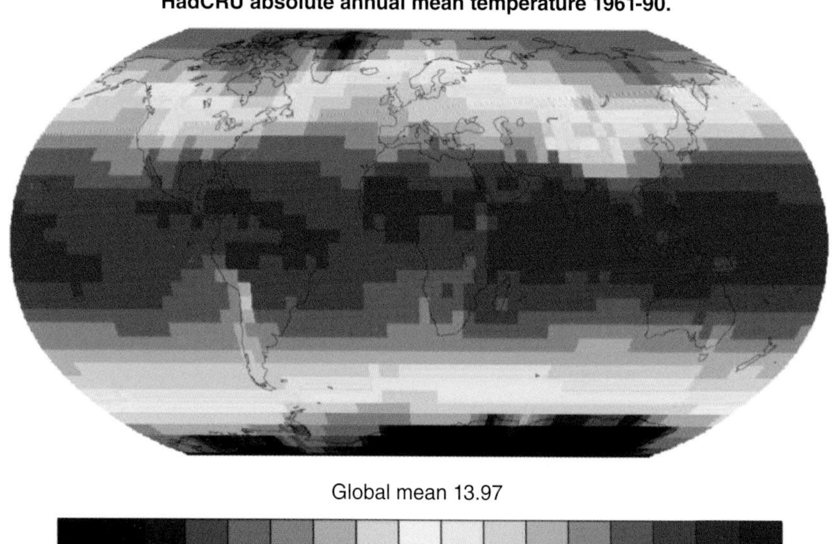

Fig. D.17 Surface air temperature climatology based on the 1961–1990 mean from NCEP/NCAR Reanalysis data. *Credit: University Corporation for Atmospheric Research (UCAR): Climate Data Guide*

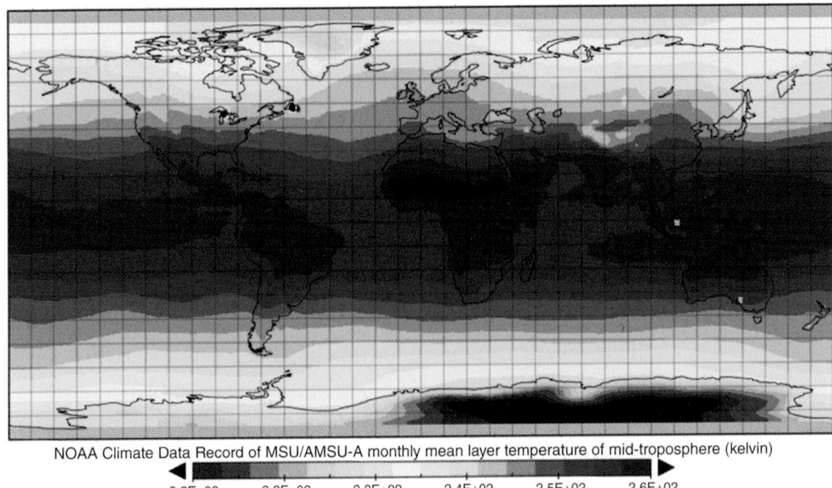

Fig. D.18 Climatology of the mid-troposphere (about 5 km altitude) temperature: the 1979–2008 mean. *Credit: NOAA NCEI, doi:10.7289/V5QF8QTK.*

is thus a straightforward generalization to multiple variables of the concept of a normal derivative for a function of one variable: dy/dx.

Temperature also has a gradient in the zonal direction, i.e., the temperature gradient in the east–west direction as longitude changes while latitude does not, denoted by $\partial T/\partial \theta$.

Long ago, the visualization of a function of latitude and longitude was mostly done using contour lines on the latitude–longitude map. The temperature would have the same value everywhere on a given contour line.

If we consider the variation of temperature with respect to elevation, then altitude z should be included: $T = g(\phi, \theta, z)$. This is a function of three variables. The visualization can be done by animation of color maps $T = g(\phi, \theta, z_i)$ with monotonically ordered altitudes z_i, or a series of color maps listed according to the order of z_i. Figure D.18 shows the mid-troposphere (about 5 km altitude) temperature. At this altitude, even the tropical temperature is below the freezing point of water.

When the variations are not very sudden, $T = g(\phi, \theta, z)$ can be visualized by a 3D color map. This technique is often applicable to ocean water temperature and salinity.

If the temperature history is considered, then time is a variable: $T = h(\phi, \theta, z, t)$. This is a function of four variables. A visualization of this 4D function can be provided by an animation in time for any given z. Animation of a 3D color map is possible as well, but this can be confusing.

D.4.2 Calculate Partial Derivatives

When walking on a mountain, we may face the choice between different paths, of which one is a shortcut but is steep (i.e., having a large slope or large derivative), and others are longer roads but are not as steep (i.e., smaller slopes or derivatives). Thus, the slope or derivative on a surface depends on both the location and direction, and it is called a

Fig. D.19 No. 6 Tianmen Mountain Zigzag Highway in the Zhangjiajie National Forest Park, Hunan Province, China: The total length is 11 km, and the elevation rises from 200 m to 1300 m.

directional derivative, or a partial derivative when along the direction of an axis, such as an x-partial derivative or y-partial derivative.

Figure D.19 shows Tianmen Mountain, China and its Zigzag Highway with an average slope of 10%, rising 1,100 m elevation over a distance of approximately 11 km. The steep winding mountainous highway is among the most dangerous roads in the world. The figure also shows gondola cables and cars. The cable length is 7,455 m, and its elevation rise is 1,277 m. The average slope is approximately 17%. In general, the gondola cable has steeper slopes than those of the winding highway. The cable and the highway have different orientations. If one chooses to climb the mountain on foot along a very steep mountain path, one may experience slopes larger than 17%. If one chooses to climb a cliff shown in the figure, the slope is close to infinity.

The calculation of partial derivatives follows the same rules as those for finding an ordinary derivative, except that we treat all the other independent variables as constants when taking the derivative with respect to one independent variable. For example, if $u = (1 - y^2)\sin x$, then

$$\frac{\partial u}{\partial x} = (1 - y^2)\frac{d\sin x}{dx} = (1 - y^2)\cos x \text{ (treat } y \text{ as constant)}, \tag{D.74}$$

$$\frac{\partial u}{\partial y} = \frac{d(1 - y^2)}{dy}\sin x = -2y\sin x \text{ (treat } x \text{ as constant)}. \tag{D.75}$$

One can calculate derivatives not only in the directions of x and y, but also along any given direction $\vec{n}$ which has an angle θ measured counterclockwise from the positive x-axis direction. This directional derivative along the direction $\vec{n}$ is

$$\frac{\partial u}{\partial \vec{n}} = \frac{\partial u}{\partial x}\cos\theta + \frac{\partial u}{\partial y}\sin\theta. \tag{D.76}$$

When $\theta = 0$, this reduces to $\frac{\partial u}{\partial x}$, and when $\theta = \pi/2$, this reduces to $\frac{\partial u}{\partial y}$.

One can use the usual R derivative command to find partial derivatives. For example,

```
D(expression(sqrt(6400^2- (x^2+y^2))),"x")
# -(0.5 * (2 * x * (6400^2 - (x^2 + y^2))^-0.5))
D(expression(sqrt(6400^2- (x^2+y^2))),"y")
# -(0.5 * (2 * y * (6400^2 - (x^2 + y^2))^-0.5))
```

D.4.3 Traveling Wave Solution to the Wave Equation

The wave equation refers to a specific type of equation involving second partial derivatives with respect to both time and space. For example, to model an idealized equatorial traveling wave, we may use the following wave equation model

$$w_{tt} - c^2 w_{xx} = 0, \tag{D.77}$$

where $w = g(x,t)$ represents a climate parameter, such as pressure, as a function of both space x and time t. Here, x is the positional coordinate along the equator pointing from west to east, w_{tt} is the second derivative of w with respect to time t, and w_{xx} is the second derivative of w with respect to the space coordinate x. Thus, the above equation can also be written in the following form:

$$\frac{\partial^2 w}{\partial t^2} - c^2\frac{\partial^2 w}{\partial x^2} = 0. \tag{D.78}$$

An equation that includes partial derivatives is called a partial differential equation (PDE). This particular PDE is usually called the wave equation, because it is a mathematical model for various kinds of waves. This equation is related to the ordinary differential equation (ODE) model $d^2x/dt^2 = F(x,t)/m$ derived from Newton's second law of motion: $F = ma$, which governs many kinds of motions, including free-fall.

Any function $w = g(x,t)$ that satisfies the wave equation is called a solution of the wave equation.

Recall that the solutions of an algebraic equation are numbers. For example, $x^2 - 1 = 0$ has $x_1 = 1$ and $x_2 = -1$ as its solutions. The solutions to an ODE are functions. For example, the solutions to $dx/dt + x = 0$ has a general solution $x(t) = C\exp(-t)$ for any constant C, the value of which is to be determined by an initial condition $x(0) = A$. This initial condition forces the constant C to be equal to A.

Similarly, the solutions of a PDE are functions of two or more variables, such as the aforementioned $w = g(x,t)$.

A traveling wave solution to the above wave equation is a simple and specific form of the solution function $g(x,t) = f(x - ct)$. This wave moves and maintains its original shape, which is why the "traveling wave" is named, as if an entity is traveling. As an example, we may represent an equatorial westerly wave as

$$w = f(x - ct) \tag{D.79}$$

where $w = f(\xi)$ is a single-variable function with $\xi = x - ct$ depending on both position x and time t, and c is the constant wave speed. When $c > 0$, then the wave travels from west to east, that is, it is a westerly wave. To understand that the wave travels from west to east, i.e., from a smaller x to a larger x, we can follow a specific observation point B on the wave profile, say $f(\xi = b)$, which means $b = x - ct$ has a given value, and thus, solving for x, we have $x = b + ct$ must become larger as t increases, because c is positive. Thus, the observation point B moves to the east.

This traveling wave function has the following partial derivatives:

$$\frac{\partial w}{\partial t} = \frac{df}{d\xi}\frac{\partial \xi}{\partial t} = -c\frac{df}{d\xi}, \tag{D.80}$$

$$\frac{\partial w}{\partial x} = \frac{df}{d\xi}\frac{\partial \xi}{\partial x} = \frac{df}{d\xi}, \tag{D.81}$$

$$\frac{\partial^2 w}{\partial t^2} = \frac{\partial}{\partial t}\left(\frac{\partial w}{\partial t}\right) = c^2\frac{d^2 f}{d\xi^2}, \tag{D.82}$$

$$\frac{\partial^2 w}{\partial x^2} = \frac{\partial}{\partial x}\left(\frac{\partial w}{\partial x}\right) = \frac{d^2 f}{d\xi^2}. \tag{D.83}$$

We thus have

$$\frac{\partial^2 w}{\partial t^2} - c^2\frac{\partial^2 w}{\partial x^2} = c^2\frac{d^2 f}{d\xi^2} - c^2\frac{d^2 f}{d\xi^2} = 0. \tag{D.84}$$

The wave equation is a PDE having second-order derivatives with respect to both space and time. The dimensional analysis of the wave equation tells us that the dimension of c must be the dimension of speed, i.e., the dimension of length or space divided by the dimension of time:

$$[c] = \frac{[space\ in\ one\ direction]}{[time]} = [speed]. \tag{D.85}$$

The function $w = f(x - ct)$ is called a traveling wave solution to the wave equation. Here, the wave refers to the shape of the wave determined by the function $f(\xi)$, which can be a sinusoidal wave, for example, or it might be a single-crest solitary wave.

The wave equation PDE can have infinitely many kinds of solutions. In particular, the wave equation can be used as an idealized mathematical model for many different types of wave phenomena, such as sound waves, water waves, and Rossby waves. A traveling wave is the simplest type of solution of the wave equation. Other solutions are possible, depending on the physical factors and on initial conditions and boundary conditions. PDEs form an important branch of mathematics, and they are used extensively in the study of many aspects of physical phenomena, including climate modeling and numerical weather prediction.

D.4.4 Diffusive Solution to a Heat Equation

Another frequently encountered PDE in climate science is the heat diffusion equation:

$$T_t = DT_{xx}, \tag{D.86}$$

where t is time and x is a one-dimensional space variable, such as the location on an idealized road. Here, T is temperature having dimension $[T] = \Theta$, and D is a constant.

We can verify that the heat diffusion function in Section 1.5,

$$T = \frac{Q}{\rho C\sqrt{2\pi Dt}} \exp\left(-\frac{x^2}{4Dt}\right) \tag{D.87}$$

is a solution of the above heat equation.

Applying the product rule and chain rule of differentiation

$$\frac{d}{dx} f(x)g(x) = f'(x)g(x) + f(x)g'(x) \quad \text{(Product rule)}, \tag{D.88}$$

$$\frac{d}{dx} f(g(x)) = \frac{df}{du}\frac{du}{dx}, \quad u = g(x) \quad \text{(Chain rule)}. \tag{D.89}$$

to the above heat diffusion function yields

$$\begin{aligned} T_t &= \frac{Q}{\rho C\sqrt{2\pi D}} \left(\frac{d}{dt} t^{-1/2}\right) \exp\left(-\frac{x^2}{4Dt}\right) + \frac{Q}{\rho C\sqrt{2\pi Dt}} \left(\frac{d}{dt} \exp\left(-\frac{x^2}{4Dt}\right)\right) \\ &= \frac{Q}{\rho C\sqrt{2\pi D}} (-1/2)t^{-3/2} \exp\left(-\frac{x^2}{4Dt}\right) + \frac{Q}{\rho C\sqrt{2\pi Dt}} \exp\left(-\frac{x^2}{4Dt}\right) \left(\frac{-x^2}{4D}(-t^{-2})\right) \\ &= \left(-\frac{1}{2t} + \frac{x^2}{4Dt^2}\right) T(x,t), \end{aligned} \tag{D.90}$$

$$\begin{aligned} T_x &= \frac{Q}{\rho C\sqrt{2\pi Dt}} \left(\frac{d}{dx} \exp\left(-\frac{x^2}{4Dt}\right)\right) \\ &= \frac{Q}{\rho C\sqrt{2\pi Dt}} \exp\left(-\frac{x^2}{4Dt}\right) \frac{d}{dx}\left(-\frac{x^2}{4Dt}\right) \\ &= -\frac{x}{2Dt} T(x,t), \end{aligned} \tag{D.91}$$

$$\begin{aligned} T_{xx} &= \frac{\partial}{\partial x}\left(-\frac{x}{2Dt} T(x,t)\right) \\ &= T(x,t)\frac{\partial}{\partial x}\left(-\frac{x}{2Dt}\right) + \left(-\frac{x}{2Dt}\right)\frac{\partial T(x,t)}{\partial x} \\ &= \left(-\frac{1}{2Dt} + \frac{x^2}{4D^2t^2}\right) T(x,t). \end{aligned} \tag{D.92}$$

Thus,

$$DT_{xx} = D\left(-\frac{1}{2Dt} + \frac{x^2}{4D^2t^2}\right) T(x,t) = \left(-\frac{1}{2t} + \frac{x^2}{4Dt^2}\right) T(x,t) = T_t. \tag{D.93}$$

This completes the verification that the diffusion function (D.87) is a solution of the heat equation (D.86).

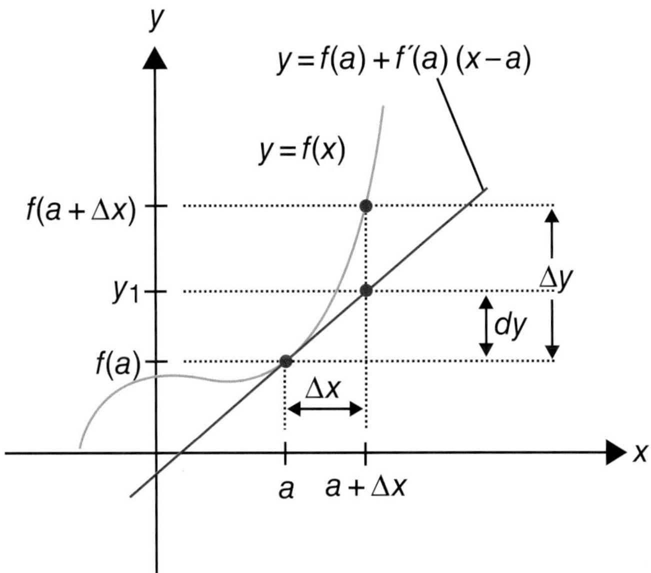

Fig. D.20 A diagram showing the geometric relationships of $dx = \Delta x$, $dy \neq \Delta y$, and slope $f'(a)$

D.4.5 Differentials and Exact Differentials

The concept of differentials is often used in atmospheric thermodynamics and other areas of climate science.

The differential of a variable is a small increment, denoted by dx, dy etc. If y is a differentiable function of x, then dy is also related to dx by

$$dy = f'(x)dx. \tag{D.94}$$

The increment of y is related to the increment of x via the slope $f'(x)$ of the function. Figure D.20 illustrates the geometric meaning of the differential dy as the exact increment for the tangent line. A straight line is called a linear curve, or linear line. Thus, dy means a linear approximation to Δy, which is the true increment of y. However, for x, the small increment dx and Δx are equal, because x is an independent variable. The tangent line

$$y = f(a) + f'(a)(x - a) \tag{D.95}$$

is a linear approximation of the function $y = f(x)$ in the neighborhood of $(a, f(a))$. In other words, it is a straight line approximation to a curve $y = f(x)$ near the point $(a, f(a))$. This concept of linear approximation is used often in developing climate science models.

The increment for functions of more than one variable can be defined in a similar way. If $z = f(x,y)$, then

$$dz = f_x(x,y)dx + f_y(x,y)dy. \tag{D.96}$$

For example, the equation of state for an ideal gas, $pV = nR^*T$, has a differential form

$$nR^*dT = pdV + Vdp. \tag{D.97}$$

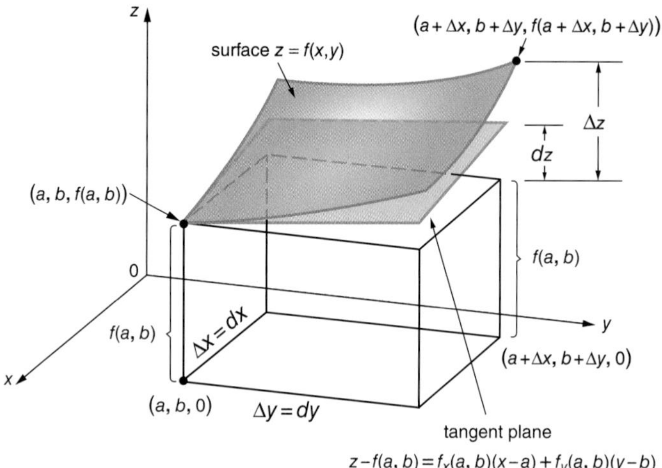

Fig. D.21 A diagram showing the geometric relationships of small increments for a function of two variables: $\Delta x = dx$, $\Delta y = dy$, but $\Delta z \neq dz$.

This means that the increment of temperature T involves two processes: one is the increment of volume V with a fixed pressure p, and another is the increment of pressure p with a fixed volume V. Atmospheric air may usually be regarded as an ideal gas for our purposes, so this equation of state is employed in climate models as well as in many other meteorological models.

Figure D.21 illustrates the geometric meaning of this differential. The equation of the tangent plane at point $(a, b, f(a,b))$ is

$$z = f(a,b) + f_x(a,b)(x-a) + f_y(a,b)(y-b). \tag{D.98}$$

This is clearly an extension of the linear approximation from the single variable case. One can, of course, extend this linear approximation to a function of many variables in a similar way.

At this point, we may ask the question: what is the function $f(x,y)$ that makes the following equation hold all the time?

$$f_x(x,y)dx + f_y(x,y)dy = 0. \tag{D.99}$$

The answer is very simple:

$$f(x,y) = C, \quad \text{(a constant)}. \tag{D.100}$$

This solution is not trivial. It defines a set of curves on the xy-plane. For example,

$$x^2 + y^2 = C \tag{D.101}$$

defines a set of circles having the center at the origin and the radius equal to $\sqrt{C}$ when C is positive.

A further question is: how can one solve the following differential equation?

$$u(x,y)dx + v(x,y)dy = 0. \tag{D.102}$$

This is equivalent to the ODE

$$v(x,y)\frac{dy}{dx} = -u(x,y). \tag{D.103}$$

The solution can be either implicit $F(x,y) = C$ or explicit $y = f(x)$.

The circle solution is a special case: $u = x, v = y$.

Another special case is called the separation of variables, in which the x terms can be placed on one side of the equation and the y terms on the other. For example, $u = (1+x)/y$ and $v = (1+y)/x$. The above circle solution is also from a separable equation.

Still another special case is called an exact differential, for which one can find a function $F(x,y)$ that satisfies the ODE

$$u(x,y)dx + v(x,y)dy = 0 \tag{D.104}$$

such that

$$F_x dx + F_y dy = 0. \tag{D.105}$$

The condition for an equation $u(x,y)dx + v(x,y)dy = 0$ to be exact is that

$$u_y = v_x. \tag{D.106}$$

D.4.6 Integral of a Multivariate Function

This subsection explores integrations of multivariate functions, that is, functions of more than one independent variable. The integral of a single variable function $y = f(x)$ in the interval $[a,b]$

$$\int_a^b f(x)dx \tag{D.107}$$

or over a domain D on the real line

$$\int_D f(x)dx \tag{D.108}$$

can be extended to higher dimensions, such as in two-dimensional space (i.e., two variables)

$$\iint_{D_2} f(x,y)dxdy, \quad \text{(Double integral)} \tag{D.109}$$

and in three-dimensional space

$$\iiint_{D_3} f(x,y,z)dxdydz \quad \text{(Triple integral).} \tag{D.110}$$

Here, D_2 and D_3 are integration domains in the xy-plane or the xyz-space. For example, D_2 can be a rectangle, a round disk, a triangle, or a region of an irregular shape. Similarly, D_3 can be a cube, a cylinder, a tetrahedron, or a body of an irregular shape.

If $f(x) = \rho(x)$ is the linear density of a rod (units: kg m^{-1}), then $\rho(x)dx$ is the mass of the small section dx. The total mass of the rod from a to b is the sum of the masses of the small sections:

$$m_1 = \int_a^b \rho(x)dx. \tag{D.111}$$

The same can be said about the double and triple integrals. If $f(x,y) = \rho(x,y)$ is the density of a unit round disk (units: kg m^{-2}), then the total mass of the disk is the sum of the masses of the small pieces, denoted by $dxdy$:

$$m_2 = \iint_{D_2} \rho(x,y)dxdy \tag{D.112}$$

where $D_2 = \{(x,y) : x^2 + y^2 \leq 1\}$.

The mass of a three-dimensional unit sphere can be expressed as

$$m_3 = \iiint_{D_3} \rho(x,y,z)dxdydz, \tag{D.113}$$

where $\rho(x,y,z)$ is the density function with units equal to kg m^{-3}, $\rho(x,y,z)dxdydz$ is the mass of a small volume element denoted by $\rho(x,y,z)dxdydz$, and $D_3 = \{(x,y,z) : x^2 + y^2 + z^2 \leq 1\}$.

In climate modeling applications, one can apply the 3D integral to calculate the total mass of seawater, for example, or the total mass of the atmosphere, when the density functions are known.

Recall that derivatives of a multivariate function can be calculated by reducing them to the derivative of a single variable. The same principle applies to multiple integrals. The calculation of multivariate integrals can be done by reducing them to a single variable integration.

A simple example is $f(x,y) = x^2 + y^2$ over a square $(x,y) \in (0,1) \times (0,1)$:

$$\begin{aligned}
I_2 &= \iint_{R_2} (x^2 + y^2)dxdy \\
&= \int_0^1 dx \int_{-1}^1 (x^2 + y^2)dy \\
&= \int_0^1 dx(x^2y + y^3/3)|_0^1 dy \\
&= \int_0^1 dx(x^2 1^2 + 1^3/3) \\
&= \left[(x^3/3 + (1/3)x)\right]_0^1 \\
&= [(1^3/3 + 1/3 \times 1)] - 0 = 2/3.
\end{aligned} \tag{D.114}$$

In climate science, the integrals $\iint_{D_2}$ and $\iiint_{D_3}$ are often encountered in mathematical models. Evaluating multivariate integrals analytically is often difficult, and it may be impossible in most applications. Numerical integrations then become necessary.

However, using polar coordinates and spherical coordinates, it is sometimes possible to integrate analytically on a 2D disk or a 3D ball. Two examples are given below.

Example D.18 Use a double integral to find the volume of a unit ball: We use polar coordinates (r, θ) with $x = r\cos\theta, y = r\sin\theta$, where θ is the longitude variable in $[0, 2\pi]$ and r is the radius variable in $[0, 1]$. The equation for the surface of a unit ball (i.e., a unit sphere) is

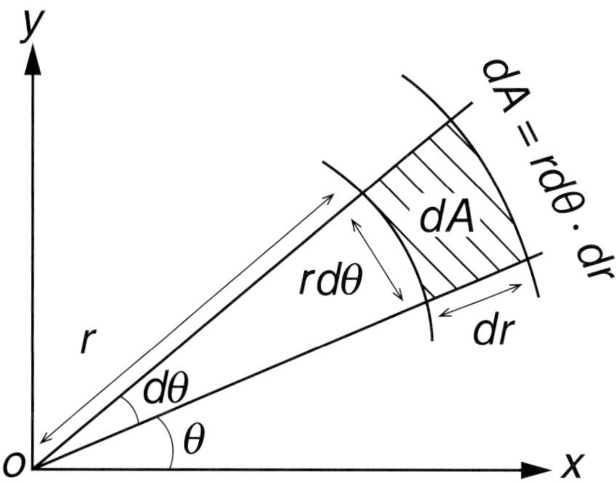

Fig. D.22 A small area in polar coordinates.

$$r^2 + z^2 = 1. \tag{D.115}$$

This can be written as

$$z = \pm\sqrt{1-r^2}. \tag{D.116}$$

Thus, the small volume of the right cylinder with the cross-sectional area $dA = rd\theta dr$ is $2\sqrt{1-r^2}dA$ (see Fig. D.22). Thus, the total volume of the unit ball is

$$V = \iiint_{D_3} dV = \iint_{D_2} 2\sqrt{1-r^2}dA. \tag{D.117}$$

This can be further reduced to

$$V = \iint_{D_2} 2\sqrt{1-r^2}\ rd\theta dr = \int_0^{2\pi} d\theta \int_0^1 dr\ 2\sqrt{1-r^2}\ r. \tag{D.118}$$

Now we can integrate over longitude θ and radius r separately. Since the integrand is independent of θ, we have

$$\int_0^{2\pi} d\theta = 2\pi. \tag{D.119}$$

Then,

$$V = 2\pi \int_0^1 2\sqrt{1-r^2}\ rdr = 2\pi(1/3)[(1-r^2)^{3/2}]_0^1 = \frac{4\pi}{3}. \tag{D.120}$$

Example D.19 Use a triple integral to find the volume of the Arctic cap of the Earth. Note that this is a partial volume of a ball, not a partial area of the surface of a sphere.

The Arctic circle is approximately at 66.5°N. We need to find the volume of the cap of the Earth north of 66.5°N. We use spherical coordinates as described in Appendix C:

$$x = r\cos\phi\cos\theta, \tag{D.121}$$

$$y = r\cos\phi\sin\theta, \tag{D.122}$$

$$z = r\sin\phi, \tag{D.123}$$

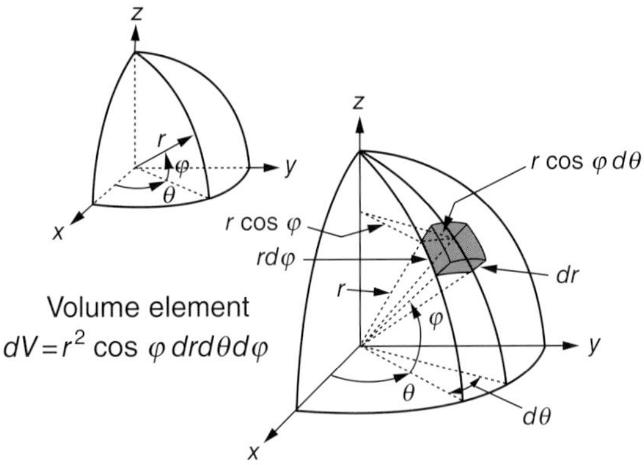

Fig. D.23 A small volume in spherical coordinates.

where ϕ is the latitude from 66.5°N to 90°N, i.e., in the interval $[0.3694\pi, \pi/2]$, θ is the longitude in $[-\pi, \pi]$, and r is the radius in $[0, R]$ with R being the Earth's radius. The small volume, also called the volume differential, in 3D expressed in spherical coordinates is

$$dV = dr \times r\cos\phi d\theta \times r d\phi = r^2 \cos\phi \; drd\theta d\phi. \tag{D.124}$$

See Fig. D.23 for this small volume, also called a volume element. The volume of the unit ball is then the summation of the volume element via a 3D integration

$$\begin{aligned} V &= \iiint_{D_3} dV \\ &= \iiint_{D_3} r^2 \cos\phi \; drd\theta d\phi \\ &= \int_0^{2\pi} d\theta \int_{0.3694\pi}^{\pi/2} \cos\phi \; d\phi \int_0^R r^2 dr \\ &= 2\pi \times \left[\sin\phi \right]_{0.3694\pi}^{\pi/2} \times \left[r^3/3 \right]_0^R \\ &= 2\pi(1 - 0.9171) \times \frac{R^3}{3} \\ &= 0.0415 \times \frac{4\pi R^3}{3}. \end{aligned} \tag{D.125}$$

Thus, the Arctic cap has only about 4% of the Earth's total volume $4\pi R^3/4$.

D.4.7 Line Integral and Work

Work is the displacement L times the force F in the direction of the displacement:

$$W = F_l L. \tag{D.126}$$

If the force and the displacement have an angle γ between them, then

$$W = FL\cos\gamma, \tag{D.127}$$

with

$$F_l = F\cos\gamma \tag{D.128}$$

as the projection of the force along the direction of the displacement. The formula in vector form is a dot product

$$W = \mathbf{F}\cdot\mathbf{L} = FL\cos\gamma, \tag{D.129}$$

where the force vector is $\mathbf{F} = (F_1, F_2)$ and the displacement vector is $\mathbf{L} = (L_1, L_2)$.

The dot product of two vectors $\mathbf{r}_1 = (a_1, b_1)$ and $\mathbf{r}_2 = (a_2, b_2)$ is defined as a scalar $\mathbf{r}_1 \cdot \mathbf{r}_2 = a_1a_2 + b_1b_2 = r_1r_2\cos\gamma$, where r_1 is $\mathbf{r}_1$'s length, r_2 is $\mathbf{r}_2$'s length, and γ is the angle between vector $\mathbf{r}_1$ and vector $\mathbf{r}_2$.

The force may depend on the spatial position, such as the atmospheric buoyancy force which may be different at different pressure levels or heights. In this case, the total work can be considered as the sum of many work elements, each of which is equal to the force at a location times a small displacement. If force is $\mathbf{F} = (u(x,y,z), v(x,y,z), w(x,y,z))$ and the small displacement resulting from the action of the force is $d\mathbf{r} = (dx, dy, dz)$ along the integral path C, the work element is then

$$dW = udx + vdy + wdz. \tag{D.130}$$

The total work done along the curve C is the sum of the work elements given by an integration

$$\int_C \mathbf{F}\cdot d\mathbf{r} = \int_C u(x,y,z)dx + v(x,y,z)dy + w(x,y,z)dy. \tag{D.131}$$

If the integral path is a closed curve in 3D space, then

$$\oint_C \mathbf{F}\cdot d\mathbf{r}. \tag{D.132}$$

If there is a function $f(x,y,z)$ such that

$$u = f_x, v = f_y, w = f_z, \tag{D.133}$$

then $f(x,y,z) = C$ is the solution of the differential equation

$$udx + vdy + wdz = 0. \tag{D.134}$$

The function $f(x,y,z)$ is called the potential of the force field (u,v,w). For example, the gravitational force $\mathbf{F} = (0,0,-mg)$ has a potential function $\phi_g = -mgz$, called the gravitational potential. Intuitively, one can conclude that the total work done by a gravitational force along an entire closed path is zero.

A special condition for the potential function f to exist in the 2D case is simple:

$$u_y = v_x. \tag{D.135}$$

Then $udx + vdy = 0$ is called an exact ordinary differential equation:

$$u(x,y) + v(x,y)\frac{dy}{dx} = 0, \tag{D.136}$$

which can define y as a function of x, implicitly or explicitly: $z = f(x,y) = C$, which implies $dz = dC = 0$. Sometimes, this function $f(x,y)$ can be found analytically.

Example D.20 Find a solution for the following differential equation:

$$(1+x^2)dx + (1+y^2)dy = 0. \tag{D.137}$$

Since $v_x = u_y = 0$, this is an exact differential. In fact, the variables of this equation are already separated:

$$(1+x^2)dx = -(1+y^2)dy. \tag{D.138}$$

Integrate this equation

$$\int (1+x^2)dx = \int -(1+y^2)dy, \tag{D.139}$$

which yields

$$x + y + \frac{x^3}{3} + \frac{y^3}{3} = C, \tag{D.140}$$

where C is an arbitrary constant.

Example D.21 When work is done on a system consisting of air through a volume compression $dV < 0$, the work element can be written as

$$dW = -pdV. \tag{D.141}$$

Here, the dimension of p is force per unit area, and the dimension of dV is volume. Thus, the dimension of pdV is force times distance, which is the dimension of work. Air pressure always acts in the direction of volume reduction, so the dot product becomes a scalar product. The total work done in an expansion or compression process of air is

$$W = -\int_P pdV, \tag{D.142}$$

where P indicates the path of the expansion or compression process.

References and Further Readings

[1] Anton, H., I. C. Bivens, and S. Davis, 2008: *Calculus: Early Transcendentals, Single Variable*, 9th Edition, John Wiley and Sons, New York.

This is one of the popular textbooks of college calculus.

[2] Cajori, F., 1985: *A History of Mathematics*, 4th Edition, Chelsea Publishing Co., New York.

> This is a classical mathematical history book and was first published in 1893.

[3] Coolidge, J. L., 1951: The story of tangents. *American Mathematical Monthly*, 58, 449–462.

> This history paper reviewed the methods of tangents developed by Pierre de Fermat in 1629 and René Descartes in 1637, before Isaac Newton was born in 1642.

[4] Curry, J. A. and P. J. Wester, 1999: *Thermodynamics of Atmospheres and Oceans*, Academic Press, New York.

> A standard textbook for an entire generation of students.

[5] Ginsburg, D., B. Groose, J. Taylor, and B. Vernescu, 1998: *The History of the Calculus and the Development of Computer Algebra Systems*. Worcester Polytechnic Institute Junior-Year Project, URL: www.math.wpi.edu/IQP/BVCalcHist/calctoc.html.

> This is a student-research paper. It discussed the methods of tangents due to Fermat and Descartes.

[6] Johnson, R. A. and G.K. Bhattacharyya, 1996: *Statistics: Principles and Methods,* 3rd Edition, John Wiley and Sons, New York.

> A popular text for college statistics.

[7] Leung, K., C. Rasmussen, S. S. P. Shen, and D. Zazkis, 2014: Calculus from a statistics perspective, *The College Mathematics Journal*, 45, 377–386.

> This and another paper by Shen and Lin (2014), are two sister articles for modernizing calculus instruction using classical concepts and contemporary technology. Also see the original manuscript at the Cornell University Library public access arXiv with URL: http://arxiv.org/abs/1406.2731 that has the same title authored by Shen, Zazkis, Leung, and Rasmussen.

[8] Lin, Q., 2010: *Calculus for High School Students: from a Perspective of Height Increment of a Curve*, People's Education Press, Beijing.

> Qun Lin (1935–) is a Chinese mathematician and an academician of the Chinese Academy of Sciences. Since the 1980s, he has been leading the effort dedicated to simplifying the instruction of calculus and linear algebra in both colleges and high schools in China.

[9] Lin, Q., 2009: *Fastfood Calculus*, Science Press, Beijing.

> This paper motivated the work of Shen and Lin (2014), who outlined a method to teach the basic concepts of calculus in two hours.

[10] Newton, I., 1736: *The Method of Fluxions and Infinite Series with Application to the Geometry of Curve-Lines*, Translated from Latin to English by J. Colson, Printed for Henry Woodfall, London. URL:

http://books.google.com/books?id=WyQOAAAAQAAJ&printsec=frontcover&source=gbs_ge_summary_r&cad=0#v=onepage&q&f=false

> This is a free online Google book: http://books.google.com. One can read Newton's original writing, ideas, and notations.

[11] Newton, I., 1729: *The Mathematical Principles of Natural Philosophy*, Translated from Latin to English by A. Motte, Printed for Benjamin Motte, London. URL:

http://books.google.com/books?id=Tm0FAAAAQAAJ&printsec=frontcover&source=gbs_ge_summary_r&cad=0#v=onepage&q&f=true

> This famous book is often known as the *Principia*, first published in 1687. One can enjoy the great master's original ideas by reading this free online Google book.

[12] Range, R. M., 2011: Where are limits needed in calculus? *American Mathematics Monthly*, 118, 404–417.

> This paper reviewed Descartes' method of tangents and showed that Descartes' approach can evaluate derivatives without the concepts of infinitesimals and differentials.

[13] Shen, S. S. P. and Q. Lin, 2014: Two hours of simplified teaching materials for direct calculus. *Mathematics Teaching and Learning*, No. 2, 2-1–2-6.

> This paper established the Descartes' direct (DD) calculus. The DD calculus method uses Descartes' methods of tangents and directly uses the concept of derivative–antiderivative (DA) pair to define a derivative and an integral. DD calculus is rigorous and avoids the procedure of limits. The English translation of this paper is freely available at Cornell University Library public access arXiv entitled "*DD Calculus*" with URL: http://arxiv.org/abs/1404.0070 Our book has adopted the DD calculus and also described the calculus method from a statistics perspective.

[14] Stewart, J., 2008: *Single Variable Calculus: Early Transcendentals*, 6th Edition, Thomson Brooks/Cole, Belmont.

This has been a very popular calculus text worldwide since its first edition in the 1980s.

[15] Suzuki, J., 2005: The lost calculus (1637–1670): Tangency and optimization without limits. *Mathematics Magazine*, 78, 339–353.

This paper reviewed the earlier ideas and methods of calculus before Isaac Newton (1642–1726) and Gottfried Leibniz (1646–1716), including the works by Renè Descartes (1637), John Wallis (1655), Jan Hudde (1657 and 1658), and Isaac Barrow (1670).

[16] Wackerly, D. D., W. Mendenhall III, and R. L. Scheaffer, 2002: *Mathematical Statistics with Applications,* 6th Edition, Duxbury.

A popular text in its area.

Exercises

Exercises D.1, D.6, D.8–D.11 require the use of a computer.

D.1 (a) Use a computer or a smartphone to calculate the first- and second-order derivatives dy/dx and d^2y/dx^2 of the function

$$y = \sqrt{9 - x^2}. \tag{D.143}$$

(b) Draw the curve for the function $y = \sqrt{9 - x^2}$ by a computer or by hand, and describe the geometric meaning of the first-order derivative dy/dx. Include the following terms: uphill slope, downhill slope, flat, and infinite slope.

(c) Describe the geometric meaning of the second-order derivative d^2y/dx^2. Include the following terms: concave-up, concave-down, and the maximum point.

D.2 Calculate the first- and second-order derivatives dy/dx and d^2y/dx^2 of the function

$$y = \sqrt{9 - x^2} \tag{D.144}$$

by hand using the chain rule, but not using a computer or a smartphone. Compare your answer with that calculated by a computer.

D.3 Consider a simple ODE model to show the exponential decay of the Earth's surface temperature to an equilibrium:

(a) Verify that the function

$$T(t) = A\exp\left(-\frac{B}{c}t\right) + \frac{Q}{B} \tag{D.145}$$

is a solution of the following ODE

$$c\frac{dT}{dt} = Q - BT. \tag{D.146}$$

Here, $T(t)$ may be regarded as the time-dependent average temperature of the Earth's surface, c may be regarded as the heat capacity of the Earth's ocean water and atmosphere, Q measures the solar radiation received by the Earth, and B helps to quantify the long-wave radiation emitted from the Earth to space.

(b) Interpret the meaning of the solution, when considering both time equal to zero and time approaching infinity. *Hint: You may use Internet resources.*

(c) Make a comprehensive dimensional analysis for the above ODE model and the solution.

D.4 Analyzing the temperature maps of Figs. D.17 and D.18, identify three locations with very large positive north-to-south directional derivatives. Use your intuition to explain the climate significance of the north-to-south directional derivatives at these locations. Do the same for the east-to-west directional derivatives.

D.5 (a) For $u = \sin(x - ct)$, find the partial derivatives u_t, u_{tt}, u_x, u_{xx}.

(b) Verify that $u = \sin(x - ct)$ is a solution of the wave equation

$$u_{tt} - 4u_{xx} = 0, \tag{D.147}$$

and find the value of c.

D.6 Examine solutions of an initial and boundary value problem of a heat equation.

(a) Verify that the following function

$$u = \sin(\pi x)\exp(-k\pi^2 t) \tag{D.148}$$

is a solution of the heat equation

$$u_t = k u_{xx}. \tag{D.149}$$

(b) How about $\sin(2\pi x)\exp(-2^2 k\pi^2 t)$? How about $u_n(x,t) = \sin(n\pi x)\exp(-n^2 k\pi^2 t)$ for a given positive integer n?

(c) What can you say about the functions $u_n(0,t)$ and $u_n(1,t)$?

(b) Use R or another computer program to plot the following curves: On the u–t plane for $y = \sin(0.2\pi)\exp(-0.1\pi^2 t)$, and on the u–x plane for $y = \sin(2\pi x)\exp(-0.1\pi^2 \times 0)$. Try to explain the physical meaning of these functions from the point of view of heat diffusion. What can you say about $u_n(x,0)$?

D.7 Find all three partial derivatives u_x, u_y, u_z of $u = 1/r$ where $r = \sqrt{x^2 + y^2 + z^2}$. Is it possible for $\nabla u = (u_x, u_y, u_z) = 0$? Think about the gravitational potential and explain your answer in physical terms.

D.8 Use the Internet to search for the density $\rho(x,y,z)$ of the Earth's atmosphere as a function of elevation, and use the integral to estimate the total mass of the Earth's atmosphere. *Hint: Use the calculus method of "cut and add." Cut the Earth's atmosphere into small pieces with volume dV and mass $\rho(x,y,z)dV$. The total mass is the integral of this $\rho(x,y,z)dV$.*

D.9 Express the Tmax January climatology of the Cuyamaca weather station (USHCN Site No. 042239) near San Diego, California, USA, as an integral when regarding Tmax as a function of time t, using the definition of an integral from the statistics perspective. The temperature data file of the Cuyamaca station is named `CA042239T.csv` and can be found from the data.zip file downloadable on the website for this book at www.cambridge.org/climatemathematics. One can also find the updated data from the Internet, such as

https://www.ncdc.noaa.gov/ushcn

D.10 Trend and derivative:

(a) Use the derivative to explain the trends of the Tmax time series in Exercise D.9.
(b) Treat the time series of the Cuyamaca January Tmin in Exercise D.9 as a smooth function from 1951 to 2010. Use the curve and its derivative to explain the instantaneous rate of change. Use the concept of derivative.
(c) Use the average rate of change for a given period of time to explain the linear trend in each of the four periods:

(i) 1951–2010
(ii) 1961 2010
(iii) 1971–2010
(iv) 1981–2010.

Try to use the concept of the mean value theorem when describing your explanation.

D.11 Use the integral concept to describe the rainfall deficit or surplus history of San Diego since January 1 of a given year according to the USHCN daily precipitation data, or do this for another location you are familiar with. You may use the integral to describe the precipitation deficit or surplus. The daily data can be found in the book data folder with filename SanDiegoDailyTandP.csv, or downloaded from

https://www.ncdc.noaa.gov/cdo-web/search

Requirements: You should use at least one figure. Your English text must be longer than 100 words.

D.12 Use the integral of temperature with respect to time to interpret the concept of cumulative degree-days in agriculture. Consider the energy needed by plants to grow. *Requirements: You must use at least one figure and one table. Your English text must be longer than 100 words.*

D.13 In 3D space, the gravitational field is $\mathbf{G} = (0,0,-mg)$, where m is mass. Along a path, a force acting against the gravitational force is $\mathbf{F} = (0,0,mg)$. This force moves a mass m along the edges of a horizontally positioned cube from $A = (0,0,0)$ to its diagonal vertex $A = (2,2,2)$. Use the line integral to calculate the total work done by this process, using at least two paths.

D.14 Calculate the integral

$$I_1 = \int_0^{\pi/2} d\theta \int_0^{\infty} dr\, [r\exp(-r^2)]. \tag{D.150}$$

Use the above integral to calculate the following double integral:

$$I_2 = \iint_{\Omega} \exp(-x^2 - y^2)\, dA, \tag{D.151}$$

where the integral domain Ω is the entire $x - y$ plane. *Hint: Use* $r^2 = x^2 + y^2$. *Formulate the integral using* dr *and* $d\theta$.

D.15 The ideal gas law is

$$nR^*T = pV, \tag{D.152}$$

where n is the amount of the gas in moles, $R^* = 8.3144598$ $[\text{J mol}^{-1}\ \text{K}^{-1}]$ is the universal gas constant, T is the absolute temperature in [K], p is the pressure of the gas, and V is the volume of the gas. Write the relationships of the differentials dT, dp, and dV for constant n and R^*, for three p and V cases: (a) p is constant, (b) V is constant, and (c) both p and V can change.

Explain the physical meaning of your results.

D.16 If T is constant and both p and V can change, from the gas law

$$nR^*T = pV, \tag{D.153}$$

find the relationship between dp and dV. Such a process, in which T is constant, is called an isothermal process. Discuss the physical significance of this process.

Appendix E Sample Solutions to the Climate Mathematics Exercises

The purpose of this appendix is to show students how to describe and present their solutions to the exercise problems. Only a few sample solutions are presented here. The updated solutions of additional problems and hints can be found on the book website www.cambridge.org/climatemathematics, while the complete solution manual is available only to instructors.

Exercise 1.4

Exercise 1.2 yields

$$h = \beta g t^2 \tag{E.1}$$

From the definition of velocity being a time derivative of distance, we have

$$v = \frac{dh}{dt} = 2\beta g t. \tag{E.2}$$

Use the energy conservation at the initial time $t = 0$ and the time $t = T$. When $t = 0$, the velocity is zero and hence there is no kinetic energy. The potential energy is set relative to the position at $t = T$ where the potential is zero since the body is on the ground. Hence, $E_P = mgH$. At $t = T$, there is only kinetic energy $K_E = \frac{1}{2}mv^2$. Thus, the $E_P + E_K$ at time zero and at time T yields

$$mgH = \frac{1}{2}mv^2. \tag{E.3}$$

i.e.,

$$mg\beta g T^2 = \frac{1}{2}m(2\beta g T)^2. \tag{E.4}$$

This leads to

$$\beta = 2\beta^2 \tag{E.5}$$

or

$$\beta(1 - 2\beta) = 0. \tag{E.6}$$

This equation has two solutions

$$\beta_1 = 0, \quad \text{and} \quad \beta_2 = \frac{1}{2}. \tag{E.7}$$

The solution $\beta_1 = 0$ means no motion, is hence unreasonable, and is discarded. Thus, we adopt the solution $\beta_2 = 1/2$.

Finally,

$$h = \frac{1}{2}gt^2, \tag{E.8}$$

which is given in many textbooks of physics and derived in traditional calculus textbooks by solving differential equations.

Exercise 1.5

Assume that

$$B = \alpha m^a l^b g^c. \tag{E.9}$$

Take the dimension of both sides of this equation:

$$[B] = [\alpha][m]^a[l]^b[g]^c. \tag{E.10}$$

or

$$[B] = [\alpha]M^aL^b(LT^{-2})^c = 1 \cdot M^aL^bL^cT^{-2c} = M^aL^{b+c}T^{-2c}. \tag{E.11}$$

Because Bt inside a sine function must be dimensionless, $[Bt] = 1$, or $[B]T = 1$, or

$$[B] = T^{-1}. \tag{E.12}$$

Equation (E.11) becomes

$$T^{-1} = M^aL^{b+c}T^{-2c}. \tag{E.13}$$

Comparing the exponents on both sides of this equation yields

$$0 = a \tag{E.14}$$

$$0 = b + c \tag{E.15}$$

$$-1 = -2c. \tag{E.16}$$

These equations have the following unique solution

$$a = 0, b = -1/2, c = 1/2. \tag{E.17}$$

Thus,

$$B = \alpha l^{-1/2} g^{1/2} = \alpha\sqrt{\frac{g}{l}}. \tag{E.18}$$

Exercise 1.6

Since we have assumed that $\theta = \theta_M$ at $t = 0$, the function $\sin(Bt + \pi/2)$ must be decreasing when t increases at first from zero, in order for the pendulum angle θ to decrease. This is possible only when B is negative, which in turn implies $\alpha < 0$. Let τ denote the period of the function $\sin(Bt + \pi/2)$. Because the period of $\sin x$ is 2π, we have

$$-B\tau = 2\pi \tag{E.19}$$

$$-\alpha m^a l^b g^c \tau = 2\pi, \tag{E.20}$$

or

$$\alpha = -\frac{2\pi}{\tau m^a l^b g^c} = -\frac{2\pi}{\tau}\sqrt{\frac{l}{g}}. \tag{E.21}$$

The average of your experimental data for the period τ, and also the values of l and g can be substituted into the above formula to obtain the value for α. The value should be approximately equal to one.

Finally, the pendulum motion can be modeled by

$$\theta = \theta_M \sin\left(\frac{\pi}{2} - \sqrt{\frac{l}{g}}\,t\right). \tag{E.22}$$

Exercise 1.7

Use the result in Exercise 1.5

$$\theta = \theta_M \sin\left(\frac{\pi}{2} + \alpha\sqrt{\frac{l}{g}}\,t\right) \tag{E.23}$$

to find the velocity formula at time t

$$v = l\frac{d\theta}{dt} = lB\theta_M \cos\left(Bt + \pi/2\right). \tag{E.24}$$

We again use the energy conservation principle. The potential energy E_P at the highest point when $t = 0$, $\theta = \theta_M$, and kinetic energy is zero is equal to the kinetic energy E_K at the lowest point where the potential energy is zero. The potential energy is with respect to the lowest point where the angle $\theta = 0$.

At $t = 0$, the pendulum is released from the highest point $\theta = \theta_M$. Its potential energy is

$$\begin{aligned} E_P &= mgh \\ &= mgl(1 - \cos\theta_M) \\ &\approx \frac{1}{2}mgl\theta_M^2. \end{aligned} \tag{E.25}$$

Here, the small angle assumption is used, because the above approximation works only when θ_M is small when expressed in radians. For example, $\theta_M \le 0.2$ [radian], equivalent to 11.46°, $1 - \cos\theta_M = 0.019933$, $\theta_M^2/2 = 0.020000$. The relative difference is less than 0.4%. The approximation is quite good even for an angle as large as 11.46°.

At the lowest point $\theta = 0$, assume $t = t_m$, then

$$\theta = \theta_M \sin\left(\frac{\pi}{2} + \alpha\sqrt{\frac{l}{g}}\,t_m\right) = 0. \tag{E.26}$$

If $\sin x = 0$, then $\cos^2 x = 1$. So,

$$\cos^2\left(\frac{\pi}{2} + \alpha\sqrt{\frac{l}{g}}\,t_m\right) = 1. \tag{E.27}$$

The kinetic energy at $t = t_m$ is

$$\begin{aligned}
E_K &= \frac{1}{2}mv^2 \\
&= \frac{1}{2}m\left[lB\theta_M \cos\left(Bt_m + \pi/2\right)\right]^2 \qquad \text{(E.28)} \\
&= \frac{1}{2}m(lB\theta_M)^2 \cos^2\left(Bt_m + \pi/2\right) \\
&= \frac{1}{2}m(lB\theta_M)^2 \\
&= \frac{1}{2}m\left(l\alpha\sqrt{\frac{g}{l}}\theta_M\right)^2 \\
&= \frac{1}{2}ml^2\alpha^2\theta_M^2\frac{g}{l} \\
&= \frac{1}{2}mgl\theta_M^2\alpha^2. \qquad \text{(E.29)}
\end{aligned}$$

From (E.25) and (E.29), the energy equation $E_P = E_K$ becomes

$$\frac{1}{2}mgl\theta_M^2 = \frac{1}{2}mgl\theta_M^2\alpha^2. \qquad \text{(E.30)}$$

This implies

$$\alpha^2 = 1. \qquad \text{(E.31)}$$

Thus,

$$\alpha = 1 \quad \text{or} \quad \alpha = -1. \qquad \text{(E.32)}$$

After choosing θ to be positive on the right and θ_M on the right, we take $\alpha = -1$ so that θ decreases after the pendulum is released at time zero. We discard $\alpha - 1$. Finally, the pendulum's angle as a function of time is

$$\theta = \theta_M \sin\left(\frac{\pi}{2} - \sqrt{\frac{l}{g}}t\right). \qquad \text{(E.33)}$$

Exercise 1.11

(a) The unit of q_{cond} is [W cm^{-2}], which means power per unit area. Thus, the dimension of heat flux should be

$$[q_{cond}] = [Work/T]/L^2 = ((M(L/T)^2)/T)L^{-2} = ML^2T^{-3}L^{-2} = MT^{-3}. \qquad \text{(E.34)}$$

This expression does not seem to have a clear physical interpretation. One may reorganize it as $(MT^{-2})/T$. Then MT^{-2} is similar to linear space acceleration LT^{-2}, and hence may be related to "mass acceleration."

The unit of D_H is cm^2 sec^{-1}, implying that the dimension of thermal molecular diffusivity D_H must be

$$[D_H] = L^2 T^{-1}. \tag{E.35}$$

This cannot be simplified further and can be understood as the rate of area change.

∇T's unit [K cm^{-1}] means $[\nabla T] = \Theta L^{-1}$. This is the simplest form and cannot be simplified further.

The heat capacity of seawater C_p[J g^{-1} K^{-1}] means that

$$[C_p] = [Energy]/(M\Theta) = M(L/T)^2/(M\Theta) = L^2 T^{-2} \Theta^{-1}. \tag{E.36}$$

When it is grouped as $[C_p] = (L/T)^2/\Theta$, it is the velocity squared per unit temperature. Since velocity squared measures the level of molecular vibration, $[C_p]$ thus measures how much molecular vibration the material can sustain per unit of temperature change. That is, how much energy the material can gain when the temperature is raised one degree. This, of course, is the original meaning or definition of heat capacity at a given pressure. For example, water will gain much more heat than air when heated by one kelvin. Thus, water has a larger heat capacity than air.

The density of seawater ρ[g cm^{-3}] implies that $[\rho] = M/L^3$, which is the dimension of density in a 3D space.

(b) Substituting the above in the following dimension equation

$$[q_{cond}] = [D_H]^a [\nabla T]^b [C_p]^c [\rho]^d \tag{E.37}$$

yields

$$MT^{-3} = (L^2 T^{-1})^a (\Theta L^{-1})^b (L^2 T^{-2} \Theta^{-1})^c (M/L^3)^d. \tag{E.38}$$

The right-hand side can be simplified, and leads to

$$L^0 T^{-3} M^1 \Theta^0 = L^{2a-b+2c-3d} T^{-a-2c} M^d \Theta^{b-c}. \tag{E.39}$$

Equating the exponents of both sides for each factor leads to the following four linear equations for a, b, c, d:

$$2a - b + 2c - 3d = 0 \tag{E.40}$$

$$-a - 2c = -3 \tag{E.41}$$

$$d = 1 \tag{E.42}$$

$$b - c = 0. \tag{E.43}$$

The last equation implies $b = c$. The first equation then becomes $2a + c = 3d = 3$, which can be coupled with the second equation for a and c. These two equations have a unique solution $a = c = 1$. Then $b = c = 1$ and $d = 1$.

The dimensional analysis thus has proved that

$$q_{cond} = D_H \nabla T C_p \rho. \tag{E.44}$$

(c) The unit

$$[\mathrm{cm}^2\ \mathrm{sec}^{-1}][\mathrm{K/cm}][\mathrm{J/(g\ K)}][\mathrm{g/cm}^3] \tag{E.45}$$

can be simplified to

$$\begin{aligned} & \text{cm}^2 \sec^{-1} \text{K} \times \text{cm}^{-1} \text{W} \times \sec \times \text{g}^{-1} \text{K}^{-1} \text{g} \times \text{cm}^{-3} \\ = & \text{cm}^{2-1-3} \times \sec^{-1+1} \times \text{g}^{-1+1} \times \text{W}^1 \times \text{K}^{1-1} \\ = & \text{cm}^{-2} \times \sec^0 \times \text{g}^0 \times \text{W} \times \text{K}^0 \\ = & \text{W} \times \text{cm}^{-2}. \end{aligned} \quad (E.46)$$

This is the unit of q_{cond}.

Exercises 2.5(a) and 2.6(a)

```
#Select Tmax, Tmean, Tmin from the Cuyamaca station and download
#Save the csv file named CA042239temp.csv into your working directory
#Users/sshen/climmath
#Manually remove the head and earlier data. Keep data from 1951-2014.
#Rename the data file as tempdata.csv.
#Open RStudio and set the R working directory to the data file folder
#setwd("/Users/sshen/climmath")

 dat=read.csv("data/tempdata.csv", header=FALSE)

#Verify the dat dimension
dim(dat)
[1] 768   6

#Extract Tmax data out as a vector
tmax=dat[,4]

#Separate the months from Jan to Dec: Year for row and Mon for column
tmaxmat=t(matrix(tmax,nrow=12))

#Add year index as the first column
 yr=seq(1951,2014)
tmaxmatyr=cbind(yr,tmaxmat)

#Now, the Tmax data matrix is in the following format
#The first column is year, then Jan, Feb, etc.
head(tmaxmatyr)
       yr
[1,] 1951 48.0 49.3 53.0 56.7 66.0 71.7 82.7 81.5 81.5 66.8 54.5 47.3
[2,] 1952 42.4 48.8 44.0 58.3 70.9 70.0 82.2 83.2 80.2 77.2 51.9 46.4
[3,] 1953 55.3 52.3 53.2 58.1 57.8 72.3 83.2 81.5 81.5 68.3 60.6 52.4
[4,] 1954 50.0 57.3 49.2 64.4 69.1 73.4 83.3 79.5 80.4 71.7 63.3 52.0
[5,] 1955 41.5 45.9 56.4 59.2 62.3 72.3 78.3 81.1 81.3 72.6 58.8 52.8
[6,] 1956 53.0 45.5 59.4 54.9 64.7 77.6 81.7 82.9 84.7 66.4 57.5 53.1

#Compute the 1961-1990 climatology for August
tmaxclimaug=mean(tmaxmatyr[11:40,9])
tmaxclimaug
[1] 83.55
```

Thus, the mean value of 1961–1990 August maximum temperature climatology of Cuyamaca station is 83.55 °F.

Exercise 9.10

```
#setwd("/Users/sshen/climmath")
k=0.2*pi
om=0.4*pi
x=seq(-20,20,len=1000)
time=c(1:10)
fileNames=time
labNames=c("x: location [cm]","Wave [cm]", "Wave Animation: Time=")
library(animation)
par(bg = "yellow")
ani.record(reset = TRUE)
for (i in 1:100){t = 0.1*(i-1)
    #fileNames[i]=paste("Wave at Time",time[i])
    #png(filename=paste(fileNames[i],".png"))
     plot(x,1.5* cos(k*x-om*t),col="red",lwd=12, type="o",
          ylim=c(-2,2),
     xlab=labNames[1], ylab=labNames[2],
     main=paste(labNames[3], t))
     ani.record() # record the current frame
}
oopts = ani.options(interval = 0.1) #Movie option
ani.replay() #Play the movie in the R plot window
saveHTML(ani.replay(), img.name = "record_plot")
#Show the movie on a website window
```

Exercise D.3

(a)

$$\begin{aligned}\frac{dT}{dt} &= \frac{d}{dt}\left(A\exp\left(-\frac{B}{c}t\right)+\frac{Q}{B}\right)\\ &= A\exp\left(-\frac{B}{c}t\right)(-B/c)\\ &= (T(t)-Q/B)(-B/c)\\ &= (-BT(t)+Q)/c.\end{aligned} \tag{E.47}$$

Multiplying both sides by c yields

$$c\frac{dT}{dt} = Q - BT. \tag{E.48}$$

This completes the verification.

(b) When $t = 0$, $T(0) = A + Q/B$ is the initial condition. When $t = \infty$, $T(\infty) = Q/B$. These mean that the perturbed initial temperature $A + Q/B$ always goes to a fixed temperature Q/B, regardless of the size of A. Thus, this is a stable system, which always reverts to the stable condition Q/B after an initial perturbation. Further, the decay to the stable condition is exponentially fast. The decay rate depends on the ratio of c/B, which has dimension time T and is the time scale of the system. This is so because $t/(c/B) = (B/c)t$ must be dimensionless inside an exponential function exp(). Thus, when the heat capacity c is large, the time scale is long. For example, water has a longer time scale than air. When B is large, the time scale is short. B measures diffusivity. When diffusion is fast, the system has a short time scale. For example, land has a shorter time scale than ocean.

(c) The dimension for the ODE model is

$$[c]\left[\frac{dT}{dt}\right] = [Q] - [B][T]. \tag{E.49}$$

We can interpret c as heat capacity. Then $[c] = [Work]/\Theta$. The left-hand side is then

$$[c]\left[\frac{dT}{dt}\right] = ([Energy]/\Theta)(\Theta/T) = [Energy]/T. \tag{E.50}$$

This is the changing rate of energy. Q must have the same dimension as this: $[Energy]/T$, which can be regarded as the heating rate, the amount of incoming energy per unit time.

It is required that all the additive terms in an equation must have the same dimension, in order to be possible to add these terms together. Thus, $[B][T] = [Energy]/T$ implies $[B] = [Energy]T^{-1}\Theta^{-1}$, meaning the energy per unit time per unit temperature. This is a diffusive process.

Exercise D.5

(a) Let

$$\xi = x - ct. \tag{E.51}$$

Then $\xi_x = 1$, and $\xi_t = -c$.

The chain rule can be applied:

$$u_t = u_\xi \xi_t = \cos(\xi)(-c) = -c\cos(x-ct), \tag{E.52}$$

$$u_{tt} = \frac{\partial u_t}{\partial t} = \frac{\partial}{\partial t}(-c\cos(x-ct)) = -c^2\sin(x-ct), \tag{E.53}$$

$$u_x = u_\xi \xi_x = \cos(\xi)(1) = \cos(x-ct), \tag{E.54}$$

$$u_{xx} = \frac{\partial u_x}{\partial x} = \frac{\partial}{\partial x}(\cos(x-ct)) = -\sin(x-ct). \tag{E.55}$$

(b) Substituting the partial derivatives of $\sin(x-ct)$ in Part (a) into the original wave equation yields

$$u_{tt} - 4u_{xx} = -c^2\sin(x-ct) - 4(-\sin(x-ct)) = -(c^2-4)\sin(x-ct) = 0. \tag{E.56}$$

This implies that

$$c^2 - 4 = 0. \tag{E.57}$$

Thus

$$c = \pm 2. \tag{E.58}$$

Exercise D.8

Let $\rho(r,\phi,\theta)$ denote the density of the atmosphere, where (r,ϕ,θ) are spherical coordinates with an origin at the Earth's center. The small volume with increments $dr, d\phi, d\theta$ is

$$dV = r^2 \cos\phi dr d\phi d\theta. \tag{E.59}$$

Thus, the total mass of the atmosphere is given by the following integral

$$M = \int_{Atmosphere\ Space} \rho dV = \int_{-\pi}^{\pi} d\theta \int_{-\pi/2}^{\pi/2} d\phi \int_{f(\phi,\theta)}^{\infty} dr \rho(r,\phi,\theta) r^2 \cos\phi, \tag{E.60}$$

where $r = f(\phi,\theta)$ describes the surface of the bottom of the atmosphere, i.e., the surface of sea and land. The r integration must be from this surface all the way to infinity.

We need to use Internet resources to find the density function: $\rho(r,\phi,\theta)$. The actual density can be a complicated function, which makes it impossible to find an analytic solution by hand. This means that one cannot calculate the integration by hand. The integral must be computed using computers. According to NASA,
http://nssdc.gsfc.nasa.gov/planetary/factsheet/earthfact.html
the total mass of the atmosphere is 5.1×10^{18} [kg]. This number also has observational uncertainties and varies, but all the reasonable values that have been calculated are in the range of 5.0–5.3×10^{18} [kg].

Glossary

alternative hypothesis – in a statistical test, the outcome that is not the null hypothesis. See hypothesis test.

Apollo program – a human space flight program of the United States which accomplished landing the first humans on the Moon from 1969 to 1972.

Argo – an international program that uses profiling floats to observe the global ocean.

Atlantic Multi-decadal Oscillation (AMO) – a climate cycle that affects the sea surface temperature of the North Atlantic Ocean involving different modes on multi-decadal timescales.

Cartesian coordinate system – a system to specify the position of any point in 2- or 3-dimensional space by 2 or 3 Cartesian coordinates, the signed distances from the point to 2 or 3 mutually perpendicular planes.

Central Limit Theorem – an important theoretical result of mathematical statistics which states that when the sample size is sufficiently large, the sample mean will be approximately normally distributed, even if the original variables themselves are not normally distributed.

centrifugal force – an inertial force directed away from the axis of rotation that appears to act on all objects when viewed in a rotating frame of reference.

Clausius–Clapeyron equation – a consequence of the first and second laws of thermodynamics which is of crucial importance in theoretical meteorology, quantifying the strong monotonic dependence of the saturation vapor pressure of water on temperature, thus explaining why a warmer atmosphere can contain significantly more water vapor than a cooler one.

climate change – any systematic change in the long-term statistics of climate elements (such as temperature or precipitation), sustained over several decades or longer. Climate change may be due to either natural or human causes, and the context in which the term is used often supplies information as to which causes are being considered.

confidence interval – a frequently used term in statistical data analysis which denotes an interval $[a, b]$ that has a given probability of containing the unknown true value of a random parameter, such as temperature. This probability is expressed in a percentage, called the confidence level, and quantifies the degree of confidence that the true parameter value is inside the interval $[a, b]$.

conservation laws – in physics, a statement that a particular measurable property of an isolated physical system, lacking sources and sinks of such properties, does not change as the system evolves over time. Exact conservation laws include conservation of energy, conservation of linear momentum, and conservation of angular momentum. Such conservation laws are the foundation of many mathematical models of physical systems, including climate.

continuity equation – in fluid mechanics, a mathematical statement that mass is conserved, a constraint which places restrictions on the velocity field.

continuum mechanics – a branch of physics that deals with the behavior of materials modeled as a continuous mass rather than as discrete particles.

Coriolis force – a force per unit mass arising from the Earth's rotation, and equal to $-2\mathbf{\Omega} \times \mathbf{u}$, where $\mathbf{\Omega}$ is the angular velocity of the Earth and $\mathbf{u}$ is the relative velocity of the particle. The Coriolis force acts as a deflecting force, perpendicular to the velocity, to the right of the motion in the Northern Hemisphere and to the left in the Southern Hemisphere. It alters the direction but cannot alter the speed of the particle.

Coriolis frequency – also called Coriolis parameter or Coriolis coefficient, is equal to twice the rotation rate Ω of the Earth multiplied by the sine of the latitude ϕ.

correlation coefficient – a measure of the strength of the linear relationship between two datasets. See also covariance.

coupled ocean–atmosphere GCM – a climate model that includes both atmospheric and oceanic components coupled interactively. Such a model may have a regional or a global domain. See general circulation model (GCM).

covariance – a parameter expressing a relationship between two random variables x and y or two datasets that measures how much the two variables vary together. See also correlation coefficient.

DA pairs – derivative and antiderivative pairs, a foundational concept in Descartes' direct (DD) calculus that describes the basic calculus concepts of derivative and integral in a direct and non-traditional way, without the need to introduce limits. This non-traditional approach is adopted as the preferred conceptual framework of calculus in this book. See Appendix D.

Descartes' direct (DD) calculus – an approach to calculus in which the basic calculus concepts of derivative and integral are defined in a direct and non-traditional way, without the need to introduce limits: The derivative of a curve is defined and computed from the slope of a tangent line to a curve, and the integral of the curve's slope is defined as the height of a curve. This direct approach to calculus has three distinct features: (i) it defines derivative and (definite) integral without using limits, (ii) it defines derivative and antiderivative simultaneously via a DA pair, and (iii) it posits the fundamental theorem of calculus as a natural consequence of the definitions of derivative and integral. See Appendix D.

eigenvalue and eigenvector – an eigenvector v of an n-by-n matrix A is a special non-zero n-by-1 vector which does not change direction under the multiplication action of the matrix A, i.e., Av has the same direction as v. For any two vectors in the same direction, there exists a scalar λ such that one vector is equal to the scalar times the other vector. Thus, $Av = \lambda v$. Here, the scalar λ is called the eigenvalue, which may be either real or complex. This is a geometric way to define the eigenvalue and eigenvector. An algebraic way of defining the eigenvalue and eigenvector is that the λ value for the equation $Av = \lambda v$ to have a non-trivial solution is called the eigenvalue, and the non-trivial solution v is called the eigenvector. An n-by-n matrix A usually has n eigenvalues and n eigenvectors.

El Niño–Southern Oscillation (ENSO) – abbreviated as ENSO, is a periodic fluctuation in sea surface temperature (El Niño) and in the air pressure of the overlying atmosphere (Southern Oscillation) across the equatorial Pacific Ocean. The Southern Oscillation describes a bimodal variation in atmospheric sea level pressure between observation stations at Darwin, Australia and at Tahiti. See Southern Oscillation Index (SOI).

emissivity – the ratio of the energy radiated from a material's surface to that radiated from a blackbody (a perfect emitter) at the same temperature and wavelength and under the same viewing conditions. It is a dimensionless number between 0 (for a perfect reflector) and 1 (for a perfect emitter).

empirical orthogonal function (EOF) – the spatial eigenvectors from the SVD decomposition of a space–time climate data matrix are known as empirical orthogonal functions. The first few EOFs often represent typical patterns of climate variability. Examples which are of great importance in climate science include the El Niño–Southern Oscillation (ENSO), the North American Oscillation (NAO), and the Pacific Decadal Oscillation (PDO). The efficient Golub–Reinsch (1970) SVD algorithm provided a technique for carrying out the SVD calculation without computing a covariance matrix and hence converted the EOF approach into a linear algebra method rather than a conventional statistical method.

energy balance model (EBM) – a class of climate models that are highly idealized and simplified representations of certain key aspects of a climate system, and in which the motions of an atmosphere and an ocean are not explicitly considered. An EBM is based on the principle of energy balance: the climate system in question is assumed to be in a balanced or equilibrium state, such that the incoming solar energy entering the system is equal to the outgoing energy leaving the system.

equation of state – a thermodynamic equation relating state variables, which describe the state of matter, such as pressure p, density ρ, and temperature T. An example is the ideal gas law, $p = \rho RT$, where R is the specific gas constant.

exact differential – in thermodynamics, a differential dQ is exact if Q is a state function of the system, also called a potential. The corresponding mathematical expression is the existence of $u(x,y)$ and $v(x,y)$ fields such that $dQ = udx + vdy$ satifying the condition $u_y = v_x$. The integral of an exact differential dQ around a closed path is identically zero. In general, neither work nor heat is a state function. For example, a cyclical process has the same initial and final states, but the total work w done by such a cyclical process depends on the path taken by the process, and thus is not necessarily zero. To evaluate the integral of dw around a closed path, the path must be specified. Therefore, work is not an exact differential, because dw cannot be determined by knowing only the initial and final states.

Fundamental Theorem of Calculus – relates differentiation and integration, showing that these two operations are inverses of one another. One part of the theorem shows that an antiderivative can be expressed as an integral. The other part of the theorem states that an integral can be computed using an antiderivative. See Appendix D.

general circulation model (GCM) – the acronym "GCM" is of mixed origin. Originally, it meant "general circulation model," where "general circulation" is an old term in meteorology that connotes prominent features of the long-term average circulation, such as the trade winds in the atmosphere or the Gulf Stream in the ocean. More recently, "GCM" has come to stand for "global climate model," a term that today is often used interchangeably with "general circulation model."

geopotential energy – the gravitational potential energy per unit mass at a given altitude z with respect to sea level. See Chapter 7.

geostrophic balance – an approximation, which may be appropriate for large-scale extratropical systems in either the atmosphere or the ocean, in which the horizontal velocity field is characterized by a balance between the Coriolis force and the pressure gradient force. See Section 8.4.

global warming – a popular rather than scientific term for the recent and current observed multi-decadal rise in the average temperature of the Earth's climate system, together with related effects. Many types of scientific evidence leave no doubt that the climate system is unequivocally warming. The terms "global warming" and "climate change" are often used interchangeably. The Fifth Assessment Report (2013) of the Intergovernmental Panel on Climate Change (IPCC) states, "It is extremely likely that human influence has been the dominant cause of the observed warming since the mid-20th century." One shortcoming of the term "global warming" is that many of the most significant aspects of the ongoing climate change are not global in character and not confined to warming.

greenhouse effect – an inaccurate but widely used term for the process by which a planet's atmosphere traps heat and thereby increases the temperature of the atmosphere and the planetary surface. It involves the fact that the atmosphere is mostly transparent to incoming sunlight but is partially opaque to outgoing infrared energy. In the words of the Fourth Assessment Report (2007) of the Intergovernmental Panel on Climate Change (IPCC), "To balance the absorbed incoming (solar) energy, the Earth must, on average, radiate the same amount of energy back to space. Because the Earth is much colder than the Sun, it radiates at much longer wavelengths, primarily in the infrared part of the spectrum. Much of this thermal radiation emitted by the land and ocean is absorbed by the atmosphere, including clouds, and reradiated back to Earth. This is called the greenhouse effect." The term is inaccurate, because actual greenhouses also warm by their glass walls and roofs blocking the wind, thereby preventing warm air from escaping.

greenhouse gases – the gases in the atmosphere which are responsible for creating the greenhouse effect, the most important of which are water vapor and carbon dioxide. Greenhouse gases have three or more atoms in each molecule, so that the gases which make up most of the atmosphere, nitrogen and oxygen, with only two atoms per molecule, do not contribute significantly to the greenhouse effect.

hypothesis test – a statistical inference to reject or accept a hypothesis based on data and their distribution. The process of hypothesis testing typically involves five steps: (a) formulating a null hypothesis, stating that the observations are the result of pure chance; (b) formulating an alternative hypothesis, stating that the observations are not from the null hypothesis scenario; (c) assigning a significance level, which is the probability expressed in percentage, often set to 5%, of the domain of rejecting the null hypothesis; (d) computing a test statistic index from data to check if the index is in the domain of rejecting or accepting the null hypothesis; and (e) reaching a test conclusion.

isohypse – a line connecting points of constant geopotential height on a chart or map of a constant pressure surface. Isohypses are sometimes called isoheights or height contours. The root term hypsos comes from the ancient Greek word for height. Isohypses play a role analogous to the role of isobars on a chart or map of pressure. For example, geostrophic winds are parallel to isohypses.

mean value theorem (MVT) of calculus – states that for a continuous function in a domain, there exists at least one point inside the domain such that the integral of the function over the domain is equal to the area or volume of the domain times the function value at the point. See Chapter 8 and Appendix D.

pressure gradient force (PGF) – the force which results when there is a difference in pressure across a surface. In general, a pressure is a force per unit area, across a surface. A difference in pressure across a surface then implies a difference in force, which can result in an acceleration according to Newton's second law of motion, if there is no additional force to balance it. The resulting force is always directed from the region of higher pressure to the region of lower pressure.

principal component (PC) – the column vectors of the temporal matrix obtained from a singular value decomposition (SVD) of a space–time climate data matrix are called principal components. The first few PCs often correspond to physical signals, such as the occurrence of El Niño events. See Chapters 4 and 10.

probability distribution, or probability density function – a function describing the likelihood of obtaining the possible values that a random variable can take on.

regression – Regression analysis is a set of statistical processes for estimating the relationships among variables. It includes many techniques for modeling and analyzing several variables, when the focus is on the relationship between a dependent variable and one or more independent variables. Regression analysis is widely used for prediction and forecasting. Regression analysis is also used to understand which among the independent variables are related to the dependent variable, and to explore the forms of these relationships.

singular value decomposition (SVD) – a method of analyzing climate data which decomposes a space–time climate data matrix into three parts: spatial patterns, temporal patterns, and energy. See Chapters 4 and 10.

Southern Oscillation Index (SOI) – a standardized difference between the sea level atmospheric pressures at Darwin, Australia and at Tahiti. An alternative index is the weighted SOI (WSOI) as the principal components of the SVD analysis of the pressure data at Darwin and Tahiti. Two other relevant indices are the cumulative SOI (CSOI) and the cumulative WSOI (CWSOI). See El Niño–Southern Oscillation (ENSO).

Stefan–Boltzmann law – the statement that the total radiant heat energy emitted from a surface is proportional to the fourth power of its absolute temperature. The law applies only to blackbodies, theoretical surfaces that absorb all incident heat radiation. The Stefan–Boltzmann law may be derived from Planck's law of radiation. See Section 7.5 for this derivation. Historically, the law was first found in 1879 by Josef Stefan (1835–1893) from experimental measurements made by John Tyndall (1820–1893) and was subsequently derived theoretically in 1884 by Ludwig Boltzmann (1844–1906).

Student's t-distribution – a probability distribution used when estimating the mean of a normally distributed variable with a small number of data points and an unknown standard deviation. William Gosset (1876–1937) published the t-distribution under the pseudonym "Student."

troposphere – the lowest layer of Earth's atmosphere. About 75 to 80% of the mass of the atmosphere is in the troposphere. The troposphere extends from the Earth's surface to an average altitude of about 18 km in the tropics but only to about 6 km in the polar regions in winter. Most clouds are found in the troposphere, and almost all weather occurs within this layer. The root word is the Greek word "tropos," which means turn, in recognition that turbulent mixing is important in the troposphere.

Index